T0317661

Daily Energy Use and Carbon Emissions

Daily Energy Use and Carbon Emissions

Fundamentals and Applications for Students and Professionals

Bruce E. Logan
State College
Pennsylvania, USA

This edition first published 2022

Registered Office
John Wiley & Sons, Inc., 111 River Street, Hoboken, NJ 07030, USA

Editorial Office
111 River Street, Hoboken, NJ 07030, USA

For details of our global editorial offices, customer services, and more information about Wiley products, visit us at www.wiley.com.

Wiley also publishes its books in a variety of electronic formats and by print-on-demand. Some content that appears in standard print versions of this book may not be available in other formats.

Library of Congress Cataloging-in-Publication Data applied for:
ISBN: 9781119831013

Cover image: © Ryoji Yoshimoto/Getty Images
Cover design by Wiley

CONTENTS

PREFACE

For over 20 y, I have been researching renewable energy solutions, particularly those that can be used for the energy sustainability of the water infrastructure. In the middle of 2019, I started to realize how few people, even well-educated scientists and engineers, understood how much energy they used. Everyone I talked to was clearly concerned about lowering CO_2 emissions and addressing climate change, but it seemed that there was a general absence in understanding about how much energy they used and how their own activities were connected to CO_2 emissions. Many could tell you how much money they spent on electricity or natural gas for their home, but not how much energy they used or the amount of CO_2 produced. About the only thing people could recall about CO_2 emissions was some vague number from an airplane trip. There was a clear lack of a connection between energy use in their daily lives and the resulting CO_2 emissions. These conversations led me to realize how limited my education had been on the magnitude of my own energy use, the energy use by an average person in the United States, or energy used by people in other countries. I knew very little about possible solutions to climate change or how much climate would be impacted by a specific reduction in energy use, despite all my years of work on developing renewable energy technologies. To tackle climate change, I realized that we would all need a better understanding of the sources and amounts of CO_2 emissions and other greenhouse gases (GHGs) that were occurring from our own activities.

In looking for ways to convey energy use or CO_2 emissions in plain language, I did not find many good examples. The best book that addressed these topics within the context of climate change was *Drawdown*, by Paul Hawken. While that book summarized the global importance of climate change and quantified possible solutions, the activities needed to reduce carbon emissions, and units given for numbers (gigatons of carbon dioxide equivalents), felt disconnected from my own life. To really become engaged in climate solutions I discovered, I needed a lot more information and a way to connect energy use to carbon emissions for specific activities. I set out to find the answers I needed, and writing this book helped to organize that material into something that could be used by me and others for the same purpose.

HOW THIS BOOK CAN HELP YOU UNDERSTAND ENERGY USE AND CLIMATE CHANGE

The purpose of this book is to provide information, examples, and calculations to address energy use and carbon emissions in our own lives. For example, how much electricity does your house use in a month? How does energy used for your car compare to that for your home? Will turning off lights in your home have a larger impact on CO_2 emissions than turning down the thermostat or driving 10 less miles a week? It has been difficult to answer these questions because you cannot directly compare these activities as different methods are used to describe the amount of energy used. Electricity for your house is likely billed in kilowatt hours (kWh), natural gas is quantified in units of therms (or CCF), and energy purchased for your car by the volume of gasoline. Even if you could put all these into the same units of kWh, how do you translate these energy uses into CO_2 emissions?

The approach used in this book was to address energy within the context of the full energy–food–climate–water nexus, in ways that are comprehensible and relevant to our daily lives. While we can say how much energy is used by the United States compared to another country in units of TWh, EJ, or quads, those energy units are all very large numbers that have no relevance to our own lives. The first step was to convert large numbers into smaller numbers that have meaning for individuals. Social math is a method to explain and compare large numbers in ways that are both understandable and compelling. For example, saying that something is equal to a penny in an Olympic-sized swimming pool provides a clear image of the enormous volume of water relative to the size of that penny. Therefore, I needed a social math approach.

To better express our personal energy and water use and CO_2 emissions relative to our lives, I decided to examine energy-consuming activities within the context of the minimum amounts of energy and other related things, that we need to be alive, which are energy/food, air/carbon emissions, and water. To address units that could be more easily understood, I redefined these activities using baseline units of D, C, and w. One D is the energy in the food we eat every day, 1 C is how much CO_2 we release from eating that food, and 1 w is the minimum water we need to drink every day. Now, we can compare the energy in a gallon of gasoline (15.3 D) to the energy in food we eat every day (1 D) or the average electricity used in a US home (13 D). Energy use becomes more meaningful when we calculate how much energy it took to put 1 D of food on the table in our homes or how many D it takes to feed people in other countries. We can never use much less energy than 1 D in the long-term, so the question is how much more energy do we use than that in our food? For example, how much energy does it take to produce that food or to travel to work and elsewhere in the world? These energy uses can now be calculated and compared on a daily basis in units of D.

Units other than D, C, and w are also used here as needed to make the amounts of energy understandable in different contexts, for example energy use by countries, but the focus in the book is on what you do and how much energy you use for your own activities. Several chapters are also devoted to considering energy and CO_2 emissions relative to our infrastructure as those amounts of energy or emissions are relevant to our lives as we add to our infrastructure to accommodate an increasing number of people on the planet.

A book like this requires research into many different technical fields and peer review to make sure the facts and calculations are correct. I was very fortunate to have help from several colleagues and friends in reviewing different sections and chapters. I would like to thank Charlie Anderson, Gahyun Baek, Vikash Gayah, Zhanzhao Li, Aleksandra Radlinska, Mohammad (Mim) Rahimi, Farshad Rajabipour, Le Shi, Erica Smithwick, and Susan Stewart. Thanks also to all of my colleagues for many lively and insightful conversations on energy use, climate change, and the environment.

State College, Pennsylvania
June 2021

Bruce E. Logan

CHAPTER 1

INTRODUCTION

1.1 A VERY BRIEF HISTORY OF ENERGY USE

For most of the history of mankind, there was no need for a unit of energy: It was all about having enough food to eat and perhaps a fire to stay warm. Animal domestication created the additional burden of needing to ensure a food supply for the animals, but the food-energy loop was small and centered on food and warmth. Around 5500 y ago, people started riding horses and mankind entered the bronze age with people learning to mine and heat metals to make better tools (and weapons). Human energy needs then became more than about getting enough food for our own energy needs, it also meant providing food for horses and other animals and having sufficient energy to craft tools.

From these early days when people first learned to make and use fire, our global population has remained bound to the concept of burning things. The history of energy use shows that most efforts to obtain energy were centered on *finding different things to burn*. Wood and other biomass were burned to provide heat from a fire, and then, coal was used as a more efficient energy source due to its greater energy density. Later on, we transitioned to oil and then natural gas, with oil being transformed into many other materials we could burn such as gasoline, kerosene, and jet fuel.

So here we are today with an energy infrastructure based primarily on distributing various kinds of fuels around the world to be burned and provide the energy we use to drive our cars, heat our homes, produce electricity, run factories, and maintain communities. We are used to an abundance of fossil fuels and thus have little direct connections to how much energy we use other than the cost at the gas pump to fill up our car or the money we use to pay our gas or electric bills. It is difficult to sum up the energy from these different sources because they mostly all have different units. One of the purposes of this book is to better connect us to the amount of energy we use to develop an appreciation of how it ties into our daily lives. Another purpose is to show that you can use this knowledge to reduce energy consumption and greatly decrease carbon emissions from fossil fuels into our environment.

In the first few chapters of this book, we will examine the vast array of energy units in our lives and gain an appreciation of where we are today in energy use and carbon emissions. In subsequent chapters, we will explore how much energy we use in our own lives and activities through the daily energy unit D, carbon emissions via the unit C, and water use using the unit w. Once we have examined our own activities using these three units, we can then examine energy use for our built infrastructure and see how much change we need for energy and water consumption to significantly reduce the amounts of carbon emissions in our lifetimes.

Daily Energy Use and Carbon Emissions: Fundamentals and Applications for Students and Professionals,
First Edition. Bruce E. Logan.

1.2 EARLY ENERGY AND POWER FOR TRANSPORTATION AND ELECTRICITY PRODUCTION

It is easy to imagine the path that led to defining the power of an engine in terms of horsepower, since engines were developed to provide an alternative power source to horses for work or transportation. The Scottish engineer James Watt is credited with term as he showed back in the late 18th century that steam engines he invented and developed were superior in the work they could do compared to one or even many horses. There was no "standard horse" in those days, and it is now considered that one horse with above average strength was used to provide the first definition of 1 horsepower (hp). The power provided by a horse was likely closer to its "maximum power" than the sustainable amount of power by a horse over a long period of time.

It does seem rather amazing that after all these years that in the United States we are still using a term such as horsepower for a car, motorcycle, or truck. One way to convey how much power is in a car engine is to relate units of horsepower to those of Watts (W) used for electricity. Thus, we can state that 1 hp equals 746 W or 0.746 kilowatts (kW). An old fashioned incandescent light bulb uses about 100 W, and a modern light emitting diode (LED) unit produces about the same light using only 14–20 W. It seems fitting to have these electrical power units named in honor of the engineer James Watt. Describing the power of a car engine in horsepower does not necessarily tell us how much of that power is being used. For example, car rated at 100 hp will not run continuously at this maximum power rating. Instead, you will use only a fraction of that total horsepower when you drive to the store or take a trip on a highway. Thus, our daily lives we are more connected to the energy used by a car in terms of gallons of gasoline consumed rather than hp or kW. For energy, many different units can be used such as Calories for food, megajoules (MJ) for gasoline, or kilowatt hours (kWh) for the energy used over a certain period of time. The amount of energy used per time is defined as power.

Example 1.1

Sarah drives to work every day in her car that uses 1 gal of gas for the 20 mi round trip that takes about 20 min each way. (a) What is the energy in kWh needed to bring her to work? Assume a gallon of gas contains 127 MJ of energy. (b) How much horsepower does it take for the 10-mi drive to work?

(a) For a single trip of 10 mi that uses a half of a gallon of gasoline, we can calculate the energy used with the conversion factor of 3.6 MJ/kWh as:

$$E = \frac{0.5\ \text{gal}}{1\ \text{trip}} \frac{127\ \text{MJ}}{\text{gal}} \frac{\text{kWh}}{3.6\ \text{MJ}} = 17.6 \frac{\text{kWh}}{\text{trip}}$$

(b) The energy that was used was expended over a period of 20 min. Thus, to calculate power, we need to divide energy by time, so the result in power (W) is as follows:

$$\frac{17.6\ \text{kWh}}{20\ \text{min}} \frac{60\ \text{min}}{\text{h}} = 52.9\ \text{kW}$$

To convert this to horsepower, we use the conversion factor that 1 hp = 746 W or 0.746 kW, or

$$52.9\ \text{kW} \frac{1\ \text{hp}}{0.746\ \text{kW}} = 71\ \text{hp}$$

From this calculation, we can see that this amount of power would have required 71 horses. Of course, even using 71 horses, it is unlikely that she could have arrived to work on time. A horse walks at about 4 mph, so whether she used 1 or 71 horses, if she moved at 4 mph, it would have taken 2.5 h to get to work.

The first engines to replace the horse were steam engines, which use an external combustion system fueled by coal or wood, to produce the steam. Internal combustion engines (ICEs) produce power by burning the fuel within the engine and have the advantage of not needing to use water to transfer the energy from the external combustion chamber to the engine.

Electricity production has not, until very recently, changed from the basic approach of steam engine in the sense that there is an external system that uses a fuel to produce steam, with the steam used drive a turbine that is then used to make electricity. Fuels for electricity production have evolved separately from the turbine systems. Therefore, steam engines that burned wood were replaced by power plants that produced steam from coal, oil, and natural gas. Electricity production has therefore greatly impacted how we use fossil fuels.

Looking back in time to where we first made a transition from a wood economy into a modern society, the transformation is best identified as the start of the Industrial Revolution, which is considered to be a period from 1760 to the early 1820s (1820–1840). This rapid rise in industrial production required increased use of steam and waterpower for manufacturing of chemicals, materials such as iron and textiles, and tools. Wood and coal sustained growth for a period of time but additional sources of energy were eventually needed to sustain our increasingly industrialized societies.

The impact of the industrial revolution on timber and wood resources was enormous in some locations. For example, in Pennsylvania and elsewhere in the northeastern United States, blast furnaces were used to produce iron materials needed for new industrial age, but these furnaces required coke, and coke production required large amounts of wood. Vast tracks of land were nearly completely leveled to run the furnaces in the 1800s leaving much of the land bare and releasing huge amounts of carbon stored in these forests into the air. Many decades were subsequently needed for the recovery and regrowth of these forests after they were cleared for this use. The use of wood used in these furnaces was eventually replaced with anthracite coal, shifting efforts from clearing the land of trees to mining coal.

The Discovery of Oil and the Next Age of Burning Things

The first oil use may have occurred as early as 600 BCE in China, but for the modern world, there were two notable events in the United States: oil first discovered in 1859 in Titusville, PA, and the subsequent operation of that oil well; and the Spindletop Hill oil discovery in Texas in 1901. Most large oil companies have their origins associated this Texas oilfield due to its enormous productivity. The high energy density of oil and its relative ease of extraction led to its prominence in the energy portfolio of the United States and the world in modern times.

Even the development of electricity production from nuclear fuels did not change this basic relationship between using a fuel to heat water, and then the steam being used to make electricity. Energy captured from nuclear fission is used no differently than that from combustion of fuels in steam power plants. Using a source of nuclear energy does have the advantage of not releasing the carbon in fossil fuels as CO_2 into the atmosphere, but the overall process remains tied heating water to produce steam. Nuclear production of electricity is also currently the most expensive way to produce electricity in the United States, and the radioactive waste has no permanent solution leaving future generations to deal with this waste product. A sufficient amount of fuel for nuclear reactors is also a big concern. The availability of uranium could enable the production of 100 TWh of electricity and thus a continuous use of 10 TW of power. However, based on using known existing supplies of uranium for nuclear fuel that rate of power generation would deplete uranium stores in less than a decade. There is also no way to quickly ramp up the use of nuclear fuel in the United States. A typical nuclear plant produces 1 gigawatt (GW), or a billion watts, and it takes 10 or more years to construct this type of plant in the United States. The last two nuclear power plants that went into operation in the United States were in 1996 and 2016 (US Energy Information Administration, 2020a). A total of 17 nuclear power plants have been shut down in the US, although two more units are anticipated to be added at an existing site in Georgia in 2021 and 2022 (US Energy Information Administration, 2020b).

Whether nuclear power plants can be economically operated in the future is not clear. For example, the Three Mile Island plant in Pennsylvania shut down due to high operating costs, and many operating nuclear power plants are 38 y old and reaching the end of their projected lifetimes. Furthermore, there is still no permanent solution for nuclear waste in the United States. The cost of a serious accident at a nuclear power plant, while remote, can be extremely high if one occurs. For example, addressing the damages of the Chernobyl nuclear power plant accident could exceed US $200 billion, and decontaminating the Fukushima site in Japan that was destroyed by a tsunami could cost US $470–$660 billion. Thus, despite the ability of nuclear power plants to produce electricity without CO_2 emissions, it does not seem reasonable that many additional nuclear power plants will be built in the US soon or that nuclear power can increase as a part of the US electrical power grid. However, nuclear power is projected to increase globally by 6% based on the analysis of the International Energy Agency (IEA) for meeting CO_2 reductions needed to avoid a global increase of more than 2°C by 2050 (IEA, 2017).

1.3 ENERGY AND THE CHALLENGE OF GLOBAL CLIMATE CHANGE

The main motivation for renewable energy today is reducing emissions of greenhouse gases (GHGs), with CO_2 released from burning fossil fuels as the main GHG of concern. The link between global warming and CO_2 emissions in the industrial age is clear and irrefutable. The main discussions now center around what can be done to curtail those emissions.

Much of the discussion on curbing GHGs focuses on power plants that make electricity and fuels used for transportation. However, electricity and heat production account for just 25% of all GHG emissions, and transportation (as a category) about 14% (Fig. 1.1). A nearly equal percentage of GHGs (24%) arise from agriculture, forestry, and other land use activities. Transportation is less (14%), with industry accounting for about 21% of all GHG emissions. Such information on emissions from these different sources is useful when we contemplate where we can effectively reduce GHG emissions from the perspective of changes that we could make in our own lives that might impact these GHG emissions.

The Paris Climate Agreement on climate change, signed by nearly 200 nations in December of 2015, focused on reducing CO_2 and other GHG emissions with the goal of not exceeding a global temperature rise of 1.5°C (2.7°F) by 2050. To work toward this goal, these nations have pledged to reduce CO_2 emissions by various amounts, but based on the analysis of Hausfather and Peters (2020), the extent of these pledges is not sufficient to achieve this 1.5°C goal (Fig. 1.2). With current pledges to reduce GHG emissions, we are on a path that might result in a 2.5–3.0°C

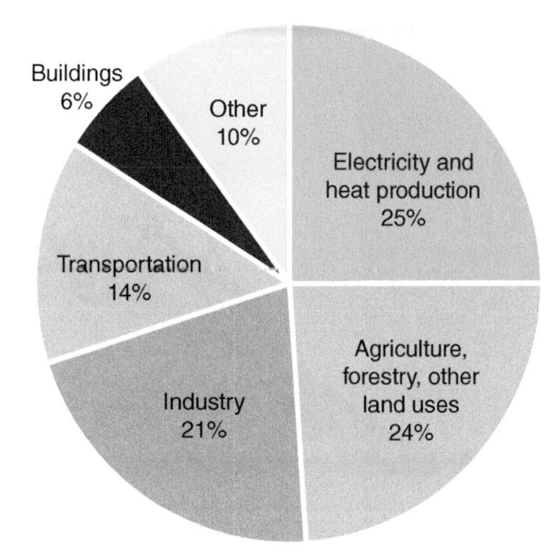

Figure 1.1 Global greenhouse gas emissions by economic sector. *Source*: Adapted from US Environmental Protection Agency (2020).

(4.5–5.4°F) increase. Therefore, not only do these nations need to adhere to these pledges going into the future, much greater changes are also needed to further reduce CO_2 and other GHG emissions. If we do not take appreciable steps toward reducing GHG emissions, a 4°C (7.2°F) scenario seems more likely on average, with the worst-case scenario reaching an increase of 5°C (9°F) or more (Fig. 1.2). Note that these averages are for increases in land and water temperatures combined. Land temperatures can average ~60% higher than these averages, raising the possibility of a rise of 8°C (>14°F) for land temperatures *on average*, and extreme event temperatures could produce

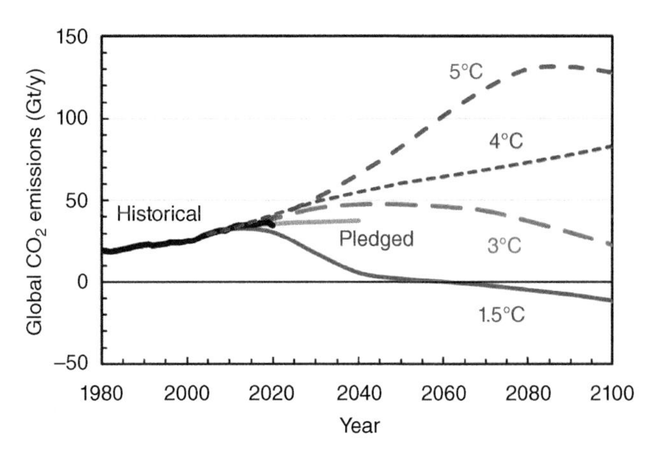

Figure 1.2 Possible scenarios for global fossil fuel emissions and their impact on global average temperature changes. Key: historical, data for global emissions from 1980 through 2020; pledged, reductions by countries; temperatures show the global rise from mitigation needed to reach the Paris agreement goal of 1.5°C, weak mitigation leading to 3°C, average predictions with no climate policies of 4°C and worst-case scenario with no climate policies of 5°C. *Source*: Adapted from Hausfather and Peters (2020).

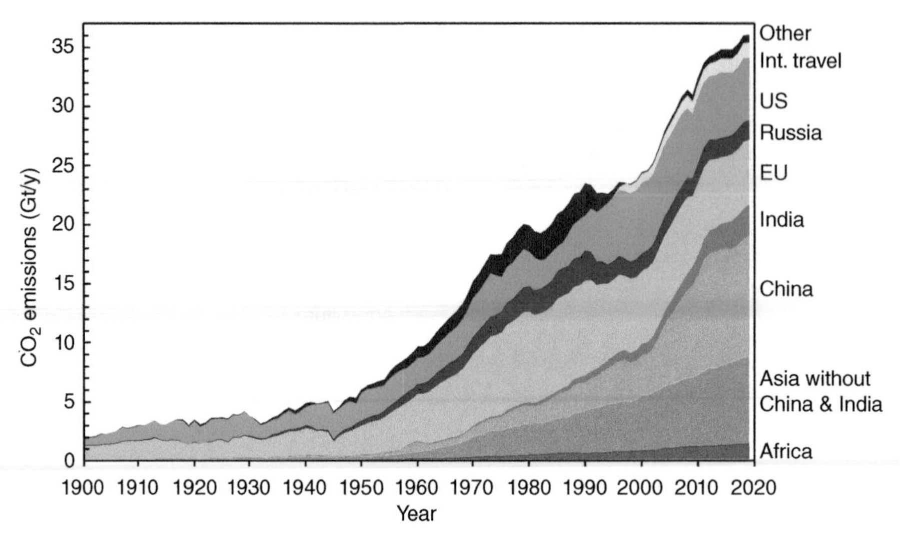

Figure 1.3 CO_2 emissions by different countries in the past century. *Source*: Adapted from Ritchie and Roser (2020).

periods of temperatures much higher than these averages (Voosen, 2021). Land temperatures are already averaging nearly 2°C higher than pre-industrial averages (Voosen, 2021). Such changes will be profound in terms of the kinds of temperatures that we experience in our daily lives for the coming decades. A more detailed assessment of CO_2 emissions and methods of carbon capture and storage are presented in Chapter 14.

There are enormous political and social challenges for reducing CO_2 emissions as the wealthier countries that generally emit higher amounts of CO_2 can greatly reduce their emissions while still enjoying a relatively comfortable lifestyle. The United States is one of the largest emitters of CO_2 per person, but in recent years, CO_2 emissions by the United States have been declining (Fig. 1.3). However, over this same period, many high CO_2 emitting industries have moved overseas, with many to China. Through rapid industrialization in China, and this country becoming a center for global manufacturing, China's CO_2 emissions have greatly accelerated and now exceed those of the United States.

Many nations in the world are considered to be "energy poor" and desire to increase the standard of living for their citizens, which would greatly increase CO_2 emissions if these standards were raised by consumption of fossil fuels. As a result, nations in Asia and other Pacific countries, along with India, are now on a path to substantially increase CO_2 emissions. Also, if demand for petroleum declines and oil prices drop, less expensive oil might become more attractive for use in many developing countries and this increased use could negate efforts by industrialized nations to reduce GHG emissions through their own reductions in CO_2 emissions. Some sources of GHG emissions are difficult to link to a specific country, such as international travel. When GHG emissions from international travel are separated into a single category, these emissions are larger than those of some countries (Fig. 1.3). Energy use and CO_2 emissions from cars and other forms of travel are further discussed in Chapters 9 and 10.

The need for energy by many people in the world is enormous, and not only because people would like a more Western lifestyle, but also because many people in the world need energy just to meet basic needs for clean drinking water and sanitation. For example, roughly 1 billion people lack sufficient access to potable water, over 2 billion lack safe drinking water in their homes, and over 4 billion people lack adequate sanitation (Osseiran, 2017). Modern sanitation is expensive. If other countries in the world follow the same path as the United States for modernizing their water infrastructure using the same technologies in the United States, for example, the increase in CO_2 emissions would be enormous based on typical energy use by the United States. An often cited number related to energy use for the water infrastructure is that approximately 3–4% of electricity production in the United States is used for the water infrastructure, which includes pumping, water treatment and distribution, and wastewater collection and treatment. However, there is much additional energy use associated with the water infrastructure that is missed in only examining electricity use, as further discussed in Chapter 5. These energy costs also do not include other costs not directly related to the electricity and energy use. One assessment of wastewater treatment of in the United States estimated that the annual cost was $25 billion (WIN, 2001). It will take the cooperation of all countries around the globe to reduce CO_2 and other GHG emissions into our shared global atmosphere, while at the same time addressing poverty and a lack of sufficient water, sanitation, and food for a large part of the world's population.

1.4 LOOKING TO THE FUTURE: THE AGE OF ELECTRO-MECHANICAL/CHEMICAL ENERGY CONVERSION AND STORAGE

The 20th century will be considered as the age of oil, or more generally, the age of "burning things" since energy use was all about extracting energy from fossil fuels. However, the 21st century needs to become the age of electro-mechanical/chemical energy based on an energy system that captures energy directly as electrical power without the need to burn fuels. Electromechanical systems, which operate based on using turbines to convert motion in one form into work or electricity are the oldest form of noncombustion-based energy conversion, but other methods have developed based on chemical catalysts and materials such as those used in solar cells. In 2019, 16.4% of energy used for electricity production was indicated by the US Energy Information Administration to be provided by renewables (wind, hydro, solar, and geothermal). As discussed in Chapter 2, these numbers are inflated based on assuming electricity production using renewables has the same efficiency as a fossil fuel power plant. Renewables were actually 6.9% of the primary energy used for electricity production and accounted for 16.7% of the electricity generated for consumers. Nuclear and biomass, categorized here as carbon neutral, accounted for 21.2% of electricity production. For a carbon-neutral energy infrastructure, we need to develop to the maximum extent possible ways to harness energy without needing to burn fuels (Fig. 1.4). The status of renewable energy use for our daily lives is addressed in subsequent chapters. In this chapter, we briefly examine the rise of these technologies over time.

Wind Energy

There were early efforts to capture wind energy using windmills around 500–900 AD, with most applications for pumping water or grinding grain. The Dutch are perhaps best known for their use of windmills dating back to the 11th century, with the number of windmills reaching over 10,000 there at one time, although estimates are that only about 1000 of these windmills remain today. The primary use of these early windmills by the Dutch was to pump water out of the lower lands to make them useful for food production. Windmills were used in the United States primarily to grind grain, pump water, and provide mechanical energy for sawmills to cut wood. In the late 1800s and early

Figure 1.4 Wood was the original fuel (left) but over time we have learned to make use of truly renewable sources of energy that do not require harvesting or burning fuels using hydropower, wind, and solar energy technologies (right of the arrow).

1900s, small wind generators (turbines) were used to produce electricity, but the electrification of the grid in the 1930s that provided easier access to electricity saw an overall decline in their use. The use of wind power to produce electricity has more recently grown substantially, reaching 8% of all electricity produced in the United States in 2019. Electricity generation using wind power has also grown around the world, with over 1.13 trillion kWh generated around the world by 129 countries.

Hydropower

The first hydropower station that produced electricity using a water turbine occurred in 1882 in Appleton, WI, and by 1889, there were about 200 hydroelectric power plants. Notably, the Hoover Dam opened in 1936 and was initially capable of producing 1.345 GW of electricity (currently rated at 2.08 GW). However, no method of producing electricity comes without other disadvantages. In the case of the Hoover Dam, the construction of a lake for the dam completely altered the river ecosystem and flowrates of the Colorado River. While hydroelectric power stations provided electricity, most of the energy demand in these earlier days was provided by coal that could be used in steam engines. The incandescent lamp, invented in 1879 by Thomas Edison, drove the wider use of electricity for lighting. In 1882, Edison built the first large scale power plant to produce electricity for lighting and other uses, and it was powered by coal. By 1900, less than 2% of the energy used to make electricity in the United States was derived from fossil fuels (coal, oil, and natural gas) compared to about 62% in 2019.

Solar Energy and Electrochemical Conversion

Solar energy has always been used as a source of heat, and as early as the 7th century BCE magnifying glasses were used to make fire and light torches. The birth of the modern electrochemical age began in 1954 at Bell Labs with the invention of the first silicon photovoltaic (PV) cell, which initially had a 4% efficiency in converting light energy into electricity. Another milestone in PV development occurred in 2001 when the Home Depot chain of stores first started selling solar power systems in a few of their stores. Fast forward to today where large solar farms can produce electricity more cheaply than that possible using fossil fuels. For example, the City of Los Angeles recently approved construction of a solar farm that will provide 7% of the electricity used by the city in 2023, at a cost of 1.997 cents/kWh. The system is a 400 megawatt (MW) array that is expected to produce 876,000 MWh of energy per year. For comparison, electricity production using natural gas is around 4 cents/kWh, with the electricity delivered to a residential customer at 13 cents/kWh. This transformation in the cost of electricity produced by solar power will revolutionize electrical power generation in the United States and the world. We currently use in 1 y the amount of energy in the sunlight (4.3×10^{20} J) that strikes the planet for 1 h. We need to harvest a greater percentage of energy from the sun to achieve a more sustainable energy infrastructure.

Biofuels

Burning biomass such as wood is the earliest form of energy harvesting from nature, and the CO_2 that is released from the wood is just being returned to the atmosphere as it was previously captured into the biomass. Using biomass in combustion processes, therefore, is a carbon-neutral method of heat generation and perhaps electricity generation, but there are other environmental issues with this or any combustion-based process. Thus, this context of electricity from biomass using combustion-based processes is not as desirable as other electro- or mechanical-conversion processes such as solar or wind energy. Burning wood, especially in uncontrolled environments, can release particulate matter, nitrogen and sulfur oxides, and carbon monoxide as well as lead, mercury, and other hazardous air pollutants. If we view biomass as solar energy storage, then it has the advantage of being able to be transformed into wood or other transportable fuels (such as ethanol or biodiesel). However, the low energy density of biomass makes it uneconomical to transport wood or other forms of biomass very far, and therefore, the point of production of fuels or electricity must be close to the biofuel source. Still, there is an enormous potential for using waste biomass for energy. An analysis by the US Department of Energy (DOE) estimated that there is more than a billion tons of waste biomass that could be converted into useful energy (Perlack et al., 2005).

There are many methods to capture biomass energy into fuels, with ethanol as the best-known example. Corn ethanol, however, has few net environmental benefits, and the separation of the ethanol from water is energy intensive. Biodiesel production is from crops is also marginally energy net positive, and both ethanol and biodiesel are used in low efficiency combustion engines. However, if cellulose rather than food crops is used to produce these biofuels, then the energy and environmental advantages become more pronounced. Alternatively, fermentation of biomass can produce methane or hydrogen gases. However, if the methane is used in an ICE for transportation, then some of the same combustion-based challenges arise with methane as the other fuels. If H_2 gas is used for transportation, it can be used in vehicles powered using fuel cells which can have about twice the energy efficiency of a typical ICE. Complete conversion of the cellulose in waste biomass could power almost all light duty vehicles if they used very efficient hydrogen fuel cells (Logan, 2019). However, only a maximum of 4 mol of H_2 can be produced from cellulose by bacterial fermentation, with the balance mostly consisting of acetic acid under optimal conditions (equivalent to 8 mol of H_2) remaining as a fermentation end product (Logan, 2004). Therefore, additional methods such as using microbial electrolysis cells are being investigated to convert acetic acid to hydrogen, but additional energy is still needed as the overall reaction is endothermic (Logan et al., 2006, 2008). The use of biomass for producing renewable energy is further addressed in Chapter 6.

Energy Storage

Intermittent production of electricity using wind and solar energy, coupled with highly variable energy demands throughout the day, require efficient energy storage methods. Batteries immediately come to mind for energy storage as they are so commonly used in toys, phones, tools, cars, and other devices. However, many batteries generate electrical power from irreversible chemical reactions. The first modern type of battery was invented by Alessandro Volta in 1800 who constructed the battery from copper and zinc electrodes. This led to the most common type of battery, the zinc-carbon battery, which are disposable batteries that are still commonly used today. Batteries that cannot be recharged, and thus cannot store electrical energy, are called primary batteries.

The first rechargeable battery was the lead acid battery that dates back to 1859. Perhaps surprisingly, lead acid batteries are still used today in applications ranging from cars to large-scale energy storage systems.

Of the many forms of energy storage available today, water storage can be the most efficient and economical method if conditions provide a sufficient elevation change of several hundred feet or more for pumping water. Pumped water storage accounts for more than 95% of energy storage installations worldwide, with 25 GW of power storage in the United States and 184 GW worldwide (in 2017). The overall efficiency of using electrical power to pump water uphill into the storage reservoir, and then produce electricity when needed by running that water back through water turbines, is about 70–80%, although some claims indicate that this efficiency can reach 87%. This approach is used to "store" energy when power plants have excess capacity and then use it when peak power production is needed. Energy storage using water is further addressed in Chapter 5.

Other methods are used to store energy in chemical form, for example by converting water to hydrogen. Water can be split to form hydrogen and oxygen gases at high efficiencies in water electrolysis systems, and then, the hydrogen gas used as a fuel in a fuel cell. However, both processes tend to be expensive and not as efficient overall as water storage. Water electrolysis and fuel cells together are estimated to be able to approach 80%, but current technologies which rely on a balance of cost and efficiency are around 50% (Grant, 2003). Water is not the only material to be moved "uphill." For example, trains can be run up a mountain to consume electricity when it is cheap and then down the mountain to generate electricity when needed (or when the electricity is worth more), or heavy loads of materials can be lifted and released to produce electricity as well. New types of batteries and other energy storage technologies continue to be researched and developed and so it is likely that the future will bring improved methods to store and recover energy depending on the specific locations and different needs for energy storage.

1.5 WHY D, C, AND w UNITS?

The single most important challenge for our future energy infrastructure is producing useable energy in forms that do not contribute to climate change. The largest driver of climate change is CO_2 emissions from fossil fuels and therefore the main challenge comes down to reducing energy use. The more fundamental challenge is understanding how we can reduce our energy use, which brings up a few central questions. How much energy do we currently use? How much of that energy is from fossil fuels? How can one person reduce energy use, and how much will that impact overall energy use compared to what industries and municipalities use?

The approach taken here in this book to answer these questions about our own energy use is to quantify energy use and climate change relative to appropriate baselines. Engineers often use dimensionless numbers to describe complex systems as a function of scale. The idea is to start with some basic relationship between two things and then to see how that relationship changes as a function of some factor such as time or flow. We also often describe changes in our social structure in terms of change relative to some point, for example percent increases in employment, inflation of money, or time spent in some activity such as screen time on your computer or phone. These approaches are used here in a way that can convey our own activities to energy and water use, and carbon emissions.

The baseline for energy used here is the energy in the food that an adult need to eat every day, or 1 D as the daily energy unit. As shown in Chapter 3, the daily energy needed by a person every day is about 2000 Cal (8.4 MJ), so we define 1 daily energy unit D as 2000 Cal (Fig. 1.5),

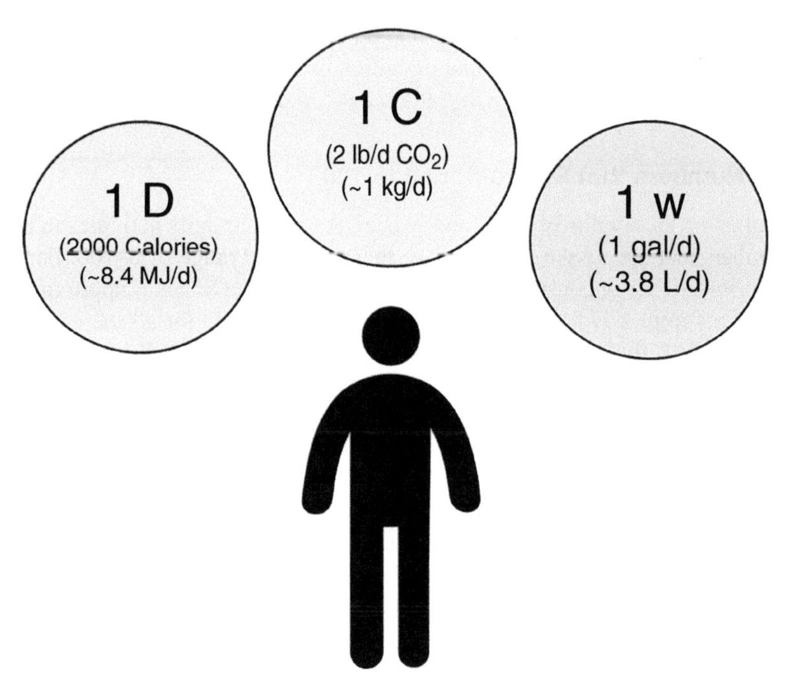

Figure 1.5 To better understand quantities in our lives related to the energy–food–climate–water nexus, we can normalize energy we use by the energy in the food we eat, CO_2 by the CO_2 we exhale every day, and water use by the amount of water we need to live for a day. *Source*: Based on Using resources from Flaticon.com. Freepik Company S.L.

or 2.32 kWh per day per person. That is roughly equal to the power continuously consumed by a 100 W lightbulb. With this definition of daily energy by one person, we can ratio all our other activities that consume energy, such as heating our home or driving our cars, all in terms of the same base unit of D. Since we have expressed energy based on food energy, the unit D provides the basis of calculations for the food–energy nexus as it directly links these two numbers.

The baseline for CO_2 emissions is the amount of CO_2 that is released back into the environment every day due to the food that we eat (Fig. 1.5). As we explain in Chapter 4, each person releases about 2 lb of CO_2 every day from eating 1 D, so we define the unit 1 C as 2 lb of CO_2 per day per person. This amount of CO_2 released is roughly equal to the CO_2 incorporated into the biomass of 15 trees. Based on that C reference point, we can express the CO_2 that is released into the environment from other activities that consumed energy in the units D. We can calculate D:C ratios for fuels such as gasoline, diesel, and natural gas. Thus, using D and C, we can facilitate our analysis of the food–energy–climate nexus. Note that C here does not refer to the element C, but instead is a defined constant of 2 lb/d of CO_2 per person. When the element carbon is used, it will have units of mass associated with it, for example 1 C = 2 lb/d CO_2 = 0.91 kg/d CO_2 = 0.23 kg-C/d (based on the molecular weights of C = 12 g/mol and CO_2 = 48 g/mol).

The baseline for water use is the amount of water that we need to drink every day to stay alive. As we show in Chapter 5 that amount of water is around 1 gal/d (about 4 L), so we define 1 w as the daily water use of 1 gal per day per person. Note that a lower-case number w is used here rather than W, as the upper case W is an abbreviation for Watts. Using w, we can ratio our actual water use to this baseline. Typically, we use around 50–170 w in our homes, although due to the very large amounts of water used for irrigation and power plant cooling, the amount normalized to the country is around 1300 w. With this final definition of w, we have the complete nexus for our food–energy–climate–water nexus using the letters D, C, and w.

The Utility of Numbers that Range from 0 to 100

Another goal of using these units of D, C, and w is to quantify our daily activities in terms of nice "rounded" numbers that are easy to use because they are in the range of 0–100. For energy use, the amount of primary energy consumed by a person in the United States is approximately 100 D, so our range of D is from 1 D for the food we eat to around 100 D for all the energy we use on average. Similarly, for C the carbon emissions normalized to a person will typically never exceed 100 C based on typical D:C ratios that are larger than 1. For water, numbers based on w can go well over a hundred, but certainly, a goal is to minimize our water use and achieve $w < 100$. The water a person uses per day in their home is around 100 w.

The unit of D is valuable for estimating electrical or fossil energy use, but also energy related to food consumption. Consider our current system in the United States where food energy is given Calories (upper case C in Calories), where 1 Cal (1 kilocalorie, kcal) equals 1000 cal (lower case c for calorie). Imagine if you were trying to evaluate the energy in the food based on eating 2,000,000 cal/d? A food item listed as 300,000 cal would certainly be a lot more confusing than the current method in the United States of listing of 300 Cal, so the simplified upper case Calorie has an advantage over the thermodynamic unit of calories. However, even with the Calories units, you still need to divide 300 by 2000 to get the percentage of Calories that you should be eating in a day. This process of evaluating food energy could therefore be improved if the item was listed in units of D, for example as 15% D, as this would indicate more directly that this food item contained 15% of your daily energy needs.

Energy Units in this Book

Energy is expressed in many different units. In Chapter 2, we examine the amount of energy used for different activities using the appropriate units for the specific energy use activities such as Btu, kWh, and Calories. In the subsequent chapters, we will analyze energy use, carbon emissions, and water use in terms of many of these characteristic units appropriate for different situations, introducing the topics using units such as quads or kWh, but thereafter also expressing energy use and other aspects of these activities in terms of the D, C, and w units. Through these analyses and presentations, it is hoped that we can all develop a clearer understanding of our energy use and how we can reduce our CO_2 and other GHG emissions to limit our own personal impact on climate change.

References

GRANT, P. M. 2003. Hydrogen lifts off—with a heavy load. *Nature,* 424, 129–130.

HAUSFATHER, Z. & PETERS, G. P. 2020. Emissions – the 'business as usual' story is misleading. *Nature,* 577, 618–620.

IEA. 2017. Energy technology perspectives 2017 – Catalysing energy technology transformations. Available: https://www.iea.org/reports/energy-technology-perspectives-2017. [Accessed February 20 2021].

LOGAN, B. E. 2004. Extracting hydrogen and electricity from renewable resources. *Environmental Science & Technology,* 38, 160A–167A.

LOGAN, B. E. 2019. Ending our hydrogen and ammonia addiction to fossil fuels. *Environmental Science & Technology Letters,* 6, 257–258.

LOGAN, B. E., AELTERMAN, P., HAMELERS, B., ROZENDAL, R., SCHRÖDER, U., KELLER, J., FREGUIAC, S., VERSTRAETE, W. & RABAEY, K. 2006. Microbial fuel cells: Methodology and technology. *Environmental Science & Technology,* 40, 5181–5192.

LOGAN, B. E., CALL, D., CHENG, S., HAMELERS, H. V. M., SLEUTELS, T. H. J. A., JEREMIASSE, A. W., & ROZENDAL, R. A. 2008. Microbial electrolysis cells for high yield hydrogen gas production from organic matter. *Environmental Science & Technology,* 42, 8630–8640

OSSEIRAN, N. 2017. *2.1 billion people lack safe drinking water at home, more than twice as many lack safe sanitation* [Online]. World Health Organization (WHO). Available: https://www.who.int/news/item/12-07-2017-2-1-billion-people-lack-safe-drinking-water-at-home-more-than-twice-as-many-lack-safe-sanitation#:~:text=lack%20safe%20sanitation-,2.1%20billion%20people%20lack%20safe%20drinking%20water%20at%20home%2C%20more,as%20many%20lack%20safe%20sanitation&text=Some%203%20in%2010%20people,report%20by%20WHO%20and%20UNICEF. [Accessed February 15 2021].

PERLACK, R. D., WRIGHT, L. L., TURHOLLOW, A., GRAHAM, R. L., STOKES, B., & ERBACH, D. C. 2005. *Biomass as feedstock for a bioenergy and bioproducts industry: The technical feasibility of a billion-ton annual supply*, Oak Ridge National Laboratory. Available: https://www1.eere.energy.gov/bioenergy/pdfs/final_billionton_vision_report2.pdf. [Accessed February 15 2021]

RITCHIE, H. & ROSER, M. 2020. *CO_2 and greenhouse gas emissions* [Online]. Available: https://ourworldindata.org/co2-and-other-greenhouse-gas-emissions [Accessed March 11 2021].

US ENERGY INFORMATION ADMINISTRATION. 2020a. *How old are U.S. nuclear power plants, and when was the newest one built?* [Online]. Available: https://www.eia.gov/tools/faqs/faq.php?id=228&t=21 [Accessed July 25 2020].

US ENERGY INFORMATION ADMINISTRATION. 2020b. *What is the status of the U.S. nuclear industry?* [Online]. Available: https://www.eia.gov/energyexplained/nuclear/us-nuclear-industry.php#:~:text=According%20to%20the%20U.S.%20Nuclear,in%20various%20stages%20of%20decommissioning. [Accessed July 25 2020].

US ENVIRONMENTAL PROTECTION AGENCY. 2020. *Global greenhouse gas emissions data* [Online]. Available: https://www.epa.gov/ghgemissions/global-greenhouse-gas-emissions-data [Accessed April 19 2020].

VOOSEN, P. 2021. Global temperatures in 2020 tied record highs. *Science,* 371, 334–335.

WIN. 2001. Water infrastructure now: recommendations for clean and safe water in the 21st century.

CHAPTER 2

ENERGY USE

2.1 UNITS OF ENERGY AND POWER

To understand energy consumption by individuals, corporations, or countries, we need to put all the different types of energy into a common language and the same units. First, we need to be careful about not using the words energy and power interchangeably. *Energy and power are not the same thing*. Energy is the amount of work done, for example, to lift a heavy rock into the back of a truck. Power is how fast you can do that, so lifting the rock can be done very quickly if you have a lot of power, otherwise you struggle to lift it up and put it in the truck, but overall, it takes the same energy whether you do it quickly or slowly. Power (P) is energy (E) per time (t), or $P = E/t$, and energy is the product of power and time, or $E = Pt$.

Energy is expressed in many different units that depend on how the energy is conveyed, for example, as a fuel for your car or electricity used in your home, and how energy is monitored, for example, based on volume of gasoline pumped into your car or the energy delivered to your home in a month. In this section, we can examine a few basic concepts: energy as a quantity of heat, energy as something we use for transportation (for example, in a car), and energy as something we pay for to various companies that sell us that energy as electricity or as a gas delivered to our home for heating.

Heating Water, in Calories

Energy units are usually expressed relative to the context for that energy used for some purpose. For example, if you want to heat water, you might use calories (cal), defined as the energy needed to heat 1 g of water by 1°C. But even for heating water, there are alternative units such as the British thermal unit (Btu) defined as the heat needed to increase the temperature of 1 lb of water by 1°F. The SI unit for heat is the Joule (J), and using this unit, we calculate the conversions of 1 cal = 4.1868 J and 1 Btu = 1055 J. The definition of 1 J, however, has nothing directly to do with heating water as 1 J is defined as the energy dissipated as heat with an electric current of 1 A passes through a 1 Ω resistor for 1 s. So, these units can be compared for something like heating water, but energy to make a pot of hot water does not relate well to other units of energy (and power) that we encounter or use.

Daily Energy Use and Carbon Emissions: Fundamentals and Applications for Students and Professionals, First Edition. Bruce E. Logan.

Your Car and Horsepower

Consider energy use in another familiar context: energy used by your car. Automobile manufacturers rate the car engine based on horsepower, but this is a unit of power, not energy. Also, the indicated horsepower is the *maximum power* of the engine and not the *average power* you will use when you drive. The dashboard in a car will provide information on how fast you are going, and more recently, the dashboard might also display the rate that you are using fuel (for example, in miles per gallon). However, neither of these sources of information tell you anything about the energy or power that you are using while driving the car. The power you use is based on the energy per some time interval. For example, if you were using a constant 20 hp in your car, after an hour, the energy use would be 5 hp times 1 h or 5 hp h. The hp h might be a unit you could calculate, but it is not a commonly used unit of energy.

What information do we need to figure out how much energy the car used? In the United States, the energy used is based on the work done by the car engine, which is evaluated in miles traveled based on the use of 1 gal of gas (mpg). For example, a car that goes 25 mi on 1 gal is 25 mpg. The mpg number is not energy used, but the energy used translated into something that we are familiar with, which in this case for a person in the United States is the volume of gas used rather than the energy. As we will see later, the fuel used by a car in Europe is reported in liters per 100 km, but this use is also not a direct report on energy consumption.

Fuels in Units of Btu

Gasoline stations do not advertise fuels based on energy content because they sell it by volume. However, a gallon of gasoline can be converted into units of energy. While the energy in a gallon of gas varies by octane and ethanol content, in general, it is considered that 1 gal of gasoline contains 127 MJ or 120 mBtu, where a capital M = mega = million and a lowercase m = 1000. If your car traveled 25 mi on 1 gal of gas, then to go 1 mi you used on average 5.1 MJ or 4.8 mBtu (see below for fuel units). I think we can all agree on these energy units are not particularly helpful in terms of understanding how much energy is in that gasoline relative to the horsepower of the engine or the energy we use for other things such as heating our home, powering electrical devices, or just energy in the food that we eat. Making these direct comparisons will require a using the same energy unit rather than all the different energy units listed in Table 2.1.

Table 2.1 Examples of some common energy units and conversions.

1 cal = 4.187 J
1 Cal = 1000 cal = 1 kcal = 4.184 KJ = 3.966 Btu
1 J = 1 Nm = 1 kgm^2/s^2 = 0.239 cal = 0.74 ft lb
1 kJ = 0.239 Cal = 0.9478 Btu
1 Btu = 1055 J = 0.252 kcal
1 kWh = 3.60 MJ = 3412 Btu
1 MWh = 3.6 GJ = 3.412 mmBtu
1 mmBtu = 10^6 Btu = 1.055 GJ
1 Quad = 10^{15} Btu = 1.055 EJ = 293.07 TWh
1 EJ = 10^9 GJ = 10^{18} J = 0.9478 quad
1 TWy = 31.5 EJ = 29.86 quad

Heating Your Home with Natural Gas, in CCF or Therms

If you use natural gas in your home, you may know what you paid for the gas you used, but do you ever pay attention to the units or compare that energy you used in the gas to your electric energy bill? Probably not! Your gas bill likely lists the gas price based on how many CCF (or therms) you used, which is fine for the vendor that sold you the gas based on the volume you used. But what is a CCF? It is a measure of the volume of gas, expressed in a mixture of metric and English units. The first C means centi, which is an SI unit used to indicate one hundred, while the following CF letters indicate cubic feet, which is an English units. The use of two letter C's that mean different things makes the CCF definition confusing. When you buy natural gas, it is monitored based on volume used, in hundreds of cubic feet. However, the density of a gas is a function of temperature, and so, it is not clear what pressure or temperature the CCF refers to in making an energy calculation. Also, like gasoline, the energy content of natural gas varies depending on the source of the gas and other factors, and thus, energy content can vary per CCF. To account for differences in energy content of natural gas, the volume of gas is converted into units called a therm, where 1 therm = 100,000 Btu = 100 mBTU. Therefore, if the energy content of the gas changes the volume of gas that you purchase could change but your bill would reflect eh amount of energy (therms) in the gas you purchased. Here, we will use 1 CCF = 103,600 Btu (US Energy Information Administration, 2019b) although the energy content of natural gas that you use could be different due to the makeup of the different gases (e.g. methane, ethane, and so forth).

Energy Use for Lightbulbs

The one energy unit many people think they know how to relate to the best is the energy for a lightbulb, as we have all had to replace lightbulbs (quite frequently as the old-style incandescent bulbs did not last very long). Lightbulbs are rated in power (Watts, W), not energy, so really you are likely to be more familiar with power than energy units for a lightbulb. For example, a 100 W lightbulb uses 100 W of power to produce light, but the energy used depends on how long the light was turned on.

A lightbulb rated at 100 W uses that amount of power for 1 h, for 24 h, or for any amount of time. As long as the lightbulb is on, it uses 100 W but the energy used requires knowing how long the lightbulb was on. To calculate the energy used, you use the time t that the lightbulb was on, with the energy calculated as $E = Pt$ and with the resulting units of energy in Watt hours (Wh). A 100 W lightbulb therefore uses 100 Wh in 1 h, 1000 Wh in 10 h, and 2400 Wh for 24 h. In units of watts (W) and kilowatt hours (kWh, or 1000 Wh), we can say a 100 W bulb uses the amounts of energy shown in Table 2.2.

Table 2.2 Comparison of power and energy at 1, 10, and 24 h for a 100 W (incandescent) lightbulb.

Basis	**1 h**	**10 h**	**24 h (1 d)**
Power	100 W	100 W	100 W
Energy	100 Wh	1000 Wh	2400 Wh
Energy	0.1 kWh	1 kWh	2.4 kWh

Table 2.3 Comparison of typical amounts of light and power produced by incandescent, CPF, and LED lightbulbs.

Incandescent lightbulbs		Alternative lightbulbs		
Light (Lu)	**Power (W)**	**Light (Lu)**	**Power, CFL (W)**	**Power, LED (W)**
450	40	400–500	8–12	6–7
800	60	650–850	13–18	7–10
1100	75	1000–1400	18–22	12–13
1600	100	1450–1700	23–30	14–20

The amount of electricity used for lighting in a house has substantially decreased over the past few decades due to the use of modern lightbulbs that use far less energy than older incandescent lamps. Most lightbulbs sold now are now LEDs (light-emitting diodes), with some older CFLs (compact fluorescent lamps) still available for purchase, or continuing to be used in homes and other locations due to their long lifetimes. LED and CFL bulbs are often rated as "60 W equivalent" or "100 W equivalent" based on the light they produce compared to the older and energy inefficient incandescent lightbulbs. This modern terminology is confusing as those W equivalent units for CFL and LED bulbs vary, and these newer lightbulbs do not actually consume that amount of power in W. Lightbulbs are more properly rated in units of lumens, abbreviated as lm, which indicates how much visible light is produced (light per area per time) by the bulbs. The old incandescent lightbulbs produced in lm and W for the lightbulb were as follows: 1600 lm (100 W), 1100 lm (75 W), 800 lm (60 W), and 450 lm (40 W). Thus, the new bulbs are rated based on producing about this much light for the older amounts of power consumed.

For the newer LED and CFL bulbs, there are ranges of lumens designated for a power equivalency to an incandescent bulb, so there is no absolute ratio between light produced and power for these lightbulbs (Table 2.3). You need to look at the writing on the side of the base of the lightbulb (in small print) or search for the information on the package to see how much power the bulbs consume to achieve a certain W equivalent claim. For example, one 60 W equivalent lightbulb uses 7 W to produce 840 lm, while another one uses 9 W to produce 800 lm, with 800 lm considered to be the standard for this lightbulb. The color spectrum of the light will impact power and the amount of overall light being used, with "daylight" bulbs generally producing more lumens for the power consumed compared to "soft light." LED lights outperform CFLs in terms of light produced relative to power consumed. There is no longer any need to use an old incandescent light (except if you want it as a heat source), so throw those out and replace them with modern LED lights to save money and energy. The important step in selecting an LED lightbulb is to carefully read the label so that you can choose the lightbulb with the lowest power that meets your lighting needs. And do not forget to turn off the lights when they are not needed.

Your home or apartment electricity use is likely billed based on units of kWh because all the things in your house use different amounts of power. The electricity provider does not bill you on the power of your devices (the energy you used per time), they bill you for the amount of energy you used. The sum of all the energy used is therefore added up over a period of time, which is usually 1 month. A typical house in the United States is about 900 kWh/mo or 30 kWh/d. To calculate the *average power* you would continuously use, you can divide this monthly use by 24 h/d to get an average power of 1.25 kW.

The Food You Eat in Calories

When you read the ingredients and energy labeling on a package in the United States, you get numbers in Calories (Cal). This is a bit confusing as 1 Cal $\neq$ 1 cal. Instead, a Calorie with a capital C is 1 kcal (1 Cal = 1 kcal), so eating 1000 C is really eating 1,000,000 calories. The energy in 1 Cal, the energy needed to heat 1000 g or 1 kg of water by 1 °C (versus 1 g of water for 1 cal).

The number of calories a person needs to eat to maintain their weight varies based on height, weight, activities, and general body metabolism. As an approximation for men and women of various heights, a reasonable estimate is that we should eat about 2000 Cal for our daily lifestyle. See more detailed information on estimating Calorie use in Chapter 3. As previously discussed, we can convert Cal to cal, Btu, J, and other units shown in Table 2.1. The most common alternative to Calories for food energy is MJ, where 1000 Cal = 1000 kcal = 4.18 MJ.

2.2 COMPARING DIFFERENT ENERGY UNITS USING kWh

All these forms of energy and power have created the use of a myriad of units, so if we want to compare them, what unit would we use? For many different energy comparisons, one preferred unit is kWh when it relates to the activities of one person. Alternatively, we will discuss in Chapter 3 a measure of energy based on a ratio of the energy use by a person to the energy in the food they eat every day, called the daily energy unit D. This unit of D is helpful as it ties our energy consumption to things that we do and can control within a reasonable range of values. The energy unit D will be the subject of Chapter 3 so here we can focus on kWh for understanding daily energy use. For very large-scale energy use, for example, for the whole United States, we will use units of exajoules (EJ).

Energy use reported in different units can be converted to kWh using the conversion factors in Table 2.2. If we convert food energy of 2000 Cal to kWh, we can see that a person needs to eat about 2.32 kWh of food every day, which is close to the energy needed to run a 100 W lightbulb all day (2.4 kWh/d) (Table 2.4). A gallon of gas (with 10% ethanol) contains 35.3 kWh (US Energy Information Administration, 2019c) or about the same amount of energy in the electricity used by an average house in the United States for one day of 30 kWh. We can also compare these numbers to the averages for energy used in the United States normalized to the population of 328.2 million

Table 2.4 Energy for different uses in various units.

Activity	Energy in typical units	Energy (kWh)
Heating 1 g of water 1°C	1 cal	1.16×10^{-6}
100 W lightbulb for 24 h	—	2.4
Daily food for a human	2000 Cal	2.32
Daily food for a horse	20,000 kcal	23
Average for a house/d	30 kWh	30
1 gal of gas	120,333 Btu	35.3
120 hp car for 24 h	2880 hp h	2150
Electricity use/d, total	0.101 quad	—
Electricity use/d/person	—	91
Total energy in USA/d	0.274 quad	—
Total energy in USA/d/person	—	245

Table 2.5 Energy content for 1 gal of three liquid fuels (gasoline, diesel, jet fuel) and natural gas for 1 CCF.

Fuel	Btu	MJ	kWh
Gasoline (1 gal)	120,333	127.0	35.3
Diesel or heating oil (1 gal)	137,381	144.9	40.3
Jet fuel (1 gal)	135,000	142.4	39.6
Natural gas (1 CCF)	103,600	109.3	30.3

Note that the energy equivalent is 1.16 CCF for 1 gal of gasoline.
References: Gasoline, diesel, and natural gas (US Energy Information Administration, 2019b), and jet fuel (US Energy Information Administration, 2016).
Source: Adapted from US Energy Information Administration, (2016) and US Energy Information Administration, (2019b).

(in 2019) for one day. For total energy use by a person, the average is 245 kWh/d, and for just electricity energy, it is 91 kWh/d. These values for the United States are based on 100.2 quads (quadrillion Btu) of energy and 37 quads of electricity used in 2019 in the Lawrence Livermore National Laboratory (LLNL) quad chart (Lawrence Livermore National Laboratory, 2019). However, as we will see below, these numbers overestimate primary energy use due to the way that renewable electricity is reported in terms of primary energy.

Energy in Fuels

The energy in fuels is given in a variety of units, most commonly Btu per gallon or barrel for liquid fuels, but sometimes, the energy in the fuel is given in units of MJ. For natural gas, typically the units are CCF (1 hundred cubic feet) or therm, as noted above. We can compare the energy in these fuels in kWh, which has a relatively narrow range of 35.3–40.3 kWh/gal. For natural gas, the energy equivalent for gasoline would be 1.16 CCF. Therefore, 1 CCF is used in Table 2.5 as it is nearly equivalent to the other fuels based on 1 gal.

Solar Energy

Another perspective on energy use is to examine the energy consumed relative to that produced by a photovoltaic panel, which is currently capable of 320 W or 0.32 kW of power, based on peak sunlight of 1000 W/m^2, or 1 kW/m^2, of panel area. However, since the sun does not shine all 24 h, the panel will not produce 7.9 kWh in one day. There are many different factors that impact the amount of kWh produced by a single panel at different locations, but the power produced in one day, averaged over the year, is usually calculated in terms of the number of hours that the panel would produce at its maximum power in a day (see Chapter 6). For example, in much of the Northeastern United States, the number of peak solar hours is 4.2 h, so the production of a 320 W panel would be 1.34 kWh. Based on this performance for the Northeastern United States, the average house using 30 kWh/d would need 22 of these photovoltaic panels to produce the *average* amount the energy used in one day. However, a system of 22 panels would produce excess energy relative to consumption during the day (when the sun is shining), but none at night, so an additional source of power would be needed when it is dark. That variation in production and use means you either need more panels and sufficient battery storage or an electrical grid that can accept excess power and provide power back to your home as needed throughout the night (or during insufficiently sunny days). Most houses are already connected to the electrical grid, and thus, it is usually possible to set up agreements with power companies so that they sell or buy your electricity throughout the year.

2.3 ENERGY USE IN THE US WITH A FOCUS ON CLIMATE CHANGE AND THE FUTURE

The energy consumption in the United States in 2019 was reported by the US Energy Information Agency (EIA) to be 100.2 quads, where 1 quad = 100 quadrillion Btu, equivalent to 29,370 terra Watt hours (TWh) of energy per year, or 3.35 TW of continuous power (quad is an abbreviation for quadrillion, or 10^{15}, and terra is an SI unit for 1 trillion, or 10^{12}). The EIA analyzes energy use in the United States on a monthly and annual basis in detailed reports that have now spanned decades. For many years, the US LLNL has used that data to prepare summaries in a format called a Sankey plot (Fig. 2.1) (Lawrence Livermore National Laboratory, 2019). Energy flows are separated into the main sources of energy for electricity production and direct uses that flow into four categories of residential, commercial, industrial, and transportation. The energy from these sources that went into useful work are called Energy Services, and energy consumed that did not accomplish a purpose are classified as rejected energy.

This energy analysis by LLNL for 2019 showed that 32.7 quad of energy in the source categories ended up as service energy, with 67.5 quad as rejected energy (overall efficiency of 32.6%). This energy efficiency is a bit lower than the average for fossil fuel power plants that have an energy efficiency of ~37.5% (32.8% for a coal power plant), primarily due to the large amount of energy used for transportation. Internal combustion engines are much less efficient in terms of conversion of energy into useful work, and so, the energy is less efficiently converted for use in transportation applications than in electricity production or uses in residential and industrial applications.

Unfortunately, there are several limitations to the way that the EIA assesses and reports energy use, including their definition of the word "primary" energy, the units used, a depiction of energy as "rejected," and a lack of perspective on presenting these data within the context of climate change. The need to reduce global CO_2 emissions is the single greatest challenge to the energy infrastructure in the coming decades. Thus, energy use data should be presented in the most direct way possible to enable decision-makers and policy analysts to view our historical and future energy use.

Use Energy Units that Make Sense Relative to Electricity Produced by Renewables

The United States remains tied to many English units, such as calories, Btu, and quad. The use of the quad unit for total energy use in the United States (Lawrence Livermore National Laboratory, 2019) makes it particularly difficult to understand our energy use as a quad is a fairly meaningless number to most people. However, the quad unit does provide for people in the US energy consumption that is currently a nice round number of ~100 (i.e. 100.2 quad in 2019) to describe energy use by different sectors, as the reported uses in quad are about the same number as percentages of the total energy.

It is advocated here that the United States and others should use energy units that are accepted standardized units of energy and not directly connected with combustion processes to heat water and produce electricity. A Btu is the energy needed to raise the temperature of 1 lb of water by 1°F and thus relies on a non-SI units of weight and temperature. Another common unit used by the International Energy Agency and the energy company BP is the Mtoe, which is defined 1 million

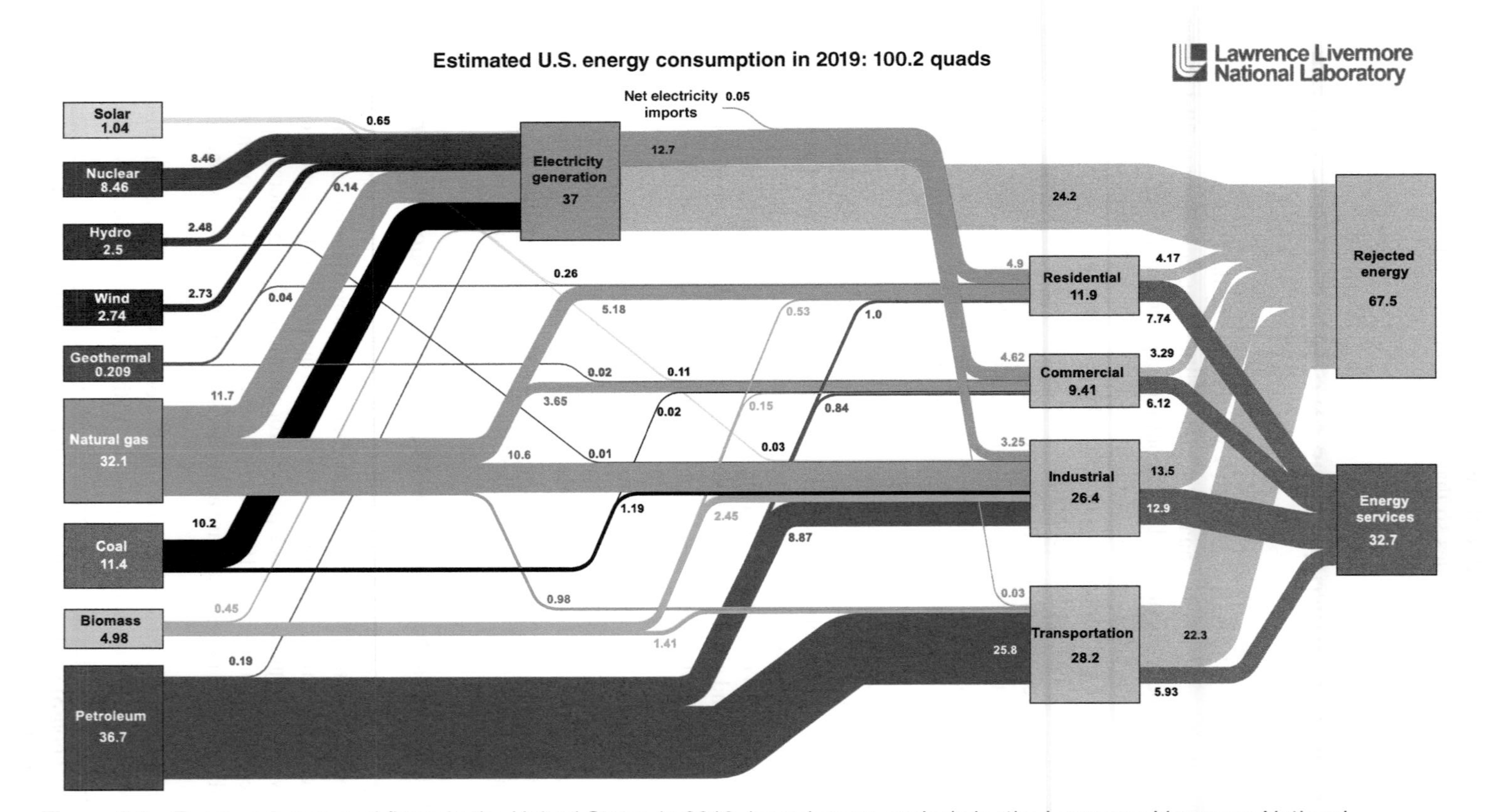

Figure 2.1 Energy sources and flows in the United States in 2019, based on an analysis by the Lawrence Livermore National Laboratory (LLNL) and the Department of Energy (Lawrence Livermore National Laboratory, 2019). The numbers indicate energy in units of quads. Numbers may not exactly sum due to rounding. *Source*: Estimated U.S. energy consumption in 2019: 100.2 quads, Lawrence Livermore National Laboratory. Reference link: https://flowcharts.llnl.gov/content/assets/docs/2019_United-States_Energy.pdf

1 Exajoule (EJ) = 0.9478 quad = 277.778 TWh
1.0551 EJ = 1 Quad = 10^{15} Btu = 293.071 TWh

Figure 2.2 Conversion factors for large-scale energy production. The recommended conventional units are EJ compared to Btu, which refer to heat, and TWh which are usually used only for electricity and not fuels. Another common unit by the EIA is bKWh, which is a billion kWh which is equal to 1 TWh.

toe, where 1 toe = 10^7 kcal or roughly the heat content of 1 tonne of oil (IEA, 2019). The use of Mtoe therefore anchors energy units to oil. For large-scale electricity production, another common unit is 1 billion kilowatt hours (1 TWh), but this is not usually used to describe the energy content of fuels even though the use of fuels and electricity are forms of energy consumption.

A more sensible energy unit, as used by some others (Holliday and Van Beurden, 2020), is the energy unit of a joule. One joule is defined using SI units as work done by a force of 1 N (Newton) acting through 1 m, or in terms of electrical energy, 1 J = 1 W s (Watt second) as the energy released in one second by a current of 1 ampere (A) through a resistance $R = 1\,\Omega$. Thus, a joule is a unit that can be used to easily relate energy in fuels and electricity, and it is not tied to heating water or combustion. Conventional units based on joules are megajoules (MJ), for example, describing the energy in the food you would eat every day (8.4 MJ/d = 2000 kcal) or exajoules (EJ, or 10^{18} J) for very large-scale energy calculations where 1 EJ = 0.948 quad = 2.39×10^7 Mtoe = 278 TWh. For energy consumption by the US in 2019, we therefore have 105.7 EJ for 2019 (Fig. 2.1). The annual energy use by the United States in units of EJ is a number that is conveniently also close to 100 and therefore useful in terms of thinking about different contributions to this total in percentages. Some convenient conversion factors related to EJ are emphasized in Figure 2.2.

Definition of Renewable, Carbon Neutral, and Fossil Fuels

Renewable energy sources are defined here as energy sources that are not derived from the use of any fuels. A fuel is generally considered as something that is oxidized to produce energy. Thus, wood, food, and coal are clearly fuels when used to produce heat energy. Wind, solar, geothermal, and hydropower can be used to do work or produce electricity, but there is no fuel consumed in the process (although certainly energy is captured). Therefore, renewables are considered here to be only these four energy sources.

Biomass is not categorized here as a "renewable," which is different classification than used by some others, for example, by the EIA and oil and other energy companies. Here, we define biomass and nuclear energy in the category of "carbon neutral" as fuels are used but there is no net release of CO_2 during the process of electricity generation. For biomass, the CO_2 in the fuel was obtained from the atmosphere by plants, and thus, when the fuel is consumed, the CO_2 is released back into the atmosphere and there is no net release of CO_2. For nuclear energy, the radioactive energy sources are not oxidized by combustion, but the decay of the uranium by nuclear fission provides heat, like a combustion process, to generate electricity. The term fuel is used here for the energy source for nuclear power generation because material must be extracted from the environment and used, and a waste is generated. However, CO_2 is not released, and therefore, this energy source is also classified as carbon neutral here.

Fossil fuels are energy sources that exist in the ground and that release large amounts of CO_2 into the atmosphere. The three main fossil fuels are coal, natural gas, and petroleum, with lesser amounts of other gases (for example propane) and minor sources of petroleum products from other processes than oil extraction.

Primary Energy Should Be Redefined for Renewables that Do Not Use Fuels

Many presentations of energy flows, such as that in Figure 2.1, define the energy going into the systems as "primary energy," with the outcomes as energy services or energy that was used for a stated purpose, and rejected energy or energy that did not contribute to a desired outcome. For example, a power plant that produces 0.375 EJ of electricity from 1 EJ of fuel would have an efficiency or service energy of 37.5% and thus would have rejected or wasted 62.5% of energy going into the process. While this concept of primary energy makes sense for fossil fuels that use combustion-based processes to convert fuels into electricity, it does not make sense for renewable energy such as solar, wind, hydro, and geothermal. The approach for renewables has been used to present the "primary energy" for these renewable energy sources on the same basis as fossil fuels, and so, the EIA and others artificially add in extra energy relative to the electrical energy produced.

The EIA reported that 2.88 EJ of wind energy in 2019 was used (Table 2.6) to produce 0.98 EJ of electricity based on back calculating primary energy assuming an average efficiency (37.5%) for a typical fossil fuel power plant. However, this primary energy was not the actual energy efficiency based on the amount of energy possible from wind turbines, but instead how much energy a fossil fuel plant would consume to produce that same electrical energy produced by wind turbines. Similarly, the efficiency of a solar panel to convert light energy to electricity is not related to energy efficiencies for fossil fuel plants, and thus, these primary energies that were reported do not reflect the actual energy conversion efficiencies of these systems. Reporting energy efficiencies for renewable energy technologies that do not consume fuels also is not useful as the sun will shine and release energy irrespective of its capture to produce electricity. Thus, inflating these numbers for renewables only places them in the context of the inefficiencies of a conventional fossil fuel power plants and not their true process efficiencies.

Here, we will define primary energy use for renewables to be the same as the electricity produced. Primary energy use based on the original data reported by the United States in units of quads,

Table 2.6 Primary energy consumed in the United States: original, including primary energy for renewables, and revised by assuming no primary energy for the renewables.

		Original primary energy		Revised primary energy	
Category	**Source**	**Quad**	**EJ**	**Quad**	**EJ**
Renewable	Solar	1.04	1.10	0.64	0.67
	Hydro	2.49	2.63	0.94	0.99
	Wind	2.73	2.88	1.03	1.08
	Geothermal	0.21	0.22	0.12	0.13
Carbon-neutral	Biomass	4.99	5.26	4.99	5.26
	Nuclear	8.46	8.93	8.46	8.93
Fossil fuels	Petroleum	36.72	38.74	36.72	38.74
	Coal	11.32	11.94	11.32	11.94
	Natural gas	32.10	33.87	32.10	33.87
Totals		100.1	105.6	96.3	101.6

Source: Adapted from US Energy Information Administration (2020b).

when converted to units of EJ, shows that 105.6 EJ of primary energy was used in 2019. However, when additional primary energy is removed for the four renewable energy sources (solar, hydro, wind, and geothermal), then only 101.6 EJ was actually used (Table 2.6). This change is only a 4.8% decrease due the relatively small amount of energy for these sources compared to the others. However, 4 EJ represents a tremendous amount of energy that was never used as a "fuel" in the traditional meaning of combustion-based processes.

Primary Fuel Use for Electricity Generation

Overall primary energy for electricity generation in the United States in 2019 was predominantly from three sources: natural gas, coal, and nuclear energy (Table 2.7). Unlike primary energy for all uses, petroleum was a very minor component of energy consumption in the United States for electricity generation.

The impact of the addition of primary energy for renewable fuels is shown in Figure 2.3 using the above data. In 2019, the reported energy used to produce electricity was 39.0 EJ, but when the additional primary energy (4 EJ) is subtracted out, there is 10% less energy consumed (35 EJ). Based on the revised primary energy values, renewable primary energy was 6.8% of total energy used for electricity production, compared to 16% using the original values. This reduction in primary energy consumption when shifting to renewables is important: As renewables are increasingly used, total energy consumption will decrease. Thus, all primary energy does not need to be replaced with the same amount of energy using renewables: Only the energy of actual electricity used would need to be produced with renewables.

Primary energy use and electricity consumption are reported differently by companies and agencies around the world. For example, the EIA reports a single number of 4401 kWh for electricity consumption in the United States for 2019, while the EIA reported 3956 TWh for the electric industry and 4118 kWh for total electricity generation for all sectors. As noted in the paragraph above, there are many sources and sinks which can impact this number.

Table 2.7 Primary energy for electricity generation in the United States in 2019: original, including primary energy for renewables, and revised by assuming no primary energy for the renewables.

		Primary electricity energy		
Category	**Source**	**Original (EJ)**	**Revised (EJ)**	**Electricity used (EJ)**
Renewable	Solar	0.69	0.26	0.26
	Hydro	2.62	0.98	0.98
	Wind	2.88	1.08	1.08
	Geothermal	0.15	0.06	0.06
Carbon-neutral	Biomass	0.47	0.47	0.10
	Nuclear	8.93	8.93	2.91
Fossil fuels	Petroleum	0.20	0.20	0.06
	Coal	10.74	10.74	3.45
	Natural gas	12.31	12.31	5.31
Totals		39.0	35.0	14.22

Electricity used is the electricity used that was produced by the different types of power plants.
Source: Adapted from US Energy Information Administration (2020b).

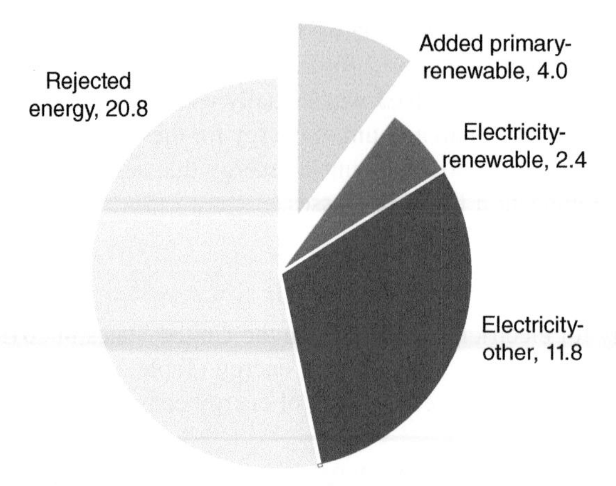

Figure 2.3 Primary energy (EJ) for electricity generation calculated for the United States in 2019. Primary energy separated into four categories based on electricity produced using renewables (solar, hydro, wind, and geothermal) or other fuels, the added energy for renewables based on the efficiency of low-efficiency power plants, and the rejected energy using other fuels. For comparisons with other low-efficiency fossil fuel plants. *Source*: Logan et al. (2020).

Other Sources and Losses for Electricity Generation

The items listed in Table 2.7 that sum to 14.22 EJ are the main sources of energy and consistent with those reported by LLNL, but they are not a complete listing of energy sources compared that are summarized in reports by the EIA, and they do not include losses after electricity generation at the power plant. There are two energy sources not included in that summary that are relevant to the electrical sector: electricity generation using other gases and other sources (Table 2.8). Other gases

Table 2.8 Additional electricity additions and losses relative to overall consumption in the United States in 2019.

Category	Electricity (EJ)
Produced by electrical sector	14.22
Electrical sector additions	
Other gases	0.016
Other sources	0.028
Subtotal	0.043
Other additions	
Commercial	0.535
Industrial	0.049
Net imports	0.140
Subtotal additions	0.72
Losses	
Pumped storage	0.019
Electrical sector direct uses	0.53
Transmission and distribution	0.94
Subtotal losses	1.47
Total with additions and losses	13.50

include blast furnace gas and other manufactured and waste gases derived from fossil fuels. Other sources include chemicals, hydrogen gas, batteries, purchased steam, pitch, sulfur, a few other miscellaneous technologies, and municipal solid waste from tire-derived fuels and non-biogenic sources. In total, these sources contribute only 0.043 EJ and are neglected here as they come from too many different sources to list, and overall, they provide a small contribution to total electrical power generation. The amount of energy in the category "other sources" is not explicitly by the EIA, and therefore, it was calculated from the difference of sum of electricity generated by specific sources and the total amount of electricity listed by the EIA.

There are other additions to electricity generation from outside the main power plants that include commercial and industrial electricity generation and imports. These three sources contribute an additional 0.72 EJ (Table 2.8). There are also losses due to consumption of electrical power at the power plants that produce electricity, so that this power is not available for retail sales. Energy losses for hydropower storage in 2019 were a total of 0.019 EJ, which was the difference between the energy needed to pump water into the storage reservoir and that recovered when the water was run back through the turbines to generate electricity. Transmission and distribution losses in the electrical grid are 0.94 EJ, or about 6.6% of the 14.22 EJ produced, or a loss of 7.0% of the 13.5 EJ available for retail sales. If electrical sector direct uses of 0.53 EJ are added to these losses, then overall losses would represent 1.47 EJ or 10.3%, without the other additions and losses. Transmission losses are very important when considering remote power generation relative to point of use applications, for example, for charging an electric vehicle. Overall, this 13.50 EJ was the amount of electricity sent for use by municipal, residential, industrial, and transportation uses in the United States with these combined additions and losses.

One of the most important electricity sources that is not included in Table 2.8 is small-scale solar electricity production which was 0.13 EJ in 2019. This small-scale solar production was equal to 50% of the commercially produced solar electricity in that year. Small-scale electricity production is not listed in Table 2.8 as only large-scale electricity sources are included in this summary. While small-scale electricity production is important relative to commercial solar electricity generation, it still accounts for <1% of overall electricity energy overall in the United States.

Efficiencies for Power Generation

Comparisons of primary energy consumption show that these are reported differently among the agencies and companies. Energy efficiencies of power plants that produce electricity are reported by the EIA based on surveys of fuel-based technologies. For non-combustible renewables, the energy efficiency is assumed to be 37.5% by the EIA, which is the average calculated for fossil fuel plants (Table 2.9). For BP, the efficiency for all renewables (non-combustible and biomass) is assumed to be 40.4%. Efficiencies were not readily apparent in the EIA report for biomass or in the BP report for fossil fuels. For calculations on fossil fuel power plant efficiencies, we will use the specific number in Table 2.9 for "primary/produced" values based on electricity use in the US or use 37.5% for a "general" case. For renewables, we use the 100% value or no additional primary energy for renewable energy sources.

Comparison of Primary Energy for Electricity Production with all Other Uses

The energy sources used for electricity generation are quite different than those used for other purposes, such as transportation and industry. Primary energy (with only the electrical energy for renewables) was 35.0 EJ, compared to 66.6 EJ for the other sources in 2019 (Fig. 2.4), and

Table 2.9 Estimated or reported energy efficiencies for electricity generation from primary fuels.

Source	EIA (%)	BP (%)	Produced/primary (%)
Renewables: solar, hydro, wind, geothermal	37.5	40.4	100
Biomass	—	40.4	21.9
Nuclear	32.6	—	32.6
Petroleum	30.8	—	31.3
Coal	32.6	—	32.2
Natural gas	43.6	—	43.1
Average	—	—	40.6

Values by the EIA are based reported data by source except for renewables. The EIA assumes non-combustible renewables of 37.5% while BP is 40.4%. The ratio of primary/produced, based on data in Table 2.7, is based on reported primary fuel use and delivered electricity and therefore is slightly different than plant efficiency.

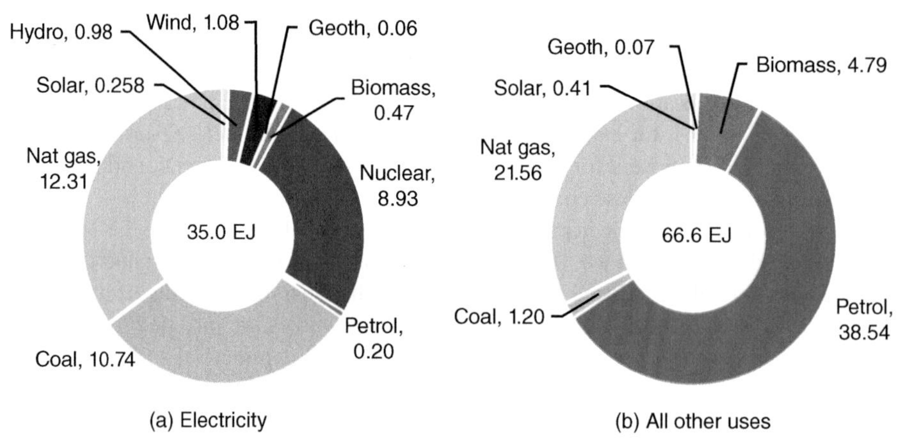

Figure 2.4 Primary energy sources for (a) electricity production and (b) all other uses for the United States in 2019. Nuclear and wind are used exclusively for electricity production, and there is negligible use of hydro for other uses. The total is 101.6 EJ. *Source*: Data from US Energy Information Administration (2020b).

petroleum completely dominated the other uses accounting for 38.54 EJ, followed by 21.56 for natural gas, and 4.79 by biomass. In comparison, little petroleum was used for electricity generation. For non-electricity uses, there was no appreciable energy consumption from nuclear, hydro, and wind energy as these energy sources were nearly completely used for electricity generation. Thus, a large shift of the United States away from fossil fuel consumption would require the largest changes in petroleum consumption.

Electricity Production

A total of 14.22 EJ of electricity were produced in 2019 for consumption (noting that some losses and additions are not included in this total as explained above). Comparisons of the relative amounts

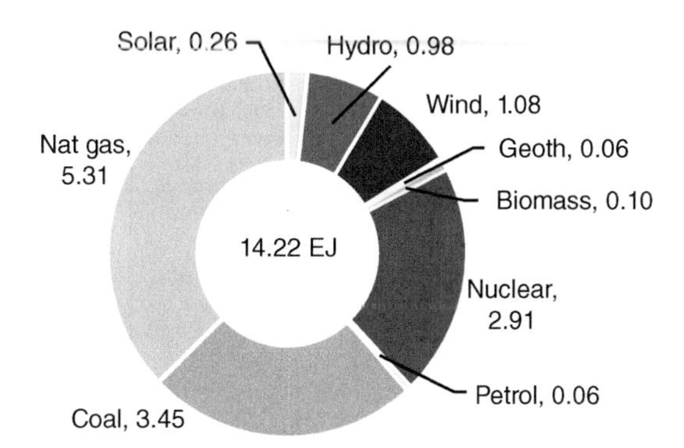

Figure 2.5 Electricity (EJ) used by the United States in 2019 based on energy source. These values do not reflect losses in transmission and do not include some other minor sources such as gases other than natural gas and hydropower storage losses. *Source*: Data from US Energy Information Administration (2020b).

of electricity produced in 2019 using different energy sources show that natural gas provided the largest source in the United States (5.31 EJ), followed by coal (3.45 EJ) and nuclear (2.91) (Fig. 2.5). Renewables provided 2.38 EJ of the total or 16.7% of the electricity. Nuclear and biomass provided 21.2% of this electricity. While the production of electricity using solar energy is rapidly increasing, it is still only about one-fourth as large as hydro or wind. The contributions of geothermal, biomass, and petroleum sources to electricity production were relatively minor. Overall, the electrical grid in the United States remains highly dependent on non-renewable energy sources.

Focus on the Amounts of Electricity and Fossil Fuels Used for Electricity Generation

From the perspective of climate change, energy use for electricity generation should be viewed in terms of two things: the amount of energy in electricity and the amount of fossil fuels consumed for that process. To decarbonize our energy infrastructure, fossil fuel use must be greatly reduced and one way to do that is to shift from combustion-based processes to electricity-dependent processes. However, if the electrical grid remains dominated by fossil fuel energy sources, then this switch will not have the desired outcome of reducing fossil fuel consumption. Therefore, it is important to reduce the amount of electricity consumed and eliminate as much as possible the use of fossil fuels in the electricity production.

To emphasize electricity use and fossil fuel consumption, a simple graphic such as the one shown in Figure 2.6 can be used. Electricity production is shown relative to the three main approaches relative to CO_2 emissions: renewables, CO_2-neutral (C-neutral) technologies, and fossil fuels. For these three areas, renewables were 2.37 EJ (16.7%), C-neutral technologies (biomass and nuclear) 3.02 EJ (21.2%), and fossil fuels made up the balance at 8.83 EJ (62.1%). Therefore, to eliminate CO_2 emissions due to fossil fuels, possible solutions are to: reduce electricity consumption to avoid the use of 8.83 EJ associated with fossil fuels (which does not seem likely); replace the 8.83 EJ derived from fossil fuels with renewables or C-neutral sources; reduce CO_2 emissions by switching to natural gas (a partial solution); or use carbon capture and sequestration technologies for all fossil fuel-related sources.

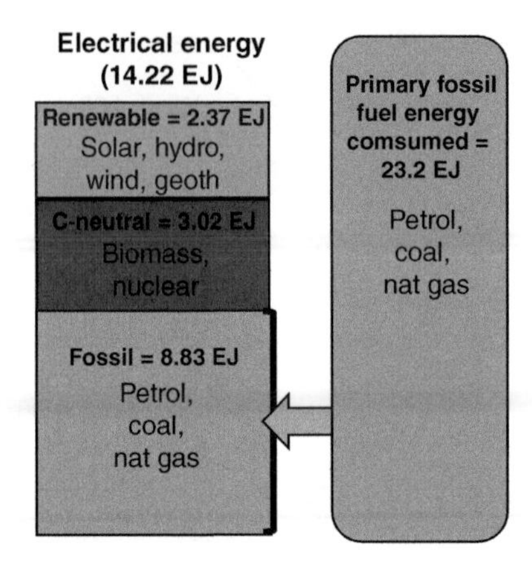

Figure 2.6 Electricity (14.22 EJ) produced for use in the United States in 2019 sorted into three categories (renewable, C-neutral, and fossil) compared to the total primary fossil fuel (PFF) energy used to produce the fossil fuel electricity (not to scale). *Source*: Reprinted with permission from Logan et al. (2020), American Chemical Society.

Replacing Coal with Natural Gas Can Rapidly Reduce CO_2 Emissions

Electricity generation using natural gas provides two advantages compared to coal. First, natural gas releases less CO_2 for the same amount of heat generated using coal. Natural gas combustion releases 50.4 metric tons of CO_2 per EJ (Mt/EJ), while coal (subbituminous) releases 1.8 times as much CO_2 for the same amount of energy (92.3 Mt/EJ) (US Energy Information Administration, 2020a). Therefore, if natural gas is used to produce steam the same way that coal is used for electricity production, then CO_2 emissions could nearly be cut in half. Second, electricity production using natural gas can be much more efficient than coal plants based on the energy used. The efficiency of a modern natural gas combined cycle plant (CCP) using commercially available technologies can reach 63.5% when operating at full load. In contrast, coal plants in 2019 had an average fuel efficiency of 32.6%, which was slightly less than that in 2010 (32.8%) due to reduced efficiencies with additional emissions controls not entirely offsetting improved process performance (US Energy Information Administration, 2020c).

Between 2010 and 2019 in the United States, there was an 8.2% reduction in annual CO_2 emissions from 5.59 to 5.13 Gt. While this decrease was helped by the growth of renewables for electricity used in 2019 (16.7%) compared to 2010 (9.3%), it was certainly aided in large part by a shift from coal to an increased use of natural gas. Between 2000 and 2010, there was a nearly equal use of coal and natural gas for all energy uses in the United States, but between 2010 and 2019, natural gas consumption increased by 55% while coal use fell by 47% (Fig. 2.7). Coal dominated as a fuel source for electricity production in 2000, but the use of natural gas has steadily increased since then, with its use for electricity generation surpassing that of coal in 2019.

In addition to the reduced use of coal during the past decade, the average efficiencies of natural gas electricity plants increased from 41.7% to 43.6% (US Energy Information Administration, 2020b), while coal plant efficiencies remained nearly constant (32.6% in 2019). These greater efficiencies of fuel to electricity conversion for natural gas resulted in 5.31 EJ of electricity being produced from natural gas compared to only 3.45 EJ using coal (Fig. 2.6) despite the nearly equal amounts of primary energy use in 2019.

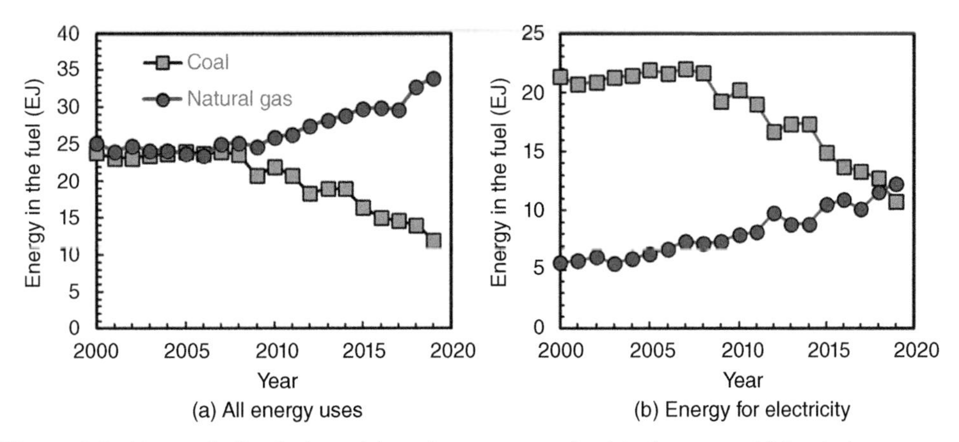

Figure 2.7 Energy in the fuel used for primary energy for (a) all uses and (b) only for electricity generation in the United States between 2010 and 2019. *Source*: Data from the US Energy Information Administration (2019a).

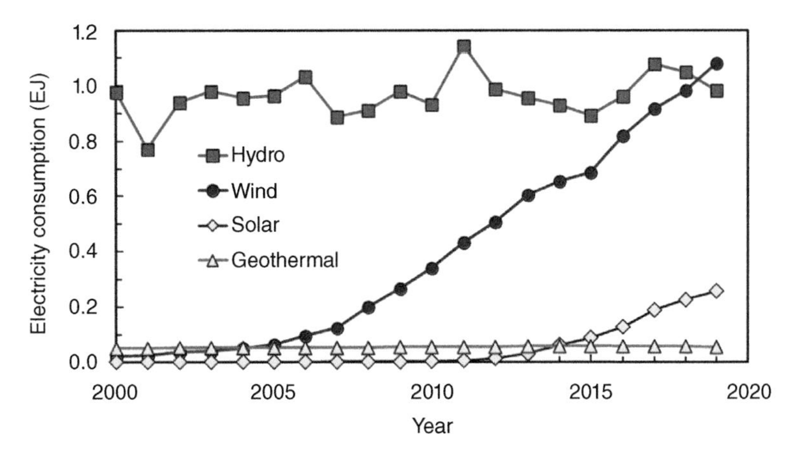

Figure 2.8 Change in renewable energy sources over time. *Source*: Data from US Energy Information Administration (2020b).

Further reductions in coal use for electricity production and increased use of natural gas CCP's can further aid in a reduction of CO_2 emissions. If electricity use was constant at the amount consumed in 2019 and coal was completely replaced by natural gas, then fossil fuel consumption would be reduced from 23.2 to 20.5 EJ. However, if natural gas plants were all operating at 63.5%, then that would reduce fossil fuel use further to 14.2 EJ and accomplish a 56% reduction in CO_2 emissions without any change in the amount of electricity generation. Therefore, natural gas will remain crucial to achieving reductions in fossil fuel consumption and CO_2 emissions in the coming years.

Electricity production using hydropower or geothermal sources has not changed appreciably changed in the past two decades (Fig. 2.8). However, wind energy started to increase rapidly in 2005 and solar began a similar climb in 2012. Solar electricity shown in Figure 2.8 is only for large-scale production and does not include that produced at smaller scales for homes. In 2019, small-scale solar production contributed 0.13 EJ of electricity, which combined with large-scale production of 0.69 EJ with be a 19% increase to 0.82 EJ. However, this amount is still relatively small compared to the 2.88 EJ in wind energy which has now reached a higher level than the historical average of hydropower over this period of 2.74 ± 0.21 EJ.

2.4 ENERGY USE AROUND THE WORLD

Energy use continues to increase globally with the Asia Pacific region having the largest energy consumption and greatest growth rate. Total energy use in the past decade grew 21% reaching 584 EJ of energy in 2019 (Fig. 2.9). Energy use grew by 39.2% for the Asia Pacific region during this 10-year period compared to 6.2% in North America. In the four years following the Paris agreement (2015–2019) to limit CO_2 emissions energy use increased by 2.4% in North America and 12.7% in the Asia Pacific Region.

One way to reduce CO_2 emissions from fossil fuels is to use electricity instead of fossil fuel with that electricity provided by renewables, such as wind, solar, hydro and geothermal, and carbon-neutral sources. However, much of the electricity produced globally is still derived from fossil fuels, and therefore, switching to electricity alone will not appreciably reduce CO_2 emissions (Fig. 2.10). Renewables and biomass energy sources provide 3.2 EJ of electricity in South and Central America, which is the highest percentage of electricity (67%) for all regions. The Asia Pacific region produces the largest amount electricity of 10.5 EJ from renewables and biomass, but due to the very high energy use for electricity in that region the overall percentage is lower (23.2%) than some other regions. Nuclear energy is used to produce about the same amount of electricity in Europe (3.3 EJ) as North America (3.5 EJ), but it is a lower percentage of total energy for North America (18%) than Europe 24%. Coal dominates energy sources in the Asia Pacific region, while petroleum is only substantially used for electricity generation in the Middle East countries. Natural gas is an important fuel for electricity production in many of these regions, and it is a preferred as an energy source to coal for electricity production due to its lower CO_2 emissions for equal energy content and the much higher efficiencies possible with natural gas compared to coal plants.

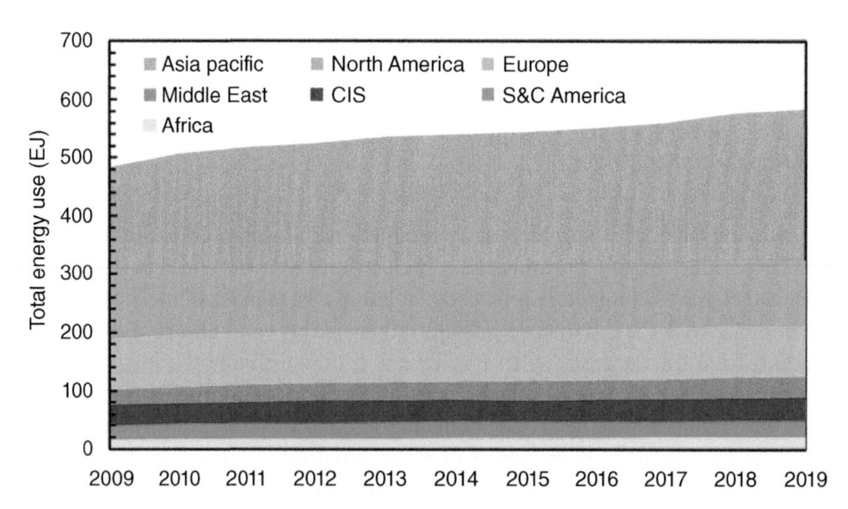

Figure 2.9 Total energy use between 2009 and 2019 separated into different regions. CIS is the Commonwealth of Independent States (including Russia). Renewable energy sources included in these totals do not include a correction for the 40.4% addition to increase primary energy for these sources. *Source*: Data from Dudley (2019).

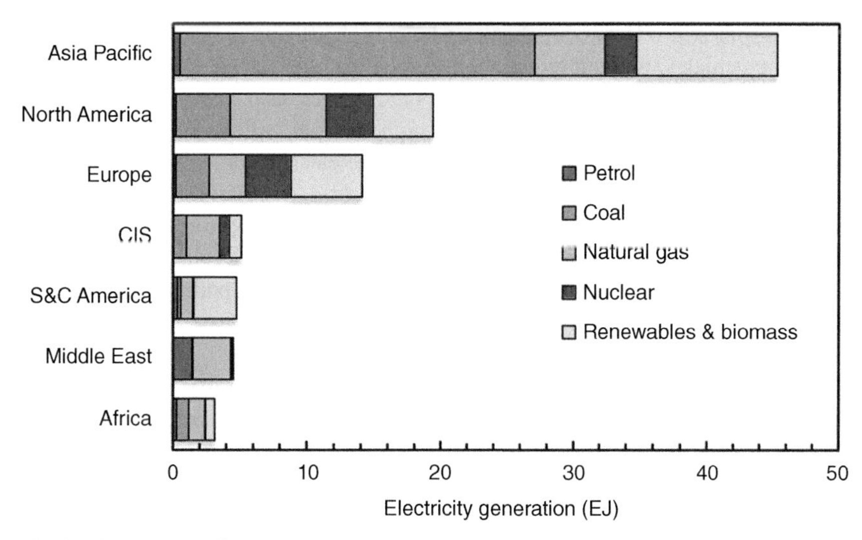

Figure 2.10 Energy use for electricity generation in 2019 separated into different regions by types of energy sources. Renewables include non-combustion sources (hydropower, solar, wind, and geothermal) assuming a 40.4% energy efficiency and carbon-neutral biomass with various efficiencies. CIS is the Commonwealth of Independent States (including Russia). *Source*: Data from (Dudley, 2019).

2.5 NEXT STEPS

In the United States, energy use grew a modest 1.8% from between 2008 and 2018, but in 2018, energy use by the United States accounted for 16.7% of total world energy use (Dudley, 2019). Energy consumption is growing faster globally than in the United States, with a growth rate of 2.9% in 2018 which was nearly double the 10-year average of 1.5% per year and the fastest growth rate since 2010. Such growth is clearly in the wrong direction relative to concerns about climate change. Energy consumption needs to decrease, and a greater percentage of energy consumption needs to obtained from renewable and carbon-neutral sources. If we consider that ~34% of energy use in the United States is for electricity generation, even if that electricity became 100% renewable energy then we would still need to address the fact that 63% of our energy consumption for other uses would be largely dependent on fossil fuels. How can we make the conversion from fossil to renewable energy?

- **Waste less energy in making electricity**: In 2019, the United States produced 14.22 EJ of electricity but consumed 35 EJ of energy to do that, wasting 20.8 EJ of raw energy. Completely eliminating coal use in favor of natural gas and then increasing the use of renewables will substantially reduce the amount of wasted energy.
- **Address energy inefficiencies in transportation**: A total of 27.2 EJ (25.8 quads) of energy from petroleum sources are used for transportation, with a small amount of other sources providing a total 29.8 EJ of energy used for transportation (Fig. 2.1). However, the actual energy used from this is only 6.26 EJ or an efficiency of only 21%.

Energy efficiency can greatly impact the amount of fuel use and carbon emissions, and therefore, improvements in efficiencies can reduce energy consumption especially for petroleum-based fuels used for transportation. However, changing from high efficiency internal combustion engine vehicles to electric vehicles, without reducing the dependency of the electrical grid on fossil fuels, does not necessarily improve energy consumption or CO_2 emissions. For example, see Chapter 9 on comparisons of energy efficiencies of cars with high mile per gallon (mpg) ratings compared to electric cars with low equivalent mpg (mpg-e). Shifting to electric cars with *renewable electricity* produced by solar panels on your own home, rather than grid electricity that currently contains a high proportion of fossil fuels, is one clear and clean path for reducing overall fossil fuel consumption that you can directly control. Similarly, a shift to hydrogen fuel cell vehicles with H_2 made from water electrolysis and renewable energy also provides a suitable path for reducing our carbon footprint.

2.6 HOW MUCH ENERGY SHOULD WE USE?

The two questions that could be asked, that do not have the same answer, are "How much energy *will we use* in the future" and "How much energy *should we use* in the future"? In thinking about predicting the future, we should first consider how well have we previously predicted energy use. In 2006, it was estimated based on population growth and energy use that energy use would increase by 94% from 426 EJ in 2001 to 820 EJ in 2050 for a global population of 9.4 billion (Lewis and Nocera, 2006). By 2017, global energy use had already reached 616 EJ or an increase in 45% after only 16 years with 23 years more to go. Thus, the projected increase was likely an underestimate if energy use continues along the historical path of 2001–2017. In 2019, the EIA estimated that by 2050, we would use 961 EJ per year or 17.2% more than the estimate made for 2050 back in 2001. Clearly, both predictions provided some guidance but they were not very accurate projections of future energy use. We cannot really know how much energy we will use in 2050 but we can expect that globally, people will want to use more energy than we use today, particularly as the population of the planet continues to increase. Unless there are large changes in energy use, it seems probable that global use could increase by ~50% compared to today.

How much energy should we use? The "2000 W" society, founded in 1998 by the Swiss Federal Institute of Technology in Zurich (ETH Zurich), suggested that power used per person should be on average 2000 W, or 2 kW, per person. This seems a reasonable goal for 2050 for already industrialized countries if it could be done without lowering the standard of living. To put that number in perspective, we can compare it to some other numbers based on primary energy use for Switzerland, the United States, and the world as shown in Table 2.10. The global predictions are made on the basis of the data from an EIA report for 2050 (Birol, 2019), and the existing energy use data for Switzerland and the United States are from a BP report (Dudley, 2019). From this comparison, we can see that global averages may approach 2 kW, but industrialized countries such as Switzerland are using much more energy. The United States, which has a very high per-capita energy use, consumes about 5 times that of the goal of 2 kW. From the perspective of needing to reduce CO_2 emissions to zero by 2050, combined with an energy use substantially less than that currently used, reversing climate change seems to be an enormous and un-precedented challenge for the world.

The 2000-W Society further suggests that 1500 W of the 2000 W, or 1.5 kW, should be obtained from renewable energy sources, with 0.5 kW therefore allowed from fossil fuels (equivalent to ~1 ton of CO_2 per person per year). However, it is not clear whether the amount of "renewable" energy for electricity would be calculated based on primary energy or the energy in the electricity (or other sources such as biofuels) that was used. As discussed in Section 2.3, added energy should

Table 2.10 Examples of power used (continuous) per person in kW, or the total annual energy use (EJ/y), for countries or the world in different years.

Case	kW	EJ/y
Goal for Swiss citizens in 2050	2	–
Global average for 7.6 billion people (2017)	2.6	623
Global predicted for 9.4 billion people (2050)	3.3	978
For 8.593 million people in Switzerland (2019)	4.24	1.13
For 328.2 million people in the US (2019)	10.2	105.6

Source: Data from Dudley (2019).

not be included for renewables when making comparisons to other energy sources as the actual energy use becomes distorted. In Table 2.7, for example, it was claimed by the United States that production of 0.26 EJ of electricity required 0.69 EJ of primary energy. Therefore, do we say that the energy use per person for solar energy based on primary energy (0.69 EJ) or the actual electricity used (0.26 EJ)? It is advocated here that getting to 1.5 kW of renewable energy should be based on the actual renewable energy consumed and not the inflated energy use based on a fossil fuel power plant.

For the Swiss citizen, the total amount of energy used in 2019 was 4.24 kW, with 1.17 kW (27.4%) electricity provided by hydropower, 0.79 kW (18.6%) from nuclear, and 0.15 (3.5%) from renewables. Therefore, a total of 2.11 kW or ~50% are from renewable and C-neutral energy sources and the balance (2.13 kW) from fossil fuels. However, these data include "added in" primary energy for renewables and hydropower sources for fossil fuel plant, which was assumed to be 40.3% in 2019 based on the BP report (Dudley, 2019). If that additional energy is removed, then the total drops to 3.45 kW, with ~39% of the total energy renewable or carbon neutral (1.8% renewable, 13.7% hydro, and 23.1% nuclear).

The inflation of primary energy for renewables was also included for the values given in Table 2.10 for the United States from Dudley (2019), so 10.2 kW for people in the United States in 2019 would be reduced to 9.82 kW when this additional energy was removed (101.6 EJ/y). With this revised energy use, the amount of renewable and C-neutral energy is 16.8% (17.1 kW) of the total, which is quite a bit smaller percentage of these energy sources than that of Switzerland (39%), but in both cases, the total energy use is much less than the goal of 0.5 kW per person. These calculations highlight the challenges of achieving the goals of the 2000-W society even for countries such as Switzerland that have relatively bountiful supplies of hydropower.

References

BIROL, F. 2019. *The future of hydrogen: seizing today's opportunities*. ISSN: 2072-5302, ISBN: 978-92-64-28230-8, https://www.iea.org/reports/the-future-of-hydrogen.

DUDLEY, B. 2019. BP Statistical Review of World Energy 2019, 68 ed. Available: https://www.bp.com/content/dam/bp/business-sites/en/global/corporate/pdfs/energy-economics/statistical-review/bp-stats-review-2019-full-report.pdf. [Accessed April 16 2021].

HOLLIDAY, C. & VAN BEURDEN, B. 2020. *Shell energy transition report* [Online]. Shell. Available: https://www.shell.com/energy-and-innovation/the-energy-future/shell-energy-transition-report/_jcr_content/par/toptasks.stream/1524757699226/3f2ad7f01e2181c302cdc453c5642c77acb48ca3/web-shell-energy-transition-report.pdf [Accessed August 7 2020].

IEA. 2019. *Projections: Energy policies of IEA countries 2019 edition* [Online]. International Energy Agency. Available: https://webstore.iea.org/ods-projections [Accessed August 7 2020].

LAWRENCE LIVERMORE NATIONAL LABORATORY. 2019. *Estimated U.S. energy consumption in 2019: 100.2 quads* [Online]. Available: https://flowcharts.llnl.gov/content/assets/docs/2019_United-States_Energy.pdf [Accessed May 9, 2020].

LEWIS, N. S. & NOCERA, D. G. 2006. Powering the planet: Chemical challenges in solar energy utilization. *Proceedings of the National Academy of Sciences of the United States of America*, 103, 15729–15735.

LOGAN, B. E., ROSSI, R., BAEK, G., SHI, L., O'CONNOR, J. & PENG, W. 2020. Energy use for electricity generation requires an assessment more directly relevant to climate change. *ACS Energy Letters*, 5, 3514–3517.

US ENERGY INFORMATION ADMINISTRATION. 2016. Annual Energy Outlook 2016, Appendix G.

US ENERGY INFORMATION ADMINISTRATION. 2019a. *U.S. electricity generation from renewables surpassed coal in April* [Online]. Available: https://www.eia.gov/todayinenergy/detail.php?id=39992 [Accessed April 22 2020].

US ENERGY INFORMATION ADMINISTRATION. 2019b. *Units and calculators explained* [Online]. Available: https://www.eia.gov/energyexplained/units-and-calculators/ [Accessed May 8, 2020].

US ENERGY INFORMATION ADMINISTRATION. 2019c. *Units and calculators explained – Energy conversion calculators* [Online]. Available: https://www.eia.gov/energyexplained/units-and-calculators/energy-conversion-calculators.php [Accessed May 9, 2020].

US ENERGY INFORMATION ADMINISTRATION. 2020a. *How much carbon dioxide is produced when different fuels are burned?* [Online]. Available: https://www.eia.gov/tools/faqs/faq.php?id=73&t=11 [Accessed August 7 2020].

US ENERGY INFORMATION ADMINISTRATION. 2020b. *July 2020 monthly energy review* [Online]. Available: https://www.eia.gov/totalenergy/data/monthly/pdf/mer.pdf [Accessed August 7 2020].

US ENERGY INFORMATION ADMINISTRATION. 2020c. *Natural gas-fired electricity conversion efficiency grows as coal remains stable* [Online]. Available: https://www.eia.gov/todayinenergy/detail.php?id=32572#:~:text=Over%20time%2C%20as%20more%20combined,rate%20of%207%2C340%20Btu%2FkWh [Accessed July 24 2020].

CHAPTER 3

DAILY ENERGY UNIT D

3.1 DEFINING THE DAILY ENERGY UNIT D

To reduce our energy use, we need to be able to relate the quantity of energy that we use to something we know. There are probably few things in our lives where we know how much energy we use in units of kilowatt hour (kWh), megajoule (MJ), or Btu, so it is quite difficult to relate our own energy use using these different units. In Chapter 2, we examined energy use using different fuels by converting the original units into the same units, for example from MJ to kWh. This approach has the advantage of putting these different energy uses into the same "language." However, units of kWh or MJ are still quite abstract as it is difficult to relate kWh, for example, to something we have direct experience with every day. The one thing that we do know is that we need energy and so we get that in the food we eat, and everyone around the world needs to eat, so food energy is the one common energy experience that we all share.

Let us assume that on average a person needs to eat around 2000 Cal a day, although that number varies quite a bit as discussed below. Then, we define 2000 Cal has the unit of one daily energy unit D. Using this, we can express other energy units in terms of daily use based on energy for our home, car, and other activities that allows us to compare those numbers to the baseline of energy use based only on food energy. As shown in Example 3.1, by defining daily energy use as 2000 Cal/d = 1 D, we also obtain the result that 2.32 kWh/d = 1 D.

Example 3.1

(a) Convert 2000 Cal/d per person to units of kWh/d and the daily energy unit D. (b) Since a unit of energy per time is power, convert kWh/d to Watts.

(a) The unit Cal is really kilocalories (kcal), so we can say 2000 Cal = 2000 kcal, and then, we can units of kcal to kWh using 860 kcal/kWh as:

$$2000\,\text{Cal}\,\frac{1\,\text{kcal}}{1\,\text{Cal}}\,\frac{1\,\text{kWh}}{860\,\text{kcal}} = 2.32\,\text{kWh}$$

Daily Energy Use and Carbon Emissions: Fundamentals and Applications for Students and Professionals,
First Edition. Bruce E. Logan.

Therefore, we have the relationship that:

$$2000\,\frac{\text{Cal}}{\text{d cap}} = 2.32\frac{\text{kWh}}{\text{d cap}} = 1\,\text{D}$$

(b) We can convert these units to Watts, using 24 h in a day, as

$$2.32\frac{\text{kWh}}{\text{d cap}}\,\frac{1\,\text{d}}{24\,\text{h}}\,\frac{10^3\,\text{W}}{\text{kW}} = 97\,\frac{\text{W}}{\text{cap}}$$

We have calculated that 1 D = 2.32 kWh/d cap = 97 W per person, or about the power used to run a 100 W lightbulb. We can use these relationships for calculations to convert other energy units into the daily energy unit D.

Is 2000 Cal/day the only reasonable baseline for food a person eats in a day? The answer is no, there are other values are possible amounts of Calories you should eat every day that varies based on height, weight, age, level of activity, and your own particular physiology (Brazier, 2018). Google how much energy you need to eat every day and you will find a range of numbers. One often quoted number for a person in the United States is 2500 Cal/d. If you look further, you can see that using that number is more typical for a male in the United States than a female. It is estimated that a 30-y-old male uses 2400 Cal/d if sedentary, and 2600 Cal/d if moderately active. However, a 30-y-old female would use 1800–2000 Cal/d for that same range of sedentary to moderately active lifestyle (Burwell and Vilsack, 2020).

To develop a better sense of Calorie needs by a person, a more useful approach is to look at calculating Calories needed based on height, weight, and level of activity for a man versus a woman. The analysis starts with the baseline measured respiration rate (BMR), which is the baseline energy needed to fuel your body. The Calories needed to sustain you can increase based on your activities. One equation used to calculate your energy use with different activity levels is (Brazier, 2018):

$$E = 10\,W + 6.25\,H - 5\,A + G \tag{3.1}$$

where W (kg) is the weight, H (cm) height, A (y) the age, and G is the gender adjustment constant which is $G = 5$ for a man and $G = -161$ for a woman. For example, for 30-y-old men and women of average height and weight in the United States, the Calories for a relatively sedentary (office) life is 2220 for the man and 1710 for the woman, or about 1970 Cal/d on average (Table 3.1). With

Table 3.1 Calories needed for an average American (Man, 90 kg, 175 cm tall; Woman, 78 kg, 153 cm tall) about 30 y old.

Activity level	**Multiplier**	**Calories (kcal)/d**		
		Man	**Women**	**Average**
BMR	1.00	1850	1430	1640
Sedentary	1.20	2220	1710	1970
Slightly active	1.375	2540	1960	2250
Moderately active	1.55	2870	2210	2540
Active	1.725	3190	2460	2830

Numbers rounded to tens place; averages may differ due to rounding.
Source: Based on Brazier (2018).

Table 3.2 Calories needed for people on average in China (Man, 66 kg, 167 cm tall; Woman, 57 kg, 156 cm tall) aged about 30 y old.

Activity level	Multiplier	Calories (kcal)/d		
		Man	Women	Average
BMR	1.00	1560	1240	1400
Sedentary	1.20	1870	1480	1680
Slightly active	1.375	2150	1700	1920
Moderately active	1.55	2420	1920	2170
Active	1.725	2690	2130	2410

Numbers rounded to tens place; averages may differ due to rounding.
Source: Based on Brazier 2018 and Chinacities.com (2015).

a slightly active lifestyle, this increases to 2250 Cal/d on average for men and women. A slightly active lifestyle could include walking 3 mi a day, which adds 300 Cal based on using 100 Cal/mi.

There are also differences on average heights and weights of people in different countries. For example, in China, the average male is 66 kg in weight and 167 cm tall, compared to 90 kg and 175 cm for an American. Thus, Calorie needs would be lower on average for a man in China than in the United States. The average for a sedentary lifestyle decreases to an average of 1680 Cal/d for men and women in average in China (Table 3.2) compared to 1970 for the American citizen.

These differences in Calorie needs in Tables 3.1 and 3.2 are due to height and weight differences for individuals that are the same age (30 y old) and having these averages for weight and height. However, Calorie needs increase with weight when people have the same height. The body mass index (BMI) is a measure of body fat based on weight and height, and it applies to adult men and women. People with a BMI of 18.5–24.9 are considered to have normal weight for their height, while overweight people would have a BMI of 25–29.9 and be considered obese with a BMI >30. An average American male, given the data in Table 3.1, would have a BMI of 29.1. If an American had a normal BMI of 22 (US National Heart, Lung, and Blood Institute, 2020) for their height (167 cm) they would weigh 68 kg (150 lb) and have an average requirement of 2080 Cal/d for a sedentary lifestyle. This weight is more like that of an average male in China who weighs 66 kg, which explains most of the differences in the calculated Calorie uses between the above two tables.

There are additional factors which impact average food use but the most important one is food waste. On average, approximately 1/3 of all food is wasted in the United States and other industrialized countries (Wilkinson, 2020). Therefore, a listing of average food consumed by a person will vary to a much greater extent than Calorie needs. For example, if the average person in the United States needs to eat 2500 Cal, then the amount of food produced for that person could be the average for a US citizen is 3750 Cal/d (1/3 wasted) to 5000 Cal/d (1/2 wasted). The energy used to provide food (and lost to food waste) is addressed further in Chapter 8. The daily energy unit used here of 1 D = 2000 Cal is therefore chosen as it the energy needs for an average US person with an affluent and sedentary lifestyle, recognizing that many Americans tend to consume around 2500 Cal/d = 1.35 D, and thus are overweight, and overall could consume as much as 2.5 D = 5000 Cal/d due to a combination of overeating and a large amount of food waste.

3.2 EXAMPLES USING D

The idea of calculating energy use based on what we use per day is to reduce energy consumption by the whole country to eliminate (as much as possible) CO_2 emissions from fossil fuels. First, we consider the average energy use by a person in the United States in 2019 based on a population of 7.6 billion people, with an energy use of 101.6 EJ. Let us convert that total energy use per person to kWh per day and then to units of D as shown in Example 3.2.

Example 3.2

The energy use for the United States in 2019, with added energy removed for renewables, was 101.6 EJ. Convert this energy use into units of (a) terra watts of power, (b) kWh per year, and (c) units of D based on a population of 328.2 million people.

(a) Based on a unit conversion of 1 EJ = 277.8 TWh, and then converting into continuous energy, we have

$$101.6\frac{\text{EJ}}{\text{y}}\,\frac{277.8\,\text{TWh}}{\text{EJ}}\,\frac{1\,\text{y}}{365\,\text{d}}\,\frac{\text{d}}{24\,\text{h}} = 3.22\,\text{TW}$$

(b) We calculate energy in units of kWh as

$$101.6\frac{\text{EJ}}{\text{y}}\,\frac{277.8\,\text{TWh}}{\text{EJ}}\,\frac{10^9\,\text{kWh}}{\text{TWh}} = 2.82\times 10^{13}\,\frac{\text{kWh}}{\text{y}}$$

This is such a large number that we have nothing could really relate this to in our lives, so that is why we try to normalize this energy use per person.

(c) By using D values, we can accomplish both a reduction in the size of the number and relate it to the population. Converting kWh to D, we have

$$2.82\times 10^{13}\,\frac{\text{kWh}}{\text{y}}\,\frac{1\,\text{y}}{365\,\text{d}}\,\frac{1\text{D}}{\frac{2.32\,\text{kWh}}{\text{d cap}}}\,\frac{1}{328.2\times 10^6\,\text{cap}} = 101.5\,\text{D}$$

Note that the number is calculated as 101.6 D if starting with 101.6 EJ as the different decimal place in part c is due to rounding.

From Example 3.2, we see that the energy use in 2019 was about 101.6 D. This result is interesting because the total energy use in EJ (101.6 EJ) is the same number as D. This is just a coincidence that results from the number of people in the United States, unit conversions, and days in a year. We will see a difference in D values based on energy use for different sized populations. Primary energy consumption for electricity was 35 EJ, or about 34% (35/101.6 × 100) of energy use, resulting in 35 D. The electricity produced (14.22 EJ) is also the same number in units of D (i.e. 14.2 D). These numbers are summarized in Table 3.3, along with some other energy uses in terms of kWh and D to show how energy use varies for these different situations.

Electricity Use for a Home in the United States

The average home in the US uses about 900 kWh/mo, or 30 kWh/d. That energy use translates to 12.9 D for the house, but on average 2.52 people live in a single home. Thus, the daily energy unit per person is on average 5.1 D, or 6.5 D with two people in the home, or 4.3 D if three people live there. This amount of electrical energy used for a home varies quite a lot in the United States (50

Table 3.3 Energy for different uses converted from kWh per day to units of D.

Activity	Energy per day (kWh/d)	D
Daily food for a human (2000 Cal)	2.32	1
100 W lightbulb for 24 h	2.4	1.03
Daily food for a horse	23	10
Average daily electricity for a house	30	12.9
With 2.52 people per house	11.9	5.1
Average daily natural gas heating for a house	73	31
With 2.52 people per house	30	12.4
1 gallon of gas	35.3	15.2
Electricity/person/d in USA	91	39
Total energy/person/d in USA	245	102

states and Washington DC), and the additional use of oil or natural gas for heating for many houses is likely not included in the electricity use numbers. Louisiana has the highest electricity use of 41.9 kWh/d, which would translate to 7.2 D for 2.52 people (Electric Choice, 2020). Louisiana also has one of the lowest electricity costs in the United States. Maine and Hawaii have the lowest per capita uses of electricity, and some of the highest electricity costs (Hawaii has the highest cost), with 18 kWh/d for Maine and 16.9 kWh/d for Hawaii. The electrical power goes into heating a home varies with location and weather, but for similar conditions also by the method of heating such as heat pumps, electrical baseboard, or natural gas.

Energy Use for Heating with Natural Gas

A house that is 2200 ft^2 that heats using natural gas uses the equivalent of 2190 kWh/mo of energy, at a cost of about \$100 per month (Energy Services Group, 2020). This is 31.1 D that is used for the home, or 12.3 D assuming 2.52 people live in that home. A comparison of the electricity use of 5.1 D and energy use with natural gas at 12.3 D shows that a much greater percentage of energy for your home could be due to energy from natural gas consumption. However, electricity is produced at a power plant with fossil fuels has about a 37.5% production efficiency, and a 93.4% transport efficiency (6.6% loss in electricity transmission and distribution), or 35.0% overall efficiency, so that the 5.1 D of electricity delivered used is about 14.6 D of primary energy (discussed further below). The average given for electricity is based on some homes that use only electricity for heating, and other homes that use primarily gas for heating. The final energy use for the sum of these two items is therefore difficult to predict. There is additional information on energy use for heating a home in Chapter 9.

Example 3.3

If a home uses 2.4 CCF (see Section 2.1) a day of natural gas for heating, (a) how much energy is used every day (kWh/d)? (b) Calculate the daily energy unit D assuming three people live in the home.

(a) To make this calculation, we need to know how much energy is in 1 CCF of natural gas. The energy content of natural gas varies, but on average is about 100,000 Btu. Gas companies report the energy content in units of Therm based on the actual heating value of the gas. Here, we will use 1 CCF = 103,600 Btu based on the Energy Information Agency (EIA) website for average in the United States. Thus, we have for kWh/d, which is power used,

$$1\ \text{CCF}\ \frac{1.036 \times 10^5\ \text{Btu}}{\text{CCF}}\ \frac{1\ \text{kWh}}{3413\ \text{Btu}} = 30.3\ \text{kWh}$$

$$P = 2.4\frac{\text{CCF}}{\text{d}}\ 30.3\frac{\text{kWh}}{\text{CCF}} = 72.7\ \frac{\text{kWh}}{\text{d}}$$

(b) To calculate D, we use the conversion 2.32 kWh/d = 1 D, and the number of people that is assumed to be 3, or

$$\text{D} = 72.7\frac{\text{kWh}}{\text{d}}\ \frac{1}{3\ \text{cap}}\ \frac{\text{D}}{2.32\ \frac{\text{kWh}}{\text{d cap}}} = 10.4\ \text{D}$$

Energy Use in the Home

Heating can be a large percentage of the electricity used in many homes, but the energy consumed for cooling and certain appliances in your home can quickly increase energy use in the home. Based on the US Energy Information Administration, energy use for cooling (air conditioning) is now slightly exceeding energy used for heating in an average home, with the sum of heating and cooling accounting for 45% of electricity use based on the use inside the home (Table 3.4). If "other uses" are considered, such as pool and spa heaters, backup electricity generators, exterior lighting, and others, then outdoor use is similar (31%) to that of space heating and cooling (31%). The next large use of electricity inside the home is the water heater, followed by refrigeration and lighting. In the past, lighting was a much larger part of the electricity bill, but the replacement of inefficient incandescent lightbulbs with compact fluorescent and then light emitting diode (LED) lightbulbs has substantially changed the distribution of energy use in the home. The percentages listed in Table 3.4 for the home (without other uses) are based on a total home annual use of 1437×10^9 kWh. Assuming 2.52 people per home and 328.2 million people, that works out to be about 30.2 kWh/d or 13.0 D, which is essentially the same as 12.9 D given in Table 3.3.

Refrigerators in a home can consume an appreciable amount of electricity, sometimes at a rate that is no longer necessary based due to increased cooling efficiencies and improved insulation. New refrigerators that have a high Energy Star rating use ~450–500 kWh/y (for 300–500 L capacity), compared to 900 kWh/y in the 1990s, and 1700 kWh/y in the 1970s (Pierce, 2003). An annual energy use of 490 kWh is 1.3 kWh/d or 0.58 D. Clearly, there have been substantial improvements in energy efficiencies of refrigerators over this period. However, given the low costs, many people are choosing to purchase larger units, which is driving up the kWh used per refrigerator. Also, many houses have more than 1 unit. Coolants used for these appliances are also a big concern as noted in Chapter 10 on cooling.

The energy use distribution shown in Table 3.4 can be converted to units of D by assuming 13 D for a home (30 kWh/d). Consider the numbers given in Table 3.4 compared to energy use for devices based on a certain amount of use per day and then convert those numbers to units of D. Some examples of D units for different devices based on 1 h of use per day are listed in Table 3.5. Energy use per day is complicated by how many people are in a home and duration of use. For

Table 3.4 Energy use in homes by percentage and D units for the house assuming a total of D = 13 and one person.

End use	Use-home with "other" (%)	Use-home (%)	D (home)
Space cooling	16	23	2.08
Space heating	15	22	1.95
Water heating	12	17	1.56
Refrigeration	6	9	0.78
Lighting	5	7	0.65
Clothes dryers	4	6	0.52
Televisions and related equipment	4	6	0.52
Computers and related equipment	2	3	0.26
Furnace fans, boiler circulation pumps	2	3	0.26
Freezers	1	1	0.13
Cooking	1	1	0.13
Clothes washers	1	1	0.13
Dishwashers (no water heating)	1	1	0.13
Other uses[a]	31	—	—

For a home with the US average, divide by 2.52 to get D per person (US Energy Information Administration, 2020a).
[a] Includes small electric devices, heating elements, exterior lights, outdoor grills, pool and spa heaters, backup electricity generators, and motors not listed above. Does not include electric vehicle charging.
Source: Adapted from US Energy Information Administration (2020a).

Table 3.5 Energy use by some appliances and devices in a house, assuming a certain amount of use each day to calculate D.

Appliance or device	Watts	Use (h/d)	D
Lightbulbs (10 at 9 W)	9	8 (×10)	0.31
Aquarium	24	24	0.25
Ceiling fan	35	12	0.18
Video game	36	6	0.09
Desktop computer	75	8	0.26
Cable box	140	24	1.45
TV (liquid crystal display)	150	6	0.39
Washer-clothes	255	1	0.11
TV-plasma	300	6	0.78
Dishwasher	330	2	0.28
Garage door opener	400	0.0167	0.0029
Coffee (1 pot)	1000	0.083	0.04
Dryer	2800	1	1.21
Total			5.34

Notes: Lightbulbs are based on 60 W equivalent assuming a use of 9 W. The average home might have 40–50 lightbulbs. The clothes washer and drier if used only once per week per person has an overestimated D; only 1 TV could be in use at a time. A single dishwasher cycle is closer to 0.43 D, but it is often used for multiple people or not used every day. Other uses vary as well.

example, a clothes washer is assumed to be used 300 times a year by a family of 4, or once or twice per week by a single person. If a single use is 0.11 D (Table 3.5), then multiplying 0.11 by 2/7 gives an average D of 0.031 per day. This can be compared to 0.13 D in Table 3.2 based on average homes, so if the average home has 2.54 people that would be 0.051 D, or 65% higher than 0.031 D. These differences arise from rounding errors, a range of existing washer efficiencies for old versus new washers, and frequencies of use.

One big change in the percent of energy that is used for different purposes is lighting. The energy used for lightbulbs has substantially been reduced with the advent of LED lightbulbs. An old incandescent bulb that used 60 W can now be replaced with an LED that uses only 9 W to provide the same light. In the example in Table 3.5, only 0.34 D would be needed for 10 LED bulbs rated at 9 W and used 8 h a day, compared to 2.07 D if they were 60 W bulbs. In either case, there is a large amount of energy to be saved just by turning them off when you are not needing that light.

3.3 PRIMARY ENERGY CONSUMPTION WHEN USING ELECTRICITY IN UNITS OF D

The amount of energy you consume in electricity is easily calculated from your home bill, and it can be divided among the number of people living in the home to calculate D. However, that electricity must be generated and delivered to your home, and so the amount of primary energy was consumed to produce that electricity is dependent on the makeup of power in your electrical grid. Power plants have a range of efficiencies depending on the technology and energy source for that power plant. Here we will develop average values based on the average sources of electricity in the United States, and distinguish energy uses based on adding a subscript, when needed, to the daily energy unit D. The D units and subscripts are defined as follows (Table 3.6):

- D_e: To specifically indicate energy for electricity, the subscript e is added.
- D_p: For the primary energy used to make electricity, a subscript p is used.

Table 3.6 Electricity production in units of D for the US: D_e = electricity, D_p = all primary energy used to produce that electricity, D_{ff} is the fossil fuel energy and D_{nC} is energy with no CO_2 emissions that was used to produce the electricity.

Source	D_e	D_p	D_{ff}	D_{nC}
Solar	0.26	0.26	0	0.26
Hydro	0.98	0.98	0	0.98
Wind	1.08	1.08	0	1.08
Geothermal	0.06	0.06	0	0.06
Biomass	0.10	0.47	0	0.47
Nuclear	2.91	8.92	0	8.92
Petroleum	0.06	0.20	0.20	0
Coal	3.45	10.74	10.74	0
NatGas	5.31	12.30	12.30	0
Total	**14.21**	**35.01**	**23.24**	**11.77**
Ratio	1	2.46	1.64	0.83

- D_{ff}: If we only consider the energy in the fossil fuel in any description of energy use, an additional subscript of ff is added to indicate the source of energy is fossil fuels.
- D_{nC}: Energy that is not associated with CO_2 emissions, such as renewables, biomass, and nuclear, is designated with a subscript nC.

The amount of energy needed to produce 14.2 EJ of electricity was shown in Table 2.5 based on the source of energy. Converting this electrical energy from units of EJ to D, we have the values shown in Table 3.6 for D_e, D_p, and D_{ff}. Based on the ratios of D_p/D_e, we can say that if 1 D of electricity is used that 2.46 D of primary energy would be needed to make that electricity, and that this use of electricity consumed 1.64 times as much fossil fuel energy.

Example 3.4

If a person uses 1.2 gallons of gasoline a day, and 30 kWh of electricity for their home, how much energy consumption is that in (a) energy used in units of D, and (b) energy based on the fossil fuels needed for both activities (D_{ff})?

(a) To calculate the total energy, we convert the energy in the gasoline into D using 35.3 kWh/gal of gasoline (Table 2.4) and then convert the electricity into D.

$$\text{D (gasoline)} = 1.2\frac{\text{gal}}{\text{d cap}}\frac{35.3\text{ kWh}}{\text{gal}}\frac{1\text{ D}}{2.32\frac{\text{kWh}}{\text{d cap}}} = 18.3\text{ D}$$

$$\text{D (electricity)} = 30\frac{\text{kWh}}{\text{d cap}}\frac{1\text{ D}}{2.32\frac{\text{kWh}}{\text{d cap}}} = 12.9\text{ D}_\text{e}$$

$$\text{D (total)} = 18.3\text{ D} + 12.9\text{ D}_\text{e} = 31.2\text{ D}$$

(b) For the fossil fuel energy, the energy in gasoline is all fossil fuel energy, and so D (gasoline) = D_{ff}. For the electricity, we first convert the energy used into fossil fuel D_{ff} using the ratio 1.64 from Table 3.6 and then add in the D from the gasoline.

$$\text{D}_\text{ff}\text{ (electricity)} = 12.9\text{ D}_\text{e}\frac{1.64\text{ D}_\text{ff}}{\text{D}_\text{e}} = 21.2\text{ D}_\text{ff}$$

$$\text{D}_\text{ff}\text{ (total)} = 18.3\text{ D}_\text{ff} + 21.2\text{ D}_\text{ff} = 39.5\text{ D}_\text{ff}$$

This comparison in Example 3.4 shows that it initially appeared that more energy was consumed by using the gasoline than the electricity. However, when the electricity energy is multiplied by the ratio of fossil fuel energy in the electricity, there is slightly more fossil fuel energy consumed by using electricity from the grid (21.2 D_{ff}) than the gasoline (18.3 D_{ff}). If the primary energy had been used, by including the energy in all the sources, the total energy for making that electricity would have been 31.7 D, bringing the total to 50 D. Thus, it is important to consider more than just the electricity energy used in your home. The primary energy used to supply that energy and energy losses delivering that electricity to your home will impact overall fossil fuel use, and therefore CO_2 emissions, for the energy use by your home.

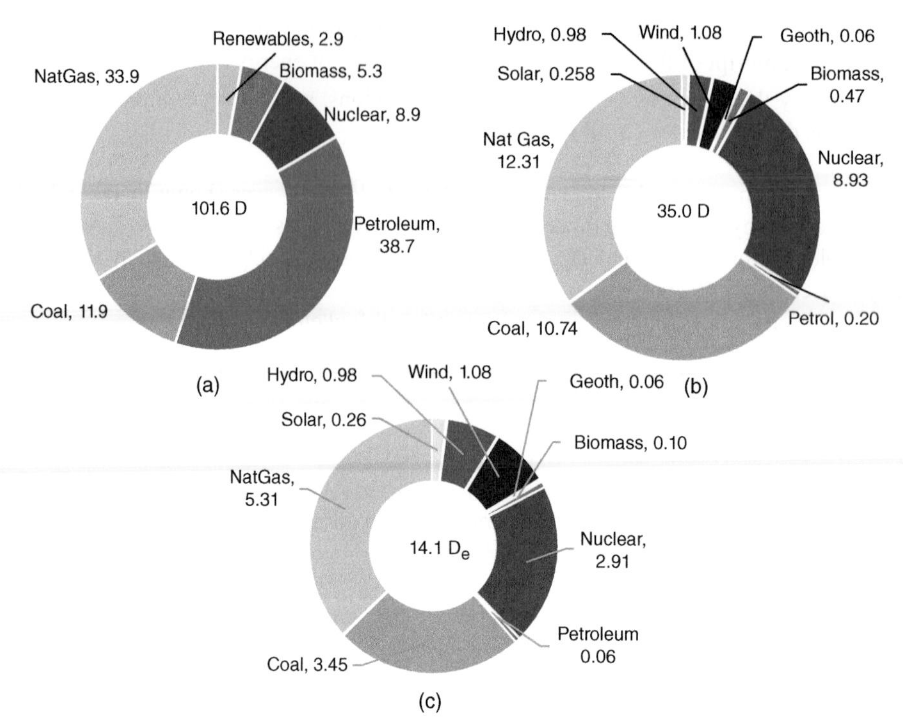

Figure 3.1 Energy use in units of D for a person the United States (2019) calculated for: (a) total energy use (b) primary energy use for electricity generation, and (c) electrical energy used (assuming a population of 328 million). *Source*: Data from US Energy Information Administration (2020b).

3.4 YOUR LIFE IN D UNITS

In Chapter 2, we examined total energy use based on 101.6 EJ of energy for the United States in 2019 with a population of 328 million. Based on conversion of this energy in units of EJ into D units, we can show this also in terms of the daily energy unit D, with 101.6 D for total energy use, 35 D for primary energy used for electricity generation, and 14.1 D the total electricity production (Fig. 3.1).

Your Own Energy Use in Units of D

It is worthwhile to consider the energy that you consume for your own lifestyle, rather than that of an "average" person in the United States, and how your daily energy use translates to fossil fuel use through fuels and electricity that you consume. While there is 101.6 D on average for a person in the United States, we do not all live in the same type of house, drive the same number of miles, and eat the same food. Therefore, how much D do you use and what part of that is directly controlled by your choices? Let us consider energy use calculated for a person with the following assumptions:

- Food consumption based on the average energy use of 13.2 D (a mix of electricity and fossil fuels, see Chapter 8) to grow, transport, and get that food on your table.

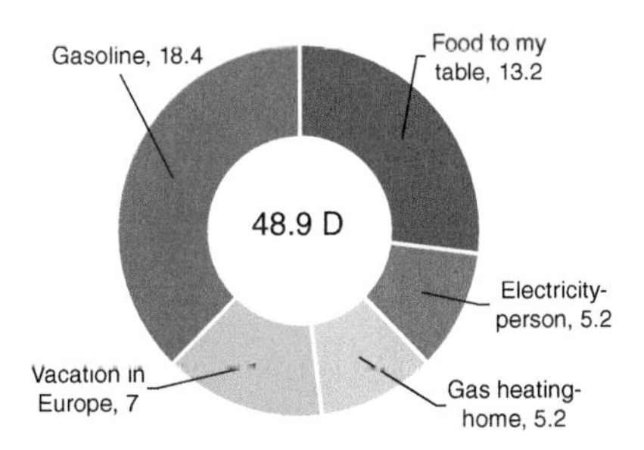

Figure 3.2 A hypothetical D footprint based on prime energy use per person: electricity and gas are normalized per 2.52 people per house, as heating is for a cold climate, and gasoline is based on 1.2 gal used per day, the national average. This analysis does not include primary energy for electricity. *Source*: Data from US Energy Information Administration (2020b).

- The average electricity use for a house of 13 D, with the average number of people (2.52 people per house), or 5.2 D
- Gas to heat a home requiring 13 D of energy (2.52 people per house)
- An annual vacation abroad, that consumes 7 D in jet fuel (round trip)
- Gasoline for your car based on using 1.21 gal/d, with 15.2 D/gal.

For these assumptions, which are mostly based on national averages for these uses, you would end up with a total of 48.9 D as summarized in Figure 3.2. How could you reduce these numbers? The energy for food is the national average, but you might be able to lower that value by food choices with less D, for example by eating a veggie burger (0.4 D) compared to a meat hamburger (6.2 D) (Chapter 6). Electricity for a home averages 13 D in the United States, and after dividing that by 2.52 people you have 5.2 D for your part of energy use for the home. The 2.52 people in a home is a national average, so that number is highly dependent on specific living conditions in each home. The overall energy use of the home is something that can be controlled through better temperature management, insulation, and lifestyles as discussed in Chapters 9 and 10. Many homes are heated with electricity, but in colder climates, natural gas, oil, or propane can be used. For Pennsylvania, about half the energy use in a home is for heating, and so estimating 13 D for natural gas heating of a single home with 2.52 people results in 5.2 D per home, and so a separate amount of energy is included for home heating using natural gas in this example.

Transportation is a very large component of energy use in the United States. Based on using 1.21 gal of gas per day (national average), that is 18.4 D per person. However, as noted in Chapter 11, the average mileage for a car is 13,500 mi would translate to 22 D. Finally, when you go on vacation, or travel for business, do you fly? One round trip from Chicago to Frankfurt could be the equivalent to 7 D averaged across 365 d in a year.

Reducing your daily energy use begins with better food choices, lower energy uses in your home by switching to high efficiency heat pumps (Chapter 9), very fuel-efficient cars (which may or may not be electric!), and more local vacations than those dependent on air travel or driving long distances. However, there are also decisions made by towns, the federal government, and industries which you may have an opportunity to participate in, so it is good to learn how you can not only modify your own daily life and home infrastructure, but also the town, state, and global infrastructure.

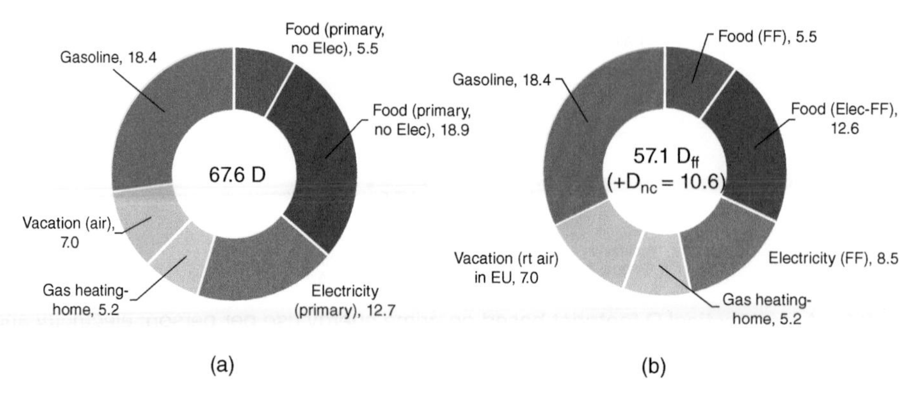

Figure 3.3 A revised footprint of energy use based on (a) the total primary energy in the electricity used for food and a home, and (b) only the primary energy for the fossil fuels used to produce that electricity. The number in parentheses shows how much electricity was consumed that did not result in CO_2 emissions (renewable and C-neutral sources). Numbers may not sum due to rounding in calculations.

Your Energy Use in Units of D_{ff} and D_{nC}

The above analysis of your daily energy use does not consider the energy that went into making the electricity. For food energy, it was estimated that 58% of the energy was electricity, with 22% using petroleum, 16% natural gas, and 4% with other sources (Canning et al., 2017). In Chapter 8, we calculate that 58% of the 13.2 D is electricity, and using the data listed in Table 3.6 for the ratio of D_{ff}:D_e, that this mixture of fuels and electricity overall required 24.4 D for food. Similarly, for the electricity use of 5.2 D per person, we have a primary energy use of 12.7 D to produce that electricity. Thus, the total amount of primary energy is now calculated to be 67.6 D (Fig. 3.3).

Not all the energy in the electricity is derived from fossil fuels, and therefore, we further consider total energy consumption from the perspective of only the primary energy derived from fossil fuels. The procedure is the same as that to calculate primary energy from the electricity, except this time we use a ratio of 1.64, resulting in 12.6 D_{ff} rather than 18.9 D for only energy in the fossil fuels for food use. For electricity used for the home, the primary energy is decreased from 12.7 D to 8.5 D_{ff}. Therefore, this results in a total energy consumption of 57.1 D_{ff} for fossil fuels, with 10.6 D_{nC} of that energy being generated by renewables or carbon-neutral power generation technologies (Fig. 3.3). As the US shifts to more renewable electricity production, our own personal carbon emissions will therefore decrease in proportion to that reduction in the use of fossil fuels.

3.5 ENERGY AND ELECTRICITY USED COMPARED TO FOSSIL FUEL USE BY DIFFERENT COUNTRIES

A shift in technologies to electrification to reduce fossil fuel use will only be successful if the electrical grid becomes less reliant on fossil fuels. The energy company BP, like some other energy companies, performs their own analyses of energy use around the world (Dudley, 2019). By using data in their reports, it is possible to calculate electricity and energy use by different countries or regions around the world. BP uses different methods and categories for summarizing energy consumption than that used here. For example, as shown in Table 2.9, they assume that on average a fossil fuel power plant has an energy efficiency of 40.4% in 2019, while the EIA calculated

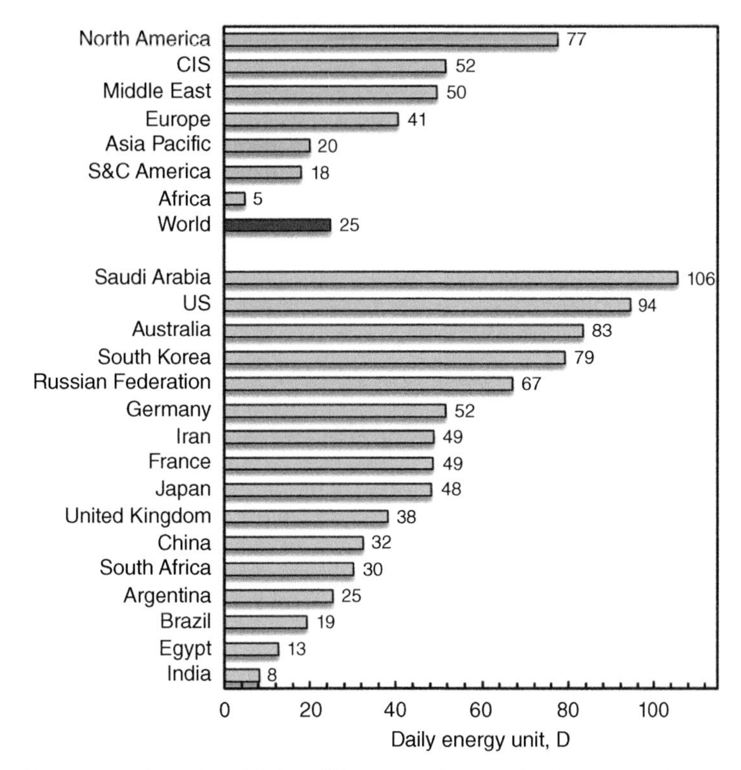

Figure 3.4 Energy use in units of D for different regions and select countries around the world based on data in Dudley (2019). *Source*: Data from Dudley (2019).

37.5%. This definition is important because primary energy for renewable sources is inflated using this efficiency.

Using this statistical data by BP on annual energy use in units of EJ, and converting into units of D based on the population of a country, in 2019 the United States had an average D of 94. This was lower than that of 101.6 D as calculated above using a different data source. D values can be compared to select countries or regions around the world to understand the regional challenges in reducing energy use (Fig. 3.4). The D for North America is the highest of all other regions with 77 D, with the lowest D of 5 for Africa. D values are calculated for other countries selected due to their high energy use in different regions, such as Saudi Arabia in the Middle East, or a being a very large country, such as India. Based on the total world population and energy use, the world average D is 25.

Electricity use can also be calculated based on data in the BP report (Dudley, 2019). Energy use for electricity generation using renewable and carbon-neutral energy sources in the BP report is sorted into three categories of renewable, nuclear, and fossil fuels. Biomass and biofuels are included in the renewable category, which is different than that assumed here (except as noted in some examples). Nuclear is separate from renewables in their report but here nuclear electricity is reported along with biofuels as carbon-neutral. The total energy use estimated for a country by BP can be different than that reported by the country. For example, in the United States, the EIA results indicated that 14.21 EJ of electricity was produced with 23.24 EJ of fossil fuels consumed as a part of that electricity production (Table 3.6). The BP report indicated that 15.85 EJ of electricity was produced by the United States with 25.92 EJ used as the portion of fossil fuel primary energy.

Example 3.5

Compare the daily energy use for electricity (D_e) and the fossil fuel energy consumed based on data in the BP report of 15.85 EJ electricity produced with primary fossil fuel energy use of 25.92 EJ. Assume a US population of 328 million.

One D is equal to 2.32 kWh/d per person, so it is a relatively direct unit conversion if we first convert the given values into TWh, with 10^9 kWh per TWh. For electricity generation, we have:

$$\frac{15.85\ \text{EJ}}{\text{y}}\ \frac{277.8\ \text{TWh}}{\text{EJ}}\ \frac{10^9\ \text{kWh}}{\text{TWh}}\ \frac{1}{328\times 10^6\ \text{cap}}\ \frac{1\ \text{y}}{365\ \text{d}}\ \frac{1\ \text{D}}{2.32\ \frac{\text{kWh}}{\text{d cap}}} = 15.9\ \text{D}_\text{e}$$

Thus, this is about 12% larger than that derived from the US EIA data of 14.21 D. Also, for both calculations, we have that the value of D is the same as that of the total EJ (both 15.9). Since the values of D align with EJ for the population of the United States, the fossil fuel use would be 25.9 D_{ff}. This estimate for fossil fuels based on the BP data is similarly about 12% larger than the US estimate.

Using data in the BP report, electricity generation can also be compared based on electrical energy source for different countries or regions. From data available in the report, comparisons were made for two regions, the Middle East and the European Union (EU), and two countries, the United States and China (Fig. 3.5). Energy use was examined based on both EJ and D for the

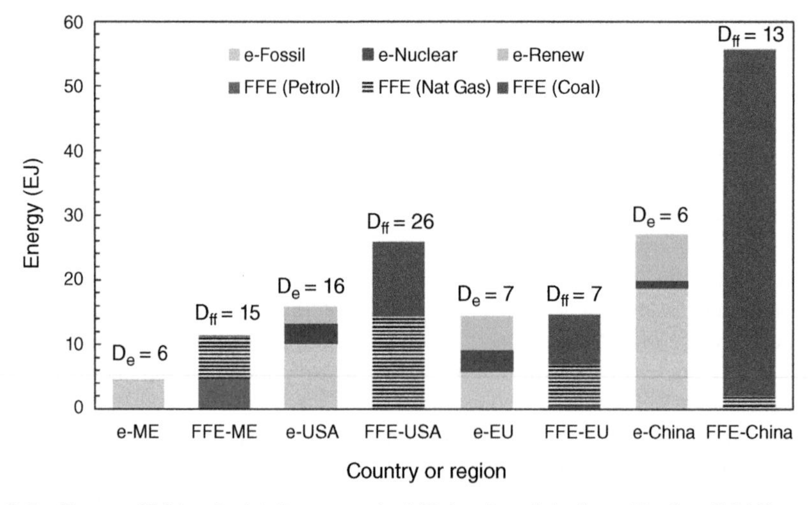

Figure 3.5 Energy (EJ) in electricity generated (D_e) using data from Dudley (2019) and fossil fuel energy (FFE, D_{ff}) used for electricity generation calculated for selected countries (USA and China) or regions (ME = Middle East, EU = European Union). Data on primary fossil fuel energy (FFE) were calculated using electricity generation efficiencies for the United States due to an absence of energy efficiencies in the report. Values here for the US different than those in Chapter 2 due to slight differences in assumptions and components of energy used for electricity generation than those in US EIA reports. Biofuels are included in the definition of renewables for these data. *Source*: Modified with permission from Logan et al. (2020).

three aggregated sources of renewables, nuclear, and fossil fuels. From this comparison, we can see that there is a reasonable balance between electricity use and fossil fuels going into that electricity for the EU, compared to a ratio of fossil energy to electricity of 1.6 for the United States and 2.1 for China. The Middle East is currently nearly completely dependent on fossil fuels for electricity generation, with slightly more than half of the electricity produced with natural gas.

The analysis based on D units shifts the focus onto per capita consumption in these countries and regions (Fig. 3.5). The variable D_e was used to indicate electricity used (not primary energy) while D_{ff} showed only the primary fossil fuel energy that went into that electricity production. For these four comparisons, the United States has both the largest electricity use per person, with $D_e = 16$, as well as the highest amount of fossil fuel used per person, with $D_{ff} = 26$. China and the other countries or regions all have similar D_e values of 6–7, indicating that there is much less electricity use per person in those countries and regions than in the United States. The total energy use by China is enormous, and it is highly dependent on coal, a fuel that has the greatest amount of CO_2 emissions based on electricity generation. China has made good progress in increasing the use of renewables as a percentage of their electricity production, but the very large population of China will make it difficult to greatly reduce CO_2 emissions without further large changes to renewables or aggressive deployment of CO_2 sequestration technologies.

3.6 CREATING GREEN D

If we are to achieve a green and renewable energy infrastructure, then electricity derived from renewable resources will play an increasingly important role. Using data in Figure 2.2, we can compare the relative amount of electricity used by the four sectors of our infrastructure in the United States compared to other sources of energy (Fig. 3.6). A good proportion of residential and commercial use is already using electricity, so it is likely that converting that electricity use over to 100% renewable electricity could greatly impact the overall D and make it green. However, the existing amount of green electricity in these categories is quite low, especially in the transportation section.

The transportation and industrial sectors provide great challenges from this perspective of moving to a green and renewable energy infrastructure. The amount of energy used by industry (27.9 EJ) is currently highly dependent fossil fuels, and a large component of total fossil fuels uses

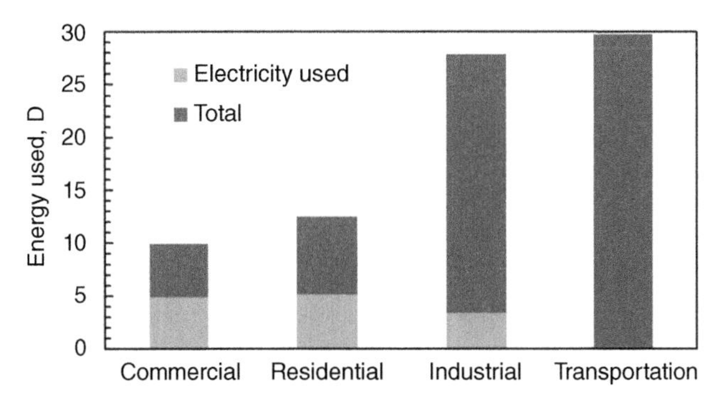

Figure 3.6 Electricity as a portion of the total energy use for four main sectors, expressed in terms of D units (based on EJ of energy, and 328.2×10^6 people in the United States in 2019). *Source*: Data from Lawrence Livermore National Laboratory (2019).

(84.6 EJ) (Fig. 2.1), suggesting that changing over to green electricity from fossil fuels will be especially challenging for this sector. It is no surprise that transportation relies heavily on fossil fuels. Currently, energy use is 29.7 D for transportation, based on total electrical energy produced normalized by the US population. The amount of electricity currently used for transportation (0.032 D) is so little it is not even visible in Figure 3.6. This amount of electricity for transportation is roughly the amount of daily energy used by every person making a pot of coffee (0.04 D, Table 3.5). From this perspective of changing from fossil fuels to electricity using renewable electricity, transportation is the greatest challenge in these four sectors, but as we shall see, also the greatest opportunity to reduce overall energy use.

Avoiding Energy Use that Goes into Rejected Energy

The energy used for work is usually based on primary energy, or the energy going into the system for use to do that work. Once used, there are two outcomes: services provided, that is the energy used to do the intended tasks; and the rejected or wasted energy. If we compare services to rejected energy for the four major use areas, and including electricity production, we can see that commercial, residential, and industrial energy use are very energy efficient compared to the other two categories (Fig. 3.7). One main reason for the apparently high efficiency is that much of the energy use goes into heating, and primary energy to heat conversion is very efficient.

Transportation is a highly inefficient process based on the very large amount of energy wasted compared to that put into services (Fig. 3.7). This inefficiency results from combustion of fuels in engines that typically operate at less than peak efficiency, so that the overall ratio of Services:Rejected ratio for transportation is 0.27, or only a 21% efficiency in energy conversion to use. Electricity production using fossil fuels is also highly inefficient when judged against the other categories, with a ratio of 0.55, with an efficiency of 34.4%.

If energy use is switched from combustion of fossil fuels to electrification based on renewable energy, then the greatest gains can be made by (i) a large reduction in energy waste due to a low amount of rejected energy based on electricity generated by renewable resources, and (ii) a greatly improved energy efficiency if transportation is switched to electrical power that is generated from renewable sources. If building-based generation of electricity using solar panels replaces the need

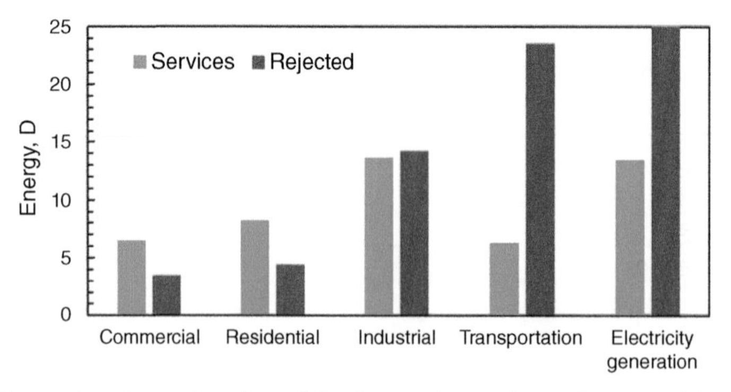

Figure 3.7 Energy input as a function of the four main sections of use, and for electricity generation, expressed in terms of D units (based on energy used and 328.2 million people in the United States in 2019). *Source*: Data from Lawrence Livermore National Laboratory (2019).

for electricity generation by large companies, there is a double benefit of reducing energy rejected in both generation and transmission to the point of use. Changing from vehicles that use very inefficient internal combustion engines to highly efficient electrical motors or fuel cells can similarly avoid a great deal of rejected energy and energy losses.

Next Steps

This analysis of energy use shows that not only do we need to focus on producing larger amounts of renewable electricity but also that transportation is the largest and most inefficient method of energy use. Therefore, while we can adjust our daily lives to reduce our personal D footprint, to have a larger impact on overall fossil fuel consumption, large changes will be needed in our energy infrastructure.

In the chapters that follow, we consider how energy is used for specific purposes, for example to power cars and other uses for transportation and international travel on airplanes, and energy and to heat and cool our homes. There are other situations which can greatly impact global CO_2 emissions that we cannot directly impact by our choices, such as carbon emissions related to our infrastructure such as consuming steel and concrete to construct buildings and other structures that make up a modern industrial society. Reducing our carbon emissions will require that the global community find solutions that can address all these energy choices that currently lead to large amounts of CO_2 emissions.

References

BRAZIER, Y. 2018. *How many calories should I eat a day?* [Online]. MedicalNewsToday. Available: https://www.medicalnewstoday.com/articles/245588 [Accessed May 4 2020].

BURWELL, S. M. & VILSACK, T. J. 2020. Appendix 2. Estimated calorie needs per day, by age, sex, and physical activity level. In: U.S. DEPARTMENT OF HEALTH AND HUMAN SERVICES AND U.S. DEPARTMENT OF AGRICULTURE (ed.) *Dietary guidlines for Americans 2015-2020*, USDA (US Department of Agriculture). 8th ed. 77–78.

CANNING, P., REHKAMP, S., WATERS, A. & ETEMADNIA, H. 2017. *The Role of Fossil Fuels in the U.S. Food System and the American Diet*, USDA, US Department of Agriculture, Economic Research Report Number 224. Access this report online: www.ers.usda.gov/publications.

CHINACITIES.COM. 2015. *Average height and weight of Chinese increase* [Online]. Available: https://www.echinacities.com/news/Average-Height-and-Weight-of-Chinese-Increase [Accessed May 5 2020].

DUDLEY, B. 2019. *BP Statistical Review of World Energy 2019.* 68 ed. BP International Limited. Available: https://www.bp.com/content/dam/bp/business-sites/en/global/corporate/pdfs/energy-economics/statistical-review/bp-stats-review-2019-full-report.pdf. [Accessed April 16 2021].

ELECTRIC CHOICE. 2020. *How much electricity on do homes in your state use?* [Online]. Available: https://www.electricchoice.com/blog/electricity-on-average-do-homes/ [Accessed April 26 2020].

ENERGY SERVICES GROUP. 2020. *The real cost of heating* [Online]. Available: https://energysvc.com/the-real-cost-of-heating [Accessed April 26 2020].

LAWRENCE LIVERMORE NATIONAL LABORATORY. 2019. *Estimated U.S. energy consumption in 2019: 100.2 quads* [Online]. Available: https://flowcharts.llnl.gov/content/assets/docs/2019_United-States_Energy.pdf [Accessed May 9 2020].

LOGAN, B. E., ROSSI, R., BAEK, G., SHI, L., O'CONNOR, J. & PENG, W. 2020. Energy use for electricity generation requires an assessment more directly relevant to climate change. *ACS Energy Letters,* 5, 3514–3517.

PIERCE, M. 2003. *Replace your old refrigerator and cut your utility bill* [Online]. Cornell Cooperative Extension. Available: http://chemung.cce.cornell.edu/resources/replace-your-older-refrigerator-to-reduce-energy-costs [Accessed April 28 2020].

US ENERGY INFORMATION ADMINISTRATION. 2020a. *How is electricity used in U.S. homes?* [Online]. Available: https://www.eia.gov/tools/faqs/faq.php?id=96&t=3 [Accessed May 10 2020].

US ENERGY INFORMATION ADMINISTRATION. 2020b. *July 2020 monthly energy review* [Online]. Available: https://www.eia.gov/totalenergy/data/monthly/pdf/mer.pdf [Accessed August 7 2020].

US NATIONAL HEART, LUNG, AND BLOOD INSTITUTE. 2020. *Calculate your body mass index* [Online]. Available: https://www.nhlbi.nih.gov/health/educational/lose_wt/BMI/bmicalc.htm [Accessed May 4 2020].

WILKINSON, K. 2020. *The drawdown review: Climate solutions for a new decade*, Project Drawdown. https://drawdown.org/sites/default/files/pdfs/Drawdown_Review_2020_march10.pdf

CHAPTER 4

DAILY CO_2 EMISSION UNIT C

4.1 DEFINING THE DAILY CARBON EMISSION UNIT C

Minimizing the impact of climate change means that we need to reduce global greenhouse gas (GHG) emissions. The mass of just CO_2 alone is a staggering number, with 33.1 billion tonnes (equal to gigatons, Gt) of CO_2 released in 2018, with CO_2 concentrations in the atmosphere reaching a record high of 417 ppm in March of 2021. Climate change is influenced by many other gases than CO_2, including methane and nitrous oxide, as well as other factors such as the rate of CO_2 sequestration in biomass and the oceans. It is difficult to really understand numbers like gigatons as such large quantities are disconnected from our lives, and so, these amounts of gases released have no particular meaning relative to the magnitude of the CO_2 release. We know we need to reduce CO_2 emissions, but most of us do not know how much CO_2 our own energy use releases every day.

One way to connect CO_2 emissions to something familiar to us all is to take the same approach used to understand energy consumption, by defining a minimum in CO_2 emissions relative to food that we eat every day. In Chapter 3, we saw that the Calories that we needed from our food every day varied for each person depending on their height, weight, activity level, and other factors. However, by choosing a reference point of 2000 Cal per person and defining that to be 1 D, we were able to ratio everything to that basic energy consumption that we all share.

We use that same normalization process for carbon emissions by defining an amount of CO_2 that is released every day based on an average from the food that we eat. While the actual CO_2 emissions from people will vary based on a lot of factors, by choosing a reference point, we can establish a baseline for all other CO_2 emissions. To ratio carbon emissions to food for each person, we set daily CO_2 emissions as the unit 1 C, where 1 C = 2 lb/d/person. If you use Google to find out how much CO_2 each person emits every day, you will probably come up with the following two numbers that are often repeated in many sites: 900 g/d based on a US Department of Agriculture (USDA) study and 2.3 lb/d. The specific USDA study, however, is never cited in these reports so it is impossible to check the accuracy of those statements. Likewise, the 2.3 lb/d of CO_2 cited in several websites never leads to a scientific source, and it seems to just be a repetition from other sites. These two numbers are somewhat different as 900 g/d = 2.0 lb/d, which is a bit less than the 2.3 lb/d.

To understand why 2 lb/d was chosen here as 1 C, we examine a range of possible definitions based on the types of food that we eat. We estimate a value for CO_2 emissions by first assuming that we eat 2000 Cal/d and that all of the carbon in our food is captured in our body. However, there

Daily Energy Use and Carbon Emissions: Fundamentals and Applications for Students and Professionals, First Edition. Bruce E. Logan.

is not any defined relationship between the energy in our food and the CO_2 emissions as the carbon and oxygen content of food varies. In Example 4.1, we calculate how CO_2 emissions vary based on the source of the Calories categorized generally as carbohydrates (sugars), fat, or protein. For each of these, we write out a stoichiometric equation based on the complete oxidation of the food (or fuel) to CO_2 and water. For a general form of each of these three categories, we can use the following: $C_6H_{12}O_6$ for a carbohydrate, based on the formula for glucose; $C_{56}H_{108}O_6$ for fat (Vitz et al., 2020); and $C_4H_{6.3}O_{1.2}N$ for a protein (Torabizadeh, 2011). On average, there are 4 Cal/g of carbohydrates and protein, while fat has 9 Cal/g.

Example 4.1

Calculate the amount of CO_2 produced for carbohydrates, fat, and protein in units of kg/d (lb/d) by assuming that a person eats 2000 Cal of only one of these food types in a day.

First, we derive a balanced equation for each of these energy sources to obtain molar ratios between the starting chemical and CO_2. Then, using the molecular weights for the chemical and 44 g/mol for CO_2, we can obtain the mass of CO_2 emitted. For carbohydrates, we balance the equation to obtain

$$C_6H_{12}O_6 + 6O_2 \rightarrow 6\,CO_2 + 6\,H_2O$$

This indicates that 6 mol of CO_2 is produced per module of glucose. Using this result, the CO_2 emission is calculated for glucose with a molecular weight of 180 g/mol as:

$$2000\,\frac{\text{Cal}}{\text{d}}\;\frac{1\text{ kg glu}}{4000\text{ Cal}}\;\frac{6\text{ mol CO}_2}{\text{mol glu}}\;\frac{44\,\frac{\text{g CO}_2}{\text{mol CO}_2}}{180\,\frac{\text{g glu}}{\text{mol glu}}} = 0.73\frac{\text{kg CO}_2}{\text{d}}\;\frac{2.2\text{ lb}}{\text{kg}} = 1.61\frac{\text{lb}}{\text{d}}$$

For fats, with a calculated molecular weight of 876 g/mol we have

$$C_{56}H_{108}O_6 + 80O_2 \rightarrow 56\,CO_2 + 54\,H_2O$$

$$2000\,\frac{\text{Cal}}{\text{d}}\;\frac{1\text{ kg fat}}{9000\text{ Cal}}\;\frac{56\text{ mol CO}_2}{\text{mol fat}}\;\frac{44\,\frac{\text{g CO}_2}{\text{mol CO}_2}}{876\,\frac{\text{g fat}}{\text{mol fat}}} = 0.63\,\frac{\text{kg CO}_2}{\text{d}}\;\frac{2.2\text{ lb}}{\text{kg}} = 1.39\frac{\text{lb}}{\text{d}}$$

For proteins, with 87.5 g/mol,

$$C_4H_{6.3}O_{1.2}N + 8.45O_2 \rightarrow 4CO_2 + 1.65H_2O + NH_3$$

$$2000\,\frac{\text{Cal}}{\text{d}}\;\frac{1\text{ kg protein}}{4000\text{ Cal}}\;\frac{4\text{ mol CO}_2}{\text{mol protein}}\;\frac{44\,\frac{\text{g CO}_2}{\text{mol CO}_2}}{87.5\,\frac{\text{g protein}}{\text{mol protein}}} = 1.00\,\frac{\text{kg CO}_2}{\text{d}}\;\frac{2.2\text{ lb}}{\text{kg}} = 2.2\,\frac{\text{lb}}{\text{d}}$$

Based on the calculation in Example 4.1, we have "narrowed" down our result to a range of 0.63–1 kg/d (1.39–2.2 lb/d) of CO_2, which is within the range of the numbers we see on the

Table 4.1 CO_2 emissions per day in units of lb/d and C for different energy input scenarios.

Assumption	Calories (d^{-1})	CO_2 (lb/d)	D	C
100% carbohydrates	2000	1.61	1	0.80
100% fat	2000	1.39	1	0.70
100% protein	2000	2.20	1	1.10
50.5% carbohydrates, 33.2% fat, and 16.4% protein	2000	1.63	1	0.83
Same, higher Calories	2500	2.04	1.25	1.02
50% carbohydrates, 20% fat, and 30% protein	2000	1.74	1	0.87
Same, higher Calories	2500	2.18	1.25	1.09

The C results are based on the definition of C = 2 lb/d of CO_2.

Internet. We can further refine these numbers by making some additional assumptions about our diet. For example, in the United States, the average diet consists of 50.5% carbohydrates, 33.2% fat, and 16.3% protein (Paddock, 2019). Using these ratios, we estimate 0.76 kg/d (1.66 lb/d) of CO_2 (Table 4.1). But what if we followed better nutritional guidelines? One recommendation is that we eat 50% carbohydrates, 20% fat, and 30% protein (Hrustic, 2017), in which case we would emit 0.79 kg/d (1.74 lb/d) of CO_2.

Some additional considerations for these calculations are that a US citizen eats on average about 2500 Cal/d, which would raise these estimates to 2.04 lb/d for the current diet proportions and 2.18 lb/d for the diet higher in protein (Table 4.1). Also, there is an enormous amount of food wasted, which can raise our effective consumption as it relates to food use to 5800 Cal/d, which would further increase our CO_2 emissions. An additional consideration is that we do not actually retain all the food calories in the food that we eat, releasing undigested amounts on average of 2% of carbohydrates, 5% of fats, and 8% of proteins (Friedrich, 2019).

So, how do we decide on a baseline for defining C? Given the uncertainty in all these numbers, we make a choice of a reference number and then just use that number for any calculation. The definition of C is just a reference point to judge other CO_2 emissions, based on reasonable assumptions of our own CO_2 emissions. Here, we choose **2 lb CO_2 per day per person = 1 C**. This number defines a reference point as something between what we are eating (2.1 lb/d based on our actual food diet and 2500 Cal or 1.25 D) and what we might eat if we had a better diet (2000 Cal/d or 1 D, with 1.8 lb/d). Units of lb/d are chosen rather than kg/d as the approximation, and rounding to kg would suggest 2 kg/d (2.2 lb/d), which would be at the upper limit of CO_2 emissions based on diet and Calorie intake by men and women around the planet. The value of 2.0 lb/d is equal to 0.90 kg/d, which is the value attributed to the uncited USDA report mentioned above.

Relating 1 C to Carbon Capture by Trees

In natural carbon cycles, plants capture CO_2 from the air to make biomass and then release it when they degrade. One way to relate the amount of CO_2 we release is to compare it to CO_2 uptake by trees. While rates of CO_2 uptake vary among different types of trees, one estimate is that young trees absorb about 13 lb of CO_2 a year, and after 10 y, they absorb around 48 lb/y (Urban Forestry Network, 2021). In the example below, calculations are shown for the number of trees needed to capture 1 C and the amount of CO_2 captured per area of forested land.

Example 4.2

(a) Assuming a tree can capture 48 lb/y of CO_2, how many trees are needed to capture 1 C? Round to an integer value. (b) If one acre of forested land is estimated to capture 2.5 tons of CO_2 per year, how many trees is that per acre? (c) If everyone in the United States (population of 326.7 million in 2018) were to plant 1 C-equivalent of trees, how much CO_2 could be captured per year? Compare that result to an annual CO_2 release from fossil fuels of 5.425 Gt/y (in 2018) where t = tonne (metric tonnes).

(a) Based on the definition of 1 C, we convert the annual uptake to a daily uptake and calculate the number of trees as:

$$\frac{2\,\frac{\text{lb CO}_2}{\text{d cap}}}{1\,\text{C}}\;\frac{\text{tree}}{48\,\frac{\text{lb}}{\text{y}}}\;\frac{365\,\text{d}}{\text{y}} = 15\,\frac{\text{trees}}{\text{cap}}$$

(b) For 2.5 tons, assuming English units (1 ton = 2000 lb), the number of trees is

$$\frac{2.5\,\text{ton}}{\text{acre y}}\;\frac{2000\,\text{lb}}{\text{ton}}\;\frac{\text{tree}}{48\,\frac{\text{lb}}{\text{y}}} = 104\,\frac{\text{trees}}{\text{acre}}$$

(c) For a population in 2018 of 326.7 million, we have

$$326.7 \times 10^6\,\text{cap}\;\frac{15\,\text{tree}}{\text{cap}}\;\frac{48\,\text{lb}}{\text{tree y}}\;\frac{\text{kg}}{2.2\,\text{lb}}\;\frac{1\,\text{t}}{1000\,\text{kg}}\;\frac{1\,\text{Gt}}{10^9\,\text{t}} = 0.11\,\frac{\text{Gt}}{\text{y}}$$

The calculated emission of 0.11 Gt/y is about 2% of the annual CO_2 annual emissions in the United States. This suggests that CO_2 emissions in the United States average around 50 times this value that was based on 1 C or 50 C. We will calculate a C value based on fossil fuel emissions for the United States below.

4.2 CO_2 EMISSIONS FROM DIFFERENT FUELS

The CO_2 emissions from various fuels can be calculated using a stoichiometric conversion of specific chemicals to CO_2 and water, as done above for food. However, the composition of oils and gasolines varies, and so instead, we rely on reported averages in the literature for combustion products from different fuels. For gasoline, one issue is whether it is pure gasoline or it contains 10% ethanol. One commonly used value for gasoline is 121 MJ/gal, although gasoline energy values tend to vary for different products and the amount of ethanol added to gasoline used for transportation fuels. The US Energy Information Agency (EIA) indicated that 1 gal of finished gasoline, which contains about 10% ethanol by volume), contained 120,333 Btu (US Energy Information Administration, 2019). Ethanol has about 34% less energy than gasoline (Wikipedia, 2020a), and thus, pure gasoline would have a higher energy value of about 124,200 Btu/gal. Since most gasoline sold in the United States now contains ~10% ethanol by volume, the blended number is used here. Also, it is estimated that carbon emissions are 19.6 lb of CO_2 per gallon of gasoline (US Energy Information Administration, 2016). Based on these numbers, we can calculate D and C values for this fuel along with the D:C ratio, as shown in the example below.

Example 4.3

Calculate the energy in a gallon of gasoline based on 120,333 Btu/gal in units of kWh and D, assuming 1 gal of gas is used per person per day. Then, calculate the D and C units for gasoline based on 19.6 lb/gal of CO_2.

Using the given information in Btu per gallon of gasoline and various unit conversions, we first convert to MJ and then further to kWh so that we have both the information in MJ and kWh, as:

$$\text{Gasoline} = \frac{120{,}333\ \text{Btu}}{\text{gal}}\frac{1\ \text{MJ}}{947.8\ \text{Btu}} = 127.0\frac{\text{MJ}}{\text{gal}}\frac{\text{kWh}}{3.6\ \text{MJ}} = 35.3\frac{\text{kWh}}{\text{gal}}$$

From this calculation, we see that the energy content of gasoline using 120,333 Btu produces 127 MJ, which is somewhat larger than the 121 MJ reported by others.

The conversion to units of D proceeds based on the use of that energy in one day as:

$$35.3\frac{\frac{\text{kWh}}{\text{d cap}}}{\text{gal}}\frac{1\ \text{D}}{\frac{2.32\ \text{kWh}}{\text{d cap}}} = 15.2\ \text{D (per gal)}$$

For conversion from lb of CO_2 emitted to C, we just use 1 C = 2 lb/d of CO_2, to obtain:

$$16.2\frac{\frac{\text{lb}}{\text{d cap}}}{\text{gal}}\frac{1\ \text{C}}{\frac{2\ \text{lb}}{\text{d cap}}} = 9.80\ \text{C (per gal)}$$

From the two results, we have the D:C ratio of 15.2/9.80 = 1.55.

The results from Example 4.3 are summarized in Table 4.2 along with other fuels For the two other liquid fuels, the values for kWh are obtained from data in Table 2.4. We see that the D:C ratios for the three different liquid fuels are all quite similar, and thus, switching among these fuels from the perspective of the finished product does not clearly impact CO_2 emission, with an average D:C of 1.6 (to two significant figures).

The situation for energy used relative to carbon emissions is quite different for natural gas compared to the other fuels, as natural gas combustion releases much less CO_2 for the same energy content of other fossil fuels. These differences are clearly shown by the D:C ratios in Table 4.2. To compare the energy in natural gas and its CO_2 emissions to the liquid fuels, we can calculate the amount of natural gas that has the same energy as a gallon of gasoline and then examine the carbon emissions. For example, for the energy content of gasoline of 15.2 D, you would need 1.165 CCF of natural gas, and the CO_2 emissions would be 13.6 lb of CO_2 or 6.82 C released using natural gas compared to 19.6 lb of CO_2 or 9.80 C for the gasoline (Table 4.2). On the basis of CO_2, the long calculation yields a ratio of 0.70 (6.82 C/9.80 C). We can obtain the same result using only the D:C ratios of the two fuels (1.55/2.23 = 0.70). This comparison clearly demonstrates the ~30% reduction in CO_2 emissions that can occur by switching from liquid fuels to natural gas if the fuels are used in the same way, for example in the same type of power plant with the same efficiency of energy conversion. We see the D:C ratio for natural gas is much higher than the other liquid fuels, indicating that more energy is available from natural gas normalized to the CO_2 emissions

Table 4.2 D and C units and ratios for different fuels, assuming 1 gal (liquid fuels) or 1 CCF (natural gas) or 1 lb (coal) are used per day; for natural gas and coal, daily unit adjusted to equal the energy in gasoline based on kWh.

Fuel	kWh	CO_2 (lb)	D	C	D:C
Natural gas (1 CCF)	30.3	11.7	13.1	5.86	2.23
(1.165 CCF)a	35.3	13.6	15.2	6.82	2.23
Diesel or heating oil (1 gal)	40.3	22.4	18.3	11.2	1.63
Jet fuel (1 gal)	39.6	21.1	17.1	10.6	1.61
Gasoline (1 gal)	35.3	19.6	15.2	9.80	1.55
Coal-bituminous (1 lb)	12.7	2.47	1.52	1.24	1.23
(10.1 lb)	35.3	24.9	15.2	12.5	1.23
Coal-subbituminous (1 lb)	9.3	1.86	1.12	0.93	1.20
(13.6 lb)	35.3	25.3	15.2	12.6	1.20

[a] natural gas volume in CCF calculated to be the same as a gallon of gasoline, with 117 lb CO_2 released from 10^6 Btu of natural gas. Due to rounding errors, the ratio for gasoline to natural gas is 1.16 based on Btu, and 1.165 using the kWh given in the table.
[b] Calculated amounts of coal needed to produce the same kWh of energy content as 1 gal of gasoline.
Source: Adapted from US Energy Information Administration (2016).

compared to the liquid fuels. While this comparison is useful to indicate inherent CO_2 emissions from using these fuels, this approach is not a full life cycle analysis. For example, the energy used to compress the natural gas and the costs for transport of these different fuels to the point of use are not considered. Gas can be efficiently transported via gas lines, but other fuels are transported mostly by ground transportation which can substantially add to their overall CO_2 emissions.

Coal

From a carbon emissions perspective, coal is the worst fuel among those examined above as all types of coal have very low D:C ratios. The four main types of coal are bituminous, subbituminous, anthracite, and lignite. Most of the coal used in the United States is subbituminous (47%) or bituminous (45%), thus making these two coals the most important relative to CO_2 emissions. The energy content of coal varies widely, but on average, subbituminous coal contains 17.7 mm Btu/ton (1 ton = 2000 lb), and bituminous coal has 24.0 mm Btu/ton. If we convert these fuels into D and C units, using the carbon emissions typical for these coals (US Energy Information Administration, 2016), we get D:C ratios of about 1.2, much lower than those of the liquid fuels (1.6), and nearly half of those for national gas. Clearly, a shift away from coal to natural gas has the largest impact on overall CO_2 emissions compared to the liquid fuels.

4.3 EMISSIONS OF CO_2 FOR DELIVERED ELECTRICITY

When you use electricity, it is important to consider how and where it was produced to understand CO_2 emissions due to its production and transport. For fossil fuels used in conventional steam-based power plants, the average energy efficiency is 37.5% according to the US EIA for fossil fuel plants (Table 2.6). In addition, 6.6% of the electricity generated is lost during transmission. Nuclear plants have lower energy efficiencies of 33–37% (Afework et al., 2019), with 32.6% reported for the United States in 2019 (US Energy Information Administration, 2020b), but they do not directly emit CO_2 during electricity generation. Based on the energy analysis in Chapter 2, we will assume plant efficiencies as given in Table 2.9 which is 37.5% for a conventional fossil fuel power plant, and assuming 6.6% for losses in transmission, this results in an overall efficiency for delivered electricity of 35%. For specific fuel sources, such as coal or natural gas, we can use the more precise efficiencies given in Table 2.9.

Combined heat and power (CHP) or cogeneration plants can have overall greater energy efficiencies than conventional plants due to the use of the heat produced along with electricity generation. When heating and cooling are combined with electricity generation, the overall efficiencies can exceed 80% and approach 85%. It is estimated that in the European Union, 11% of electricity generation uses CHP, with Finland having nearly 82% of power plants using CHP technologies.

If natural gas is used in a conventional steam-type power plant, the overall efficiency is ~37.5% for electricity generation. However, most (93.7%) natural gas fueled power plants now use gas turbines for electricity generation, with 84.7% using combined cycle gas turbine, 9% using simple cycle turbines, and only 6.3% using other technologies (US Energy Information Administration, 2020a). With a simple cycle gas turbine, the energy for electricity generation is in the range of 40–50%, but high efficiency combined cycle gas-fired plants can reach 63.5% at full load, although this efficiency is lowered without operation at full load. Here, we assume an overall efficiency of 43.6% for electricity generated, with 40.7% as delivered (6.6% transmission losses) using natural gas in a combined cycle natural gas plant (CCP) and the average efficiency reported by the EIA (US Energy Information Administration, 2020b). The maximum efficiencies for a CCP are 63.5% for electricity generated, with 59.3% as delivered.

Example 4.4

Calculate the primary energy (D_p) and carbon emissions (C) for producing 1 D_e of electricity (2.32 kWh) from subbituminous coal.

Assuming coal is used in a power plant with a combined $\eta = 32.2\%$ efficiency (Table 2.9, for ratio of primary energy to electricity produced), the D_p that would be needed to produce 1 D_e of electricity is:

$$D_p = \frac{D_e}{\eta} = \frac{1\ D_e}{0.322\ \frac{D_e}{D_p}} = 3.11$$

This calculated energy does not include losses for transmission to the point of use. Assuming a 6.6% transmission loss (or a 93.4% transmission efficiency, see Table 2.8), the primary energy needed to deliver 1 D_e is $D_p = 3.33$ (3.11/0.934). There are other gains and losses for electricity production that are not included in this calculation, such as onsite electricity use at the power plant.

The carbon emissions resulting from 1 D of electricity can be calculated from the primary energy use assuming a D:C fuel to carbon ratio of 1.20 for (Table 4.2), to obtain

$$C = \frac{1\ D_e}{C} = \frac{D_p}{D_p : C} = \frac{(3.11)}{(1.20)} = 2.59$$

Thus, if electricity produced has the same energy of D = 15.2 as a gallon of gasoline (Table 4.2), and the only fuel for the power plant was subbituminous coal, then the amount of energy in the fuel used is $D_p = 47.3$, and the carbon emissions from that fuel would be $C_{fuel} = 39.4$. These would be increased by a factor of 1.07 (1/0.934) to account for an average transmission loss of 6.6%.

Table 4.3 Fuel energy (D) and carbon emissions (C) needed to produce 1 D_e of electricity (2.32 kWh) assuming overall production efficiencies (η) for the different energy sources.

Fuel	D:C	C	η	D_p	C
Solar (photovoltaics)	0	0	1.00	1.00	—
Natural gas-CCP	2.23	0.448	0.431	2.32	1.04
Natural gas-steam	2.23	0.448	0.406	2.46	1.10
Petroleum	1.63	0.613	0.313	3.19	1.96
Coal-bituminous	1.23	0.813	0.322	3.11	2.52
Coal-subbituminous	1.20	0.833	0.322	3.11	2.59

An additional loss of 6.6% occurs on average due to electricity transmission losses.

The energy in the fuel used to produce electricity can be important for evaluating the overall carbon emissions for devices powered by electricity compared to directly using a fuel, for example for electric versus gasoline vehicles (see Chapter 9). The D values in Table 4.2 show the energy of the different fuels. If we divide that D value by the overall powerplant efficiency for converting the energy in the fuel into electricity, then we can calculate the primary energy, D_p, or the total amount of fuel needed to produce that amount of electricity. Similarly, we can calculate the C emissions based on the amount of fuel to produce the electricity and the D:C ratio for the fuel. An example calculation is shown in Example 4.3 for a power plant using subbituminous coal. For the other fuels listed in Table 4.2, we can calculate the amount of primary energy and carbon emissions to obtain the values as shown in Table 4.3. For natural gas in a CCP with a high overall energy efficiency, the carbon emission value is only 40% of that calculated for a coal plant producing electricity using subbituminous coal.

These values do not include transmission losses or other energy gains and losses for the US electrical grid. As listed in Table 2.8, the electricity produced at power plants in the United States is 14.22 EJ or 14.22 D_e but gains and losses result in delivered electricity of 13.5 D_e. Transmission losses account for 0.94 D_e of the electricity produced or a 6.6% loss (transmission efficiency of 93.4%). Thus, large-scale production of solar photovoltaic electricity would take 1.07 D_e at the source to transmit to the point of use 1 D_e. If there is no transmission, for example for home rooftop systems, the D_p would be unity.

4.4 CARBON EMISSIONS FOR PEOPLE IN UNITS OF C

In Chapter 3, we calculated that all the energy used in the United States normalized by the population (328.2 million) was equivalent to 101.6 D for 2019. A C value can be similarly calculated based on total energy use with appropriate assumptions for C values specific to the energy sources. For both renewable (defined here as solar, wind, hydro, and geothermal) and carbon-neutral energy sources (biomass and nuclear), we will assume there are no net CO_2 carbon emissions. Obviously, all these energy sources will have carbon emissions associated with their deployment and use to some smaller or greater extent. For example, getting biomass from source to point of fuel production will likely consume petroleum for the vehicles used for transportation, and corn ethanol is widely considered to be unsustainable due to the high energy demands of producing ethanol (Keeney, 2009). Some amount of energy would be consumed for the other energy sources in terms of energy needed to obtain the fuel (if one is used), construct the system or power plant, or maintain the system over time. Ultimately, some amount of the fossil fuels being used in the United States

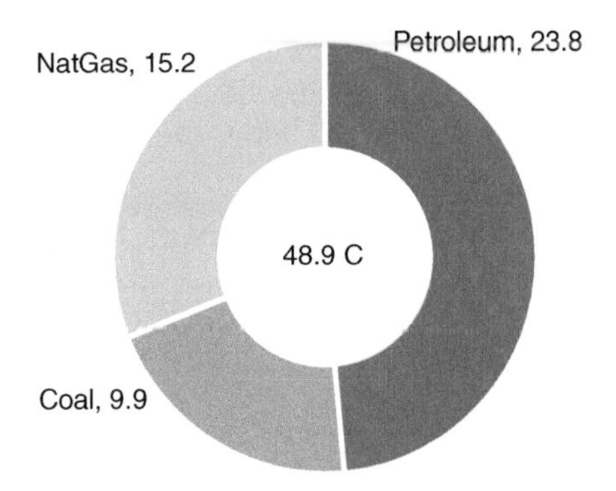

Figure 4.1 Daily carbon emissions in units of C for a person in the United States assuming only CO_2 emissions associated with fossil fuels (total of 84.5 D out of 101.6 D for all energy sources). The assumed D:C ratios are from Table 4.2, with 2.23 for natural gas, 1.2 for coal, and 1.63 for petroleum. For energy sources, see Figure 3.1.

will go into these processes. However, a more accurate analysis of emissions associated with each of these renewable technologies would require a more refined analysis than that possible here where the goal is to highlight the burden of directly using fossil fuels.

Based on the assumption of carbon emissions considering only fossil fuels, we have energy use in terms of D for the United States in 2019 (Fig. 3.1) of: natural gas, 33.9 D, coal 11.9 D, and petroleum products, 38.7 D. The sum of these three fuels is 84.5 D, out of 101.6 D used in the United States. Using the D:C values in Table 4.2, we then calculate the following distributions of C for a total of 48.9 C as shown in Figure 4.1. Based on this analysis, the D:C ratio based only on fossil fuels is 1.73, and for total energy, it would be 2.08.

With this analysis for C for total CO_2 emissions normalized to one person, we can estimate the total number of trees that would be needed to capture and sequester this amount of CO_2. In Example 4.2, we calculated ~15 trees per 1 C, and so, a total of 734 trees would be needed for each person to naturally capture and sequester the CO_2 released from the fossil fuels used in the United States of 48.9 C.

Calculating Your Own C Footprint

The carbon emissions for a person on average reflect the activities of the nation, but the real question is what your own carbon footprint might be based on your own energy use. For Figure 4.1, the D:C ratios used were previously calculated for the individual fuels. To calculate our own energy use and daily carbon emissions, we need to consider the fuels that go into the food system. To estimate a value of C based on energy for the food system and for electricity, we need to average D:C ratios based on a mix of energy inputs. These ratios can be calculated as shown in the example below.

Example 4.5

Calculate the D:C ratios in 2019 using data in Table 2.7 for (a) electricity, based on only energy going into electricity production, and (b) the total US energy system which has a different mix of fuel sources than electricity.

(a) For electricity production, the energy used is: 12.31 EJ for natural gas, 10.74 for coal, and 0.20 for petroleum. First, we calculate D and C for each fuel. Starting with natural gas, which has a D:C ratio of 2.23, we have:

$$12.31 \frac{\text{EJ}}{\text{y}} \frac{1}{328.2 \times 10^6 \text{ cap}} \frac{277.8 \times 10^9 \text{ kWh}}{\text{EJ}} \frac{1 \text{ y}}{365 \text{ d}} \frac{1 \text{ D}}{2.32 \frac{\text{kWh}}{\text{d-cap}}} = 12.3 \text{ D}$$

$$12.3 \text{ D} \frac{\text{C}}{2.23 \text{ D}} = 5.52 \text{ C}$$

The same calculation must also be made for the other two fuels. For coal, we obtain 10.7 D and 8.9 C assuming a D:C ratio of 1.2, and for petroleum, we calculate 0.2 D and 0.12 C assuming a D:C ratio of 1.63. For the complete fuel input into electricity generation, there is a total of 35.0 EJ, so the renewable fuels (which are assumed to have no CO_2 emissions) are 14.5 D. The D:C ratio for electricity based on all inputs is therefore

$$\frac{\sum \text{D}}{\sum \text{C}} = \frac{(12.3 \text{ D} + 10.7 \text{ D} + 0.2 \text{ D}) + (11.8 \text{ D})}{(5.52 \text{ C} + 8.95 \text{ C} + 0.12 \text{ C})} = \frac{35.0 \text{ D}}{14.59 \text{ C}} = 2.40 \text{ D} : \text{C}$$

Just based on the fossil fuels, the D:C ratio is 1.59, but with the carbon-neutral fuels, it increases to 2.40 showing the increased amount of energy possible per C.

(b) For a total energy use of 101.6 EJ, the calculation is made using the same procedure as above, but this time, we use total primary energy consumption (101.6 EJ) rather than the primary energy only for electricity production. For natural gas using 33.9 EJ, we obtain 33.9 D and 15.2 C. The combined C for all fossil fuels is 48.9 C. Using the appropriate D:C ratios, the average D:C ratio for all fuels is

$$\frac{\sum \text{D}}{\sum \text{C}} = \frac{101.6 \text{ D}}{48.9 \text{ C}} = 2.08 \text{ D} : \text{C}$$

This D:C ratio is for all energy. If we consider only the D in the fossil fuels (84.5 D, Table 2.6), then the D:C ratio is 1.73. These ratios are different than those for electricity due to the different amounts of the fossil fuels used for all energy production compared to electricity.

In Chapter 3 (Fig. 3.2), we calculated 48.9 D of energy based on national averages and the various energy sources that went into food processing for one person, the average national gasoline use per person; averages for a home for electricity and gas heating normalized to 2.52 people per home; and jet fuel used to travel by plane from the United States to the European Union (EU) (round trip). Next, we calculate the primary energy that went into these activities which increased the energy use to 67.6 D (Fig. 3.3). We have differentiated between the electricity that was carbon-neutral and that produced using fossil fuels, with a total of 57.1 D_{ff} of primary energy using fossil fuels and 10.6 D_{nC} that was obtained from carbon-neutral sources. Using these fuels and energy sources that went into the 57.1 D, we calculate C units with the assumptions shown in Figure 4.2 that indicated 30.5 C as the carbon emissions based on typical activities for one person. This is an overall ratio of 2.22 (67.6/30.5) for total energy to carbon emissions. In this analysis, gasoline is the largest contributor to the CO_2 emissions accounting for 40% of C.

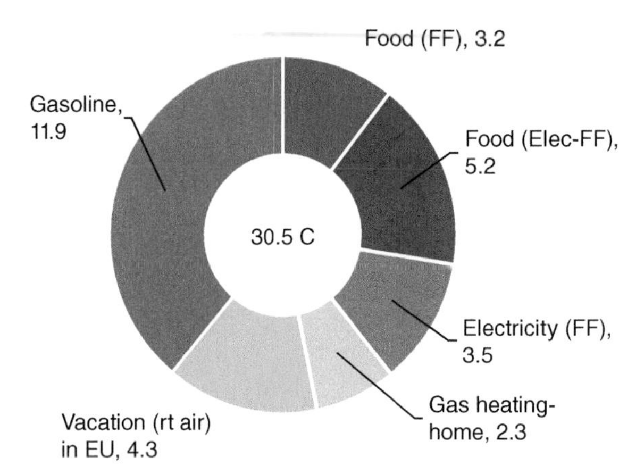

Figure 4.2 Daily carbon emissions in units of C with the assumptions of D:C ratios of: food (FF) = 1.73, food electricity (2.4), electricity = 2.4 (see Example 4.4); gas heating = 2.23 (natural gas), jet fuel for vacation = 1.61 (rt = round trip), and gasoline = 1.55 (see Table 4.2). For energy use, see Figure 3.3.

4.5 REDUCING GLOBAL CO_2 AND OTHER GHG EMISSIONS

The U.S. Environmental Protection Agency (EPA) tracks GHG emissions and has a data explorer on their site that allows calculation of these gases in any year and by multiple categories (US Environmental Protection Agency, 2020). For 2018, it was estimated that the United States emitted 5.425 Gt of CO_2 (EPA, 2020b). This calculation compares well with that estimated for fuel consumption and an average C as shown in the example below.

Example 4.6

Compare the CO_2 emissions of 5.425 Gt for 2018 reported by the EPA (2020b) *with that based on C = 48.9 C in* Figure 4.1 *assuming a US population of 326.7 million.*

From the definition of C, we know that this is the CO_2 emitted in lb/d per person. Therefore, multiplying C by the population should give us a comparable number for a population in 2018 of 327.2 million.

$$48.9\,\mathrm{C}\,(326.7\times 10^{6}\,\mathrm{cap})\,\frac{\frac{2\,\mathrm{lb\,CO_2}}{\mathrm{d\,cap}}}{\mathrm{C}}\,\frac{1\,\mathrm{kg}}{2.2\,\mathrm{lb}}\,\frac{1\,\mathrm{Gt}}{10^{9}\,\mathrm{tonne}}\,\frac{1\,\mathrm{tonne}}{10^{3}\,\mathrm{kg}}\,\frac{365\,\mathrm{d}}{\mathrm{y}} = 5.3\,\frac{\mathrm{Gt}}{\mathrm{y}}$$

This result compares well with the EPA number reasonably well given the significant figures of the other numbers and that data in Figure 4.1 were derived for 2019 with a slightly higher population.

The above analysis that produced a value of 48.9 C, calculated per person in the United States, is based only on CO_2 emissions. However, there are other GHGs, primarily methane, nitrous oxide,

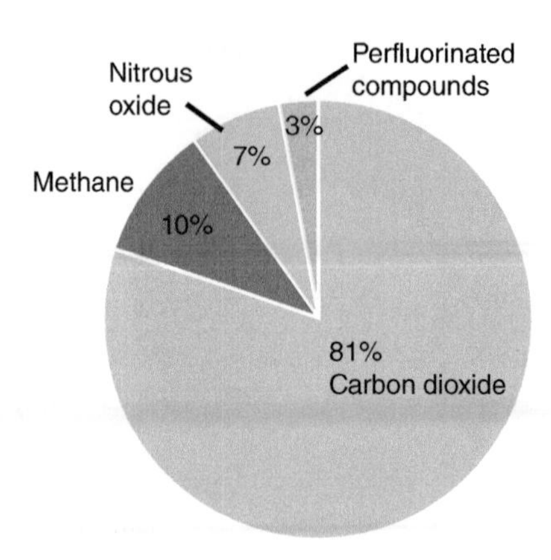

Figure 4.3 Sources of major GHGs in the United States in 2018. *Source*: Adapted from US Environmental Protection Agency (2020).

and perfluorinated gases from air conditioners and other cooling units. Total CO_2 emissions were 81% of total GHG emissions in 2018 in the United States, with the other three gases accounting for 1.252 Gt in the equivalent CO_2 emissions based on their mass releases in terms of an impact that would be the same as CO_2 (Fig. 4.3). Thus, with all these gases that is a total of 6.501 Gt/y of CO_2 equivalent emissions. Since the above calculation of $C = 48.9$ included only CO_2, if we include these other three gases in a total C value for all four of these gases, that would be equivalent to a $C_{GHG} = 60$.

Annual CO_2 and Other Greenhouse Gas Emissions Vary Based on Data from Different Sources

The amount of CO_2 estimated to be emitted by a country (or the world) can vary depending on the source of the data. Even for the same organization, the amount of CO_2 and other GHG emitted for a specific year can change over time. For example, the U.S. EPA published in their annual report in 2020 that 5.425 Gt of CO_2 was released in 2018 (EPA, 2020a), but in the 2021 report, this was revised to 5.397 Gt (EPA, 2021) (Table 4.4). For the same year, other estimates were 5.117 Gt (Dudley, 2019) and 5.281 Gt (Statistica, 2020). In general, the trends are similar over time for these different sources but given the variations in the amounts the use of four significant figures often used in these reports does not seem valid. A more extensive listing of CO_2 and other GHG emissions is in the Appendix 5.

Global CO_2 and other GHG emissions also vary by source. For example, global CO_2 emissions were estimated for 2018 to be 37.5 Gt in a UN report (United Nations Environment Programme, 2019) and about 9% less or 34.0 Gt based on an analysis by BP (Dudley, 2019). Both estimates increased for 2019 but again differed with 38.0 Gt UN report (United Nations Environment Programme, 2020) and 34.2 Gt by BP. Based on the 2019 UN report, fossil CO_2 emissions were about 65% of the total GHG emissions that included emissions associated with land use changes of 59.1 ± 5.9 Gt CO_2e. The notation CO_2e is used to account for impacts of the different gas emissions being converted to an equivalent amount of CO_2 emissions. Land use emissions include logging, fires, and changes in use (for example from forest to food crops). Without land use, the total GHG emissions were 52.4 ± 5.2 Gt CO_2e with fossil fuel emissions accounting for around 73% of the total.

Table 4.4 CO_2 emissions reported for the United States based on different sources (Gt/y).

Year	EPA 2020	EPA 2021	Dudley 2019	Statistica 2020
1990	5.128	5.117	—	5.040
2005	6.132	6.141	—	5.999
2015	5.412	5.384	5.141	5.263
2016	5.292	5.263	5.042	5.170
2017	5.254	5.219	4.984	5.131
2018	5.425	5.397	5.117	5.281
2019	—	5.274	4.965	5.130

Ranking of CO_2 Emissions Globally

The countries with the largest energy use, and therefore likely the largest carbon emissions, have the greatest responsibility for producing reductions in GHGs. Total global fossil CO_2 emissions were 37.1 Gt in 2017, which for a world population of 7.55 billion amounts to 4.9 t/y per person or ***a global average of 15 C***. Thus, the ***United States with 49 C*** (47 C in 2017) has a larger role to fill in reducing global CO_2 emissions than many other nations. That greater responsibility for achieving more reductions in carbon emissions is true for most industrialized nations compared to developing nations. The top countries, as well as some other contrasting numbers, such as the whole EU, world international shipping and world international aviation, are summarized in Table 4.5. Oil-producing nations such as Saudi Arabia can have very large C values, with the highest C of 196 calculated for Palau which has a large percentage of its income derived from tourism. Data for other years selected countries and regions are included in Appendix 5.

Table 4.5 Top countries in CO_2 emissions along with some contrasting other emissions from fossil fuels (not including all greenhouse gases) based on 2017 data.

Country	Total CO_2 (Gt/y)	CO_2 per person (tonne/y)	C
China	10.88	7.7	23
United States	5.11	15.7	47
European Union	3.55	7.0	21
India	2.45	1.8	5
Russia	1.76	12.3	37
Japan	1.32	10.4	31
Germany	0.80	9.7	29
International shipping	0.68	—	—
S. Korea	0.67	13.2	40
Iran	0.67	8.3	25
Saudi Arabia	0.64	19.4	58
Canada	0.62	16.9	51
International aviation	0.54	—	—
Palau	0.0014	64.9	196
Total for the world	37.1	4.9	15

Source: Data from Wikipedia (2020b).

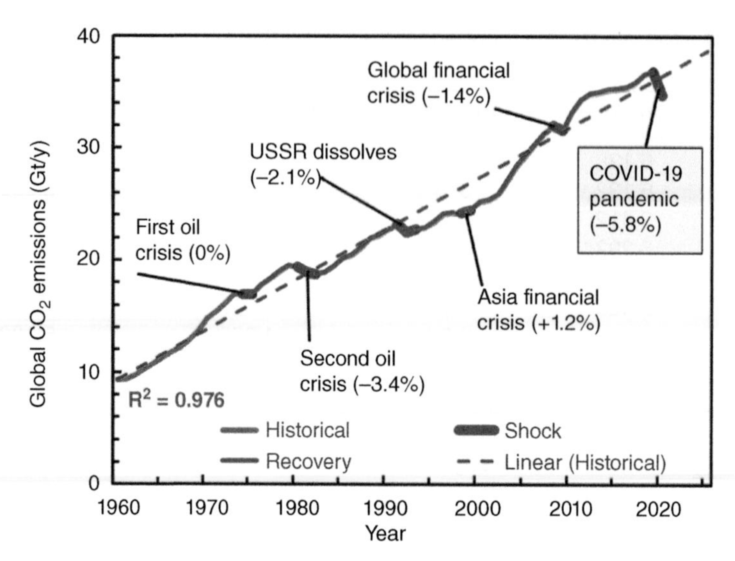

Figure 4.4 Impact of different global events on CO_2 emissions over several decades (*Source*: adapted from data in Hanna et al. (2020) between 1960 and 2019, showing CO_2 emissions and global disruptions and CO_2 emissions in 2020 from the IEA (2021)). The decline in CO_2 emissions due to the COVID-19 pandemic are the largest recorded to date, but it is expected there will again be a rebound as the economies of nations rebound. *Source*: Adapted from Hanna et al. (2020) and IEA 2021.

Impact of Global Disruptions on Carbon Emissions

There have been several major impacts on the rate of CO_2 emissions over the past few decades, including two oil crises, the dissolution of the Soviet Union, and two financial crises (Fig. 4.4). These events resulted in decreases in global CO_2 emissions for each of these events, but in each case, these emissions effectively rebounded over time so that between 1960 and 2020 there was a nearly continuous overall increase in CO_2 emissions (Hanna et al., 2020). The COVID-19 pandemic is the newest crisis that has reduced CO_2 emissions, with estimates in the spring of 2020 that emissions had decreased by 17% up to a period in early April 2020 compared to the same time period in 2019 (Le Quéré et al., 2020). In 2021, it was reported that emissions averaged over the year resulted in a global decrease of 5.8%, making it the largest decrease over the period of 1960–2021. While there is no certainty how CO_2 emissions will look going into 2021 and subsequent years, it seems likely that they will again ramp back up to where they were before the pandemic.

The events shown in Figure 4.4 may have temporarily decreased global CO_2 emissions, but we can see that overall emissions have followed a nearly linear increase over this period of time (shown by an $R^2 = 0.976$). These temporary changes in emissions have also had no impact on CO_2 concentrations measured in the atmosphere (Helm, 2020). There are other sources of CO_2 other than fossil fuels, for example those due to land use changes, that make these relatively small changes imperceptible when examining the net global impact of CO_2 and other GHG emissions. These other sources and sinks of CO_2 are further discussed in Chapter 14. Only large global efforts and a focus on reducing fossil fuels can change this rebound effect. Thus, the pandemic is an opportunity to greatly modify CO_2 emissions and keep them from increasing, but only if the energy infrastructure of the planet rapidly changes in the coming decade.

Project Drawdown

The reduction in GHGs will require not only that we discontinue CO_2 emissions from fossil fuels, but also that we change industrial and agricultural practices, personal lifestyles, and many other things that include how we grow and process food, choose and use materials for our buildings and civic infrastructure, and design and operate our transportation systems. As energy demands increase, we will need to sequester CO_2 into forests, land, oceans, and below the ground. Project Drawdown is an organization that is dedicated to identifying solutions to reduce or drawdown CO_2 and other GHGs. The first book on this subject, by Paul Hawkin in 2017, identified 100 ways for reducing CHG emissions (Hawken, 2017). Three years later, this assessment was updated to place activities into 76 categories, with changes in terms of both the ranking of different solutions as well as the amount of CO_2 that could be reduced by each solution (Wilkinson, 2020).

The top 10 most impactful efforts for reducing CO_2 emissions based on the newest assessment by Project Drawdown are shown in Table 4.6. In the original 2017 study, the #1 most important topic was identified as reducing refrigerants (fluorinated gases) that were leaking into the environment, with solutions that could lead to a reduction in −89.74 Gt of CO_2 equivalent over the 30-y period from 2020 to 2050. In the 2020 study, this topic was revised and separated into two categories with −57.7 Gt listed for refrigerant management and a ranking of #4. The use of alternative refrigerants was then listed to potentially contribute another −43.5 Gt (rank #7). Taken together, these two related items represent −101.2 Gt, which would still make them collectively larger than any other single item. Note that a minus sign is used here to distinguish reductions from other units which are emissions into the environment and thus have a positive sign.

Reducing food waste is now the #1 topic listed by Project Drawdown (Table 4.5) with a potential reduction of −87.4 Gt over the same 30-y period. The total amount by which GHGs could be reduced by the 76 solutions currently is listed as −997.2 Gt of CO_2 equivalent. If we compare the absolute current amount of GHG emissions by the United States of 6.50 Gt/y, if that US rate is not changed, it represents a total mass addition of 195 Gt over the same 30-y period as the global drawdown.

Table 4.6 The top 10 most impactful solutions to climate change in terms of equivalent CO_2 reduction.

No.	Solution	CO_2 (10^9 tonnes)	−C
1	Reduced food waste	87.4	0.93
2	Health and education	85.4	0.91
3	Plant-rich diets	65.0	0.69
4	Refrigerant management	57.7	0.62
5	Tropical forest restoration	54.5	0.58
6	Onshore wind turbines	47.2	0.50
7	Alternative refrigerants	43.5	0.46
8	Utility-scale solar photovoltaics	42.3	0.45
9	Improved clean cookstoves	31.3	0.33
10	Distributed solar photovoltaics	28.0	0.30
	Total	542.3	5.80

The value of C is shown as a negative sign as these are carbon reductions normalized to a global population of 9.4 billion people in 2050.

Source: Adapted from Wilkinson (2020).

Since units of Gt are difficult to process in terms of things in our own lives, we convert the GHG reductions identified by Project Drawdown into units of C in Table 4.5 using calculates as shown in the example below. Since we are projecting these reductions into the future, and there is no certainty of the population in 2050, this calculation is instructive just to place the enormous numbers of global CO_2 reductions on the same basis as the CO_2 emissions in our lives.

Example 4.7

Convert the –997.2 Gt of reductions in GHGs to units of C by assuming a total global population of 9.4 billion in 2050.

A Gt is defined as 10^9 tonne (metric ton, or mt), and 1 mt is 10^3 kg, so using these and the definition of C, we have

$$-997.2\ \text{Gt}\frac{1}{30\ \text{y}}\ \frac{1}{9.4\times10^9\ \text{cap}}\ \frac{10^9\ \text{tonne}}{\text{Gt}}\ \frac{10^3\ \text{kg}}{\text{tonne}}\ \frac{1\ \text{y}}{365\ \text{d}}\ \frac{2.2\ \text{lb}}{\text{kg}}\ \frac{\text{C}}{\frac{2\ \text{lb}}{\text{d cap}}} = -10.7\ \text{C}$$

Since the rate at which the carbon can be reduced is not clear, we have used the estimated population in 2050 here. The calculation spans the period of 2020 (population of 7.795 billion) to 2050 (estimated population of 9.4 billion) so an intermediate population could have been used instead of the final year estimate. If the 2020 global population was used, then the result would be –12.8 C.

The results in Example 4.7 show that all the drawdown amounts, if normalized to all the people that will be living on the planet, are smaller than our own current daily C unit of C = 49, based only on CO_2, or $C_{GHG} = 60$ if all GHGs are considered. This shows that everyone in the United States has the potential to greatly reduce global CO_2 emissions by reducing their own personal emissions. For example, using 1 gal of gasoline in a day represents a D = 15.2 and C = 9.8, so everyone reducing their gasoline use could have a large impact on global carbon emissions.

References

AFEWORK, B., HANANIA, J., STENHOUSE, K. & DONEV, J. 2019. *Energy education: Nuclear power plant* [Online]. Available: https://energyeducation.ca/encyclopedia/Nuclear_power_plant [Accessed May 15 2020].

DUDLEY, B. 2019. BP statistical review of World Energy 2019. 68 ed.

EPA. 2020a. *Greenhouse gas inventory data explorer* [Online]. Available: https://cfpub.epa.gov/ghgdata/inventoryexplorer [Accessed June 11 2020].

EPA. 2020b. Inventory of U.S. greenhouse gas emissions and sinks, 1990-2018.

EPA. 2021. Inventory of U.S. greenhouse gas emissions and sinks, 1990-2019.

FRIEDRICH, C. 2019. *Do you absorb all of the Calories you eat?* [Online]. Available: https://cathe.com/do-you-absorb-all-of-the-calories-you-eat [Accessed May 5, 2020].

HANNA, R., XU, Y. & VICTOR, D. G. 2020. After COVID-19, green investment must deliver jobs to get political traction. *Nature,* 582, 178–180.

HAWKEN, P. 2017. *Drawdown: The most comprehensive plan ever proposed to reverse global warming*, Penguin Books.

HELM, D. 2020. *Net Zero: How we stop causing climate change*, HarperCollins Publishers.

HRUSTIC, A. 2017. *How much fat, protein, and carbs should you be eating?* [Online]. Men's Health. Available: https://www.menshealth.com/nutrition/a19537348/how-much-fat-and-carbs-should-you-eat [Accessed May 5 2020].

IEA. 2021. *Global energy review: CO_2 Emissions in 2020* [Online]. Available: https://www.iea.org/articles/global-energy-review-co2-emissions-in-2020 [Accessed March 14 2021].

KEENEY, D. 2009. Ethanol USA. *Environmental Science & Technology,* 43, 8–11.

LE QUÉRÉ, C., JACKSON, R. B., JONES, M. W., SMITH, A. J. P., ABERNETHY, S., ANDREW, R. M., DE-GOL, A. J., WILLIS, D. R., SHAN, Y., CANADELL, J. G., FRIEDLINGSTEIN, P., CREUTZIG, F. & PETERS, G. P. 2020. Temporary reduction in daily global CO_2 emissions during the COVID-19 forced confinement. *Nature Climate Change*, 10, 647–653.

PADDOCK, C. 2019. *US diet still contains too many low quality carbs* [Online]. Medical News Today. Available: https://www.medicalnewstoday.com/articles/326456 [Accessed May 5 2020].

STATISTICA. 2020. *Carbon dioxide emissions from energy consumption in the U.S. from 1975 to 2020 (in million metric tons of carbon dioxide*) [Online]. Available: https://www.statista.com/statistics/183943/us-carbon-dioxide-emissions-from-1999 [Accessed April 3 2021].

TORABIZADEH, H. 2011. All proteins have a basic molecular formula *Chemical and Molecular Engineering,* 5, 501–505.

UNITED NATIONS ENVIRONMENT PROGRAMME 2019. Emissions gap report 2018, https://www.unep.org/resources/emissions-gap-report-2019.

UNITED NATIONS ENVIRONMENT PROGRAMME 2020. Emissions gap report 2020, https://www.unep.org/emissions-gap-report-2020.

URBAN FORESTRY NETWORK. 2021. *Trees improve our air quality* [Online]. Available: http://urbanforestrynetwork.org/benefits/air%20quality.htm [Accessed January 14 2021].

US ENERGY INFORMATION ADMINISTRATION. 2016. *Carbon dioxide emissions coefficients by fuel* [Online]. Available: https://www.eia.gov/environment/emissions/co2_vol_mass.php [Accessed May 20 2020].

US ENERGY INFORMATION ADMINISTRATION. 2019. *Units and calculators explained* [Online]. Available: https://www.eia.gov/energyexplained/units-and-calculators [Accessed May 8 2020].

US ENERGY INFORMATION ADMINISTRATION. 2020a. *Electricity: Form EIA-923 detailed data with previous form data (EIA-906/920)* [Online]. Available: https://www.eia.gov/electricity/data/eia923 [Accessed June 24 2020].

US ENERGY INFORMATION ADMINISTRATION. 2020b. *July 2020 monthly energy review* [Online]. Available: https://www.eia.gov/totalenergy/data/monthly/pdf/mer.pdf [Accessed August 7 2020].

US ENVIRONMENTAL PROTECTION AGENCY. 2020. *Global greenhouse gas emissions data* [Online]. Available: https://www.epa.gov/ghgemissions/global-greenhouse-gas-emissions-data [Accessed April 19 2020].

VITZ, E., MOORE, J. W., SHORB, J., PRAT-RESINA, X., WENDORFF, T. & HAHN, A. 2020. *Foods: Burning or metabolizing fats and sugars* [Online]. LibreTexts. Available: https://chem.libretexts.org/Bookshelves/General_Chemistry/Book%3A_ChemPRIME_(Moore_et_al.)/03Using_Chemical_Equations_in_Calculations/3.04%3A_Analysis_of_Compounds/Foods%3A_Burning_or_Metabolizing_Fats_and_Sugars [Accessed May 5 2020].

WIKIPEDIA. 2020a. *Ethanol fuel* [Online]. Available: https://en.wikipedia.org/wiki/Ethanol_fuel [Accessed May 12 2020].

WIKIPEDIA. 2020b. *List of coutnries by carbon dioxide emissions* [Online]. Available: https://en.wikipedia.org/wiki/List_of_countries_by_carbon_dioxide_emissions [Accessed June 12 2020].

WILKINSON, K. 2020. *The drawdown review: Climate solutions for a new decade*, Project Drawdown. https://drawdown.org/sites/default/files/pdfs/Drawdown_Review_2020_march10.pdf.

CHAPTER 5

DAILY WATER UNIT w

5.1 ENGINEERED AND NATURAL WATER SYSTEMS

Water is the single most important resource on the planet, and we use more of it than any other material. There have been several evolutions of water use over the ages, beginning with providing water to cities, progressing to water treatment to stem the spread of water-borne diseases and infusion of sanitation into modern cities in the form of used water treatment, and most recently addressing concerns about low concentrations of pollutants in the water while maintaining a sufficient supply of potable water for everyone on the planet (Sedlak, 2014). The next step in the evolution of water is the link between water and energy, as we try to reduce the use of fossil fuels and thus move to make our water infrastructure sustainable in terms of energy consumption.

In the natural environment, we seek to avoid extremes in water availability: too much in the wrong place, for example, can produce floods, and too little rainfall or storage can result in draughts. Large cities can have water availability challenges because of the volume of water needed to supply the city and typically that water must be pumped over long distances. The energy used for water pumping is particularly large for cities in the Western United States. The energy for collection of used water for treatment, and its discharge back into the environment, usually is much less energy-intensive but the release to a single location creates a "point source" with potentially large environmental impacts on ecosystems due to the very large flows at a single site. A large point source can be an environmental challenge due to, for example, a large influx of nutrients that can stimulate algal growth in receiving water bodies. However, having a point source of treated used water can also be an opportunity, providing an ability to recharge aquifers that have become depleted due to excessive pumping for use as potable uses such as drinking water, or as a source of water for irrigation.

In this chapter, we will examine the amount of water that we use and the energy needed for collecting, treating it for potable use, then distributing it to consumers or for other specific uses. Once water is used for residential and most commercial and industrial applications, it must again be collected, treated, and discharged back into the environment. Many cities and countries are already experiencing severe potable water shortages, and around 1 billion people lack sufficient access to potable water, and over 2 billion lack safe drinking water in their homes (Osseiran, 2017). Adequate used water treatment is also a health concern, and over 4 billion do not have safely managed sanitation, with around half not having access to basic sanitation. Intensive water use relative to available surface waters has resulted in some aquifers being drawn down so low that the water is either becoming unavailable as the aquifer is drained or nearly dry, or the water is so deep that it

Daily Energy Use and Carbon Emissions: Fundamentals and Applications for Students and Professionals,
First Edition. Bruce E. Logan.

requires nearly the same energy to pump it to the surface as that used to desalinate seawater. The amount of water derived from groundwater aquifers and released into the environment has been substantial. It is estimated that waters released from aquifer pumping which go into streams, rivers, and ultimately the ocean, combined with artificial reservoirs and water storage and losses of water in closed basins, have contributed 0.77 mm/y or 42% of the measured rise in sea level between 1961 and 2003 cm (Pokhrel et al., 2012).

Providing potable water and treating wastewater can place huge burdens on the energy infrastructure, and thus, providing adequate water supplies and sanitation to everyone in the world using the same technologies used today could further increase carbon emissions. Therefore, solutions are needed that are both effective and not energy- or technology-intensive.

5.2 WATER USE AND THE DAILY WATER USE UNIT w

We have discussed our daily energy needs in terms of a unit D and carbon emissions in terms of the unit C, and so, it makes sense to develop a **daily water unit w** to link the energy–CO_2–water units together as a baseline for daily energy and water consumption and CO_2 emissions. The unit for daily water use is designated here with a lowercase w since the uppercase W is used for energy units of Watts. While the unit w does not directly translate to energy use or carbon emissions, it does establish a minimum or baseline for our daily use.

What is the quantity of water for 1 w or the minimum volume of water needed per day for 1 person? One estimate is that we need about 15.5 cups (0.97 gal) of water a day to survive based on an average male of typical height and weight and 11.5 cups (0.72 gal) for a female (Mayo Clinic, 2020). Thus, **we define 1 w = 1 gal of water per day per person** or 1 gal/d/cap = 1 gpdc = 1 w, as a minimum for both men and women (because we would not want to go below that minimum for anyone). Since the quantities of w and gpd are the same, the unit of w does necessarily provide greater insight into water use than just using gpd in the United States. However, in most of the world, water and other volumes of liquids are reported in units of liters or cubic meters, so 1 w = 3.78 L/d = 0.0037 m^3/d. This w definition is therefore useful globally due to the different ratio of liters and gallons to 1 w. Using a new unit of w also serves the main purpose of focusing the amount of water used to a reference point that establishes all water use relative to the water we need to use every day just to stay alive. This baseline of 1 w then can be compared to water used to provide food we need to eat every day or the amount of water needed for a power plant and the carbon dioxide emissions that result from those activities.

Water use per person in the industrialized nations is much larger than 1 w so water use numbers can be quite large. The amount of water used per person is variable, and there is no generally accepted "average" for a person in the United States. Reported averages for water use per person range from about 52 gpcd, or 52 w, to 171 w depending on the specific conditions used in the study, year of the study, location in the United States, different water uses included in the study (i.e. water use in the home or averaged across a community), and even factors such as the weather (Wikipedia, 2020c) (Table 5.1). A study by the United States Geological Survey (USGS) on single- and multi-family homes showed an average of 89 w in 2010 and 83 w in 2015 (Wikipedia, 2020c). A USGS website indicated overall that the average person uses 80–100 w (USGS, 2020). A survey by the Water Research Foundation (WRF) of 762 households in the United States and Canada in 2016 concluded 138 w per house or 52 w based on 2.65 people per household in the survey (Wikipedia, 2020c). This amount of daily water use in the WRF study in 2016 was reported to be a 22% decrease since the 1990s (Wikipedia, 2020c), which is in reasonable agreement with 74 w in another study in 1998 (Viessman and Hammer, 2005).

Table 5.1 Average water use reported in different studies based on year and agency for household use, and one study based on normalized use across the United States.

Study	Water use, w (gpcd)	References
Water Research Foundation in 2016	52	Wikipedia (2020c)
Textbook (American Water Works Association [AWWA] study in Denver)	52	Viessman and Hammer (2005)
USGS study in 2010	83	Wikipedia (2020c)
USGS study in 2013	89	Wikipedia (2020c)
USGS website (2020)	80–100	USGS (2020)
DOE study in 2014	142	Bauer et al. (2014)
EPRI study in 2013	171	Pabi et al. (2013)
All water uses in 2013	1300	Bauer et al. (2014)

The number of people in a household is important. Average water use declined with the number of people in the household, with 97 w reported for 1 person in a house compared to 50 w with 6 people in a home (Viessman and Hammer, 2005).

A far greater amount of water use occurs outside our residences. Based on a report in 2014 using data from 2011 on water use tied to energy consumption, the public water supply, defined here as both residential and commercial sectors, consumed 11% of the total water use of 405 billion gallons per day, equivalent to 141 w (Fig. 5.1) (Bauer et al., 2014). Using the total water flow normalized to the population in that year (311.6 million), the average for all uses was 1300 w. The data in Figure 5.1 show that 48% of the water used was for thermoelectric cooling, defined as the amount of water used in power plants for cooling. The net amount of water used for cooling is a function of the type of power plant and other factors, but it is mostly a function of whether water use is used for once-through cooling (used once and then discharged into the environment) or reused many times (water is recirculated many times through a cooling tower). When water is reused in

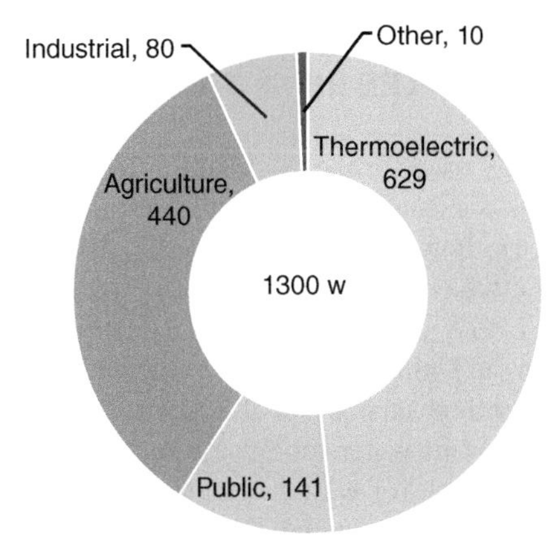

Figure 5.1 Water use in the United States in 2011 data normalized to the total population. Note that the unit w equals one gallon of water per person per day. *Source*: Data from Bauer et al. (2014).

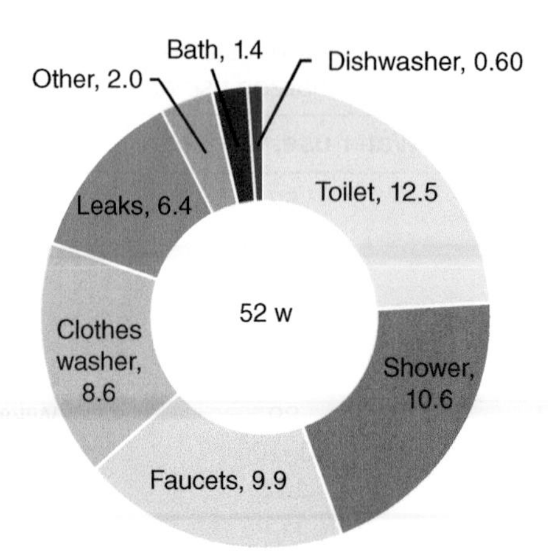

Figure 5.2 Typical water use in the home separate by uses, with all numbers in units of w, defined as one gallon per day per person. *Source*: Data from Wikipedia (2020c).

a cooling tower, a portion of that water evaporates with each use (cycle) and thus salts build up in the water which can lead to salt precipitation. Therefore, some volume of salty water in the cooling tower (called blowdown) must be replaced each cycle with additional fresh water that have a much lower concentration of salts to make up for this lost water. As a result, the power plant will still use a net amount of water for cooling even with water reuse.

Water use in the home provides some insight into its connection with energy consumption. Toilets and showers use the most water, accounting for 44% of water use in the home (Fig. 5.2). Water for showers, baths, clothes washers, faucets, and some other uses all can use hot water, with previous estimates that on average the water heater uses 17% of the electricity in a home (US Energy Information Administration, 2020a). Water heaters can use natural gas, propane, or electricity (direct heating or using a heat pump), so the exact amount of energy consumed by the water heater will depend both on the number of users and the type of system installed in a home.

5.3 ENERGY USE FOR OUR WATER INFRASTRUCTURE

Estimates of the amount of energy used for the US water infrastructure vary, with most estimates that 3% or more is used in terms of electricity. One of the first reports on energy use for the water infrastructure was by the Electric Power Research Institute (EPRI) in 2002 where it was stated that about 4% of all electricity produced in the United States was used for the water infrastructure based on 0.44 EJ (123 TWh or 14 GW of power) of electrical energy used and 12.1 EJ (3360 TWh) of total electricity production in the United States (3.7% calculated here based on stated numbers) (Goldstein and Smith, 2002). Based on an analysis of Bauer et al. (2014), energy for water use was primarily electricity used for the public water supply, estimated to be 0.105 EJ (29.3 TWh/y = 3.3 GW), with the same amount of energy for wastewater treatment (0.101 EJ). The energy uses for water activities within the residential, commercial, and industrial categories are difficult to extract as the total energy use is shown in their flow chart. Based on just the water supply and wastewater treatment only 0.21 EJ (58 TWh), this would amount to ~0.2% of electricity produced in the United States.

The focus on electricity use for water consumption misses two important aspects of energy use for the water infrastructure. First, only electricity use is considered, not other energy inputs, for example, using natural gas to heat water in homes. Second, when the energy use is given only as electricity, the *primary energy* is not presented, and so, the total energy required is not reported. If the primary fuel use is not stated that makes it more difficult to calculate fossil fuel use and thus to make better estimates of CO_2 emissions from the water infrastructure.

The analysis by Sanders and Webber (2012) provided a more complete picture of energy use in the United States for the water infrastructure as it considered all sources of energy used rather than only electricity. This assessment of water and energy use in 2010 concluded that 12.6% of primary energy consumption was used for the water-related purposes, with 5.3% of this energy used as water services (Sanders and Webber, 2012). This assessment is summarized in a Sankey plot, shown in Figure 5.3, given in units of trillion Btu. Conversion of Btu into EJ shows that a total of 13.0 EJ (12 318 trillion Btu) were used by the water sector. Of this total energy, 43.5% went into electricity generation (5.66 EJ), resulting in 2.21 EJ (2092 trillion Btu) produced for the water industry (Fig. 5.3). This analysis makes it clear that electricity as a source is only 16.9% of the total energy going into the water infrastructure. Thus, previous estimates based only on electricity energy (not primary energy) substantially underestimated energy use by a factor of 5.9 times.

The "primary energy" listed in Figure 5.3 includes 672 trillion Btu (0.71 EJ) in the category of renewable energy. However, as previously noted, this energy use is exaggerated based on a typical fossil fuel plant energy efficiency which was defined as 38.5% in 2010 (Lawrence Livermore National Laboratory, 2020). Thus, the actual energy use of 13.0 EJ would have been slightly reduced to 12.6 EJ. Energy losses of ~7% due to transmission to the point of use are also not considered.

Example 5.1

Calculate the CO_2 emissions in units of C for the water sector using data in Figure 5.3 *and the renewable energy reduced to remove the additional energy added for a fossil plant efficiency of electricity generation, for a US population in 2010 of 309.3 million.*

To calculate the units of C, we first need to convert the energy by source into units of D, and then using the D : C ratio for the different fuels, we can calculate C. For example, the coal use was 2874 trillion Btu or 3.03 EJ.

$$\text{Coal: } 2874 \times 10^{12}\text{ Btu}\,\frac{1\text{ quad}}{10^{15}\text{ Btu}}\,\frac{1.0551\text{EJ}}{\text{quad}} = 3.03\text{ EJ}$$

This energy in EJ is converted into D units as:

$$\text{D}:\ \frac{3.03\,\dfrac{\text{EJ}}{\text{y}}}{309.3\times 10^{6}\text{ cap}}\,\frac{277.8\text{ TWh}}{\text{EJ}}\,\frac{10^{9}\text{ kWh}}{\text{TWh}}\,\frac{1\text{ y}}{365\text{ d}}\,\frac{1\text{ D}}{2.32\,\dfrac{\text{kWh}}{\text{d cap}}} = 3.21\text{ D}$$

For coal, we use a D:C ratio of 1.20 for subbituminous coal and calculate C as:

$$\text{C}:\ 3.21\text{ D}\,\frac{1\text{ C}}{1.20\text{ D}} = 2.68\text{ C}$$

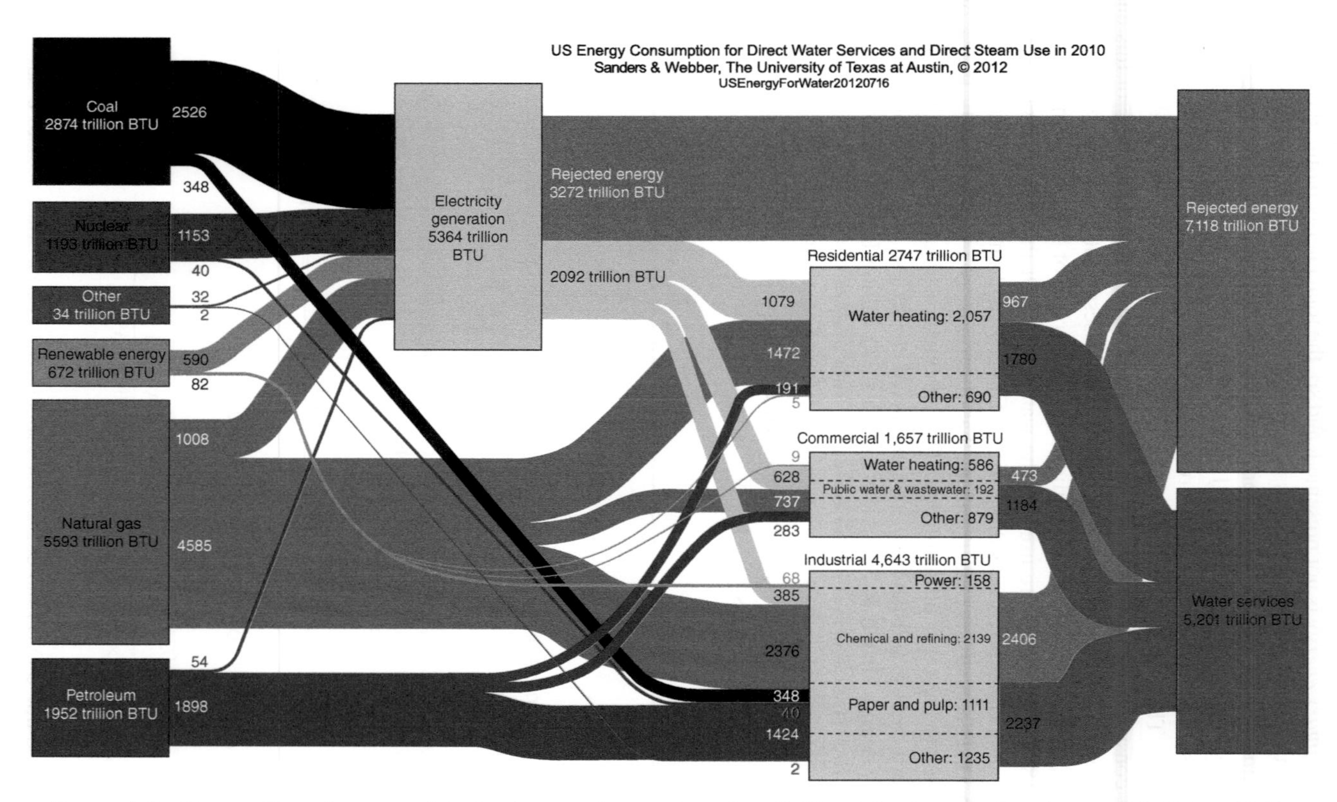

Figure 5.3 Energy flow diagram related only to water use in units of trillion Btu. *Source*: Sanders and Webber, 2012.

For the other fuels, we use the D:C ratios given in Table 4.2 (2.23 for natural gas, and 1.63 for petroleum) and assume that the "Other fuels" has a D:C value similar to that of diesel fuel. The results of the calculations are summarized in Figure 5.4. We see that the water infrastructure consumes 13.3 D of energy and produces 6.85 C of CO_2 emissions, for an average D:C ratio of 1.94.

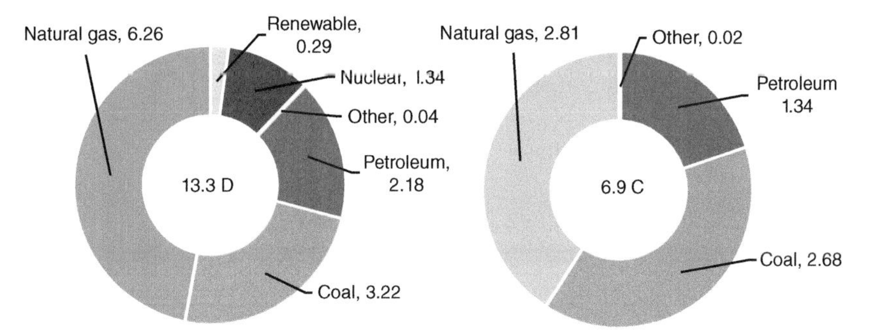

Figure 5.4 Energy use for the water infrastructure in units of D, and the CO_2 emissions in terms of C based on calculations in Example 5.1. *Source*: Data from Sanders and Webber (2012).

Energy Use for the Water Infrastructure in the Western 17 States

Detailed studies on energy use for the water infrastructure in 17 states in the western portion of the United States provide further insights into how dependent some states can be on energy for water. Drinking water and wastewater utilities used about 2% of electricity produced on average over the year (Tidwell et al., 2014), in reasonable agreement with averages reported in other studies. In addition, large-scale water conveyance consumed another 1.2–1.6% of the electricity with a further 2.6–3.7% by agricultural pumping, bringing the total to 5.8–7.4% based on electricity use. The study by Tidwell reported ranges of water use for different activities, but for brevity, only the averages of these numbers will be used here. For these 17 states, the total energy use was 1192 TWh/y, with 6.6% (78.3 TWh used for the water infrastructure.

Example 5.2

Assuming a US population of 105 million (Tidwell et al., 2014) *for the western 17 states, (a) what is the average D based on an electricity use of 78.3 TWh? (b) In 2010, the average electricity efficiency was reported as 32.2% in the Lawrence Livermore National Laboratory (LLNL) chart for that year (similar to chart for 2019 shown in Fig. 2.2). What is the primary energy in the fuel, designated here as D_p? (c) The D : C ratio for electricity generation for 2010 is estimated to be 2.0 (compared to 2.40 in Example 4.4 for 2019). What is the C value based on this ratio?*

(a) Using the given information and using a subscript e to emphasize that the is the energy in the electricity only, we have

$$\mathrm{D_e}: \frac{78.3\ \frac{\mathrm{TWh}}{\mathrm{y}}}{105 \times 10^6\ \mathrm{cap}}\ \frac{10^9\ \mathrm{kWh}}{\mathrm{TWh}}\ \frac{1\ \mathrm{y}}{365\ \mathrm{d}}\ \frac{1\ \mathrm{D}}{2.32\ \frac{\mathrm{kWh}}{\mathrm{d\ cap}}} = 0.88\ \mathrm{D_e}$$

(b) The energy efficiency can be used to calculate the primary fuel use as

$$D_p : 0.88\, D_e \frac{D_p}{0.322\, D_e} = 2.73\, D_p$$

Note that the electrical plant efficiency in 2010 of 32.2% was lower than that in 2019 of 34.3%, indicating that modern electricity production is more efficient than in previous years due to the use of natural gas (more efficient) and renewables such as solar panels (no waste heat associated with electricity generation).

(c) Using the given D:C ratio, we have

$$2.73\, D_p \frac{C}{2.0\, D_p} = 1.36\, C$$

The above example shows an important point: Evaluating our water infrastructure based on electricity used ignores the primary energy that is needed to produce that electricity. For example, we can see that we use slightly less than 1 D of electricity on average across those 17 western states, but converting that back to primary energy we obtain 2.73 D which indicates our much larger energy use due to the need to produce that electricity with other fuels (Fig. 5.5). Recalculating the data on 17 states, we can see a wide range in D values from 0.95 (Oklahoma) to 23 (Idaho). Carbon emissions will also follow the use of fossil versus renewable energy, with the C values decreasing as we shift from coal to natural gas and solar or wind electricity. The values used in Example 5.1 assumed national averages, but California is working to obtain more of its energy from renewables so its C emissions will be lower than implied by the national average D:C ratios. The amount of electricity used varies depending on the specific study, for example, with 19% estimated by others for electricity consumed by California (Stokes and Horvath, 2009) compared to 10% by Tidwell (2014).

The study by Tidwell et al. (2014) also provides important information on the distribution of energy use. The percentage of electricity used for water services by certain states is quite large relative to the averages. For example, Idaho uses up to nearly half its electricity (estimates of 34–49%) for water, with about half of that used for surface water pumping. Other high users based on the percentage of electricity used in the state are Montana (16%), Arizona (14%), and California (10%). To balance off these high percentage users, states like North Dakota and Oklahoma use <2% of electricity for their water services. In arid regions, especially California and Arizona, water is pumped over large distances. As a result, California uses 8 TWh or about 3% of its electricity just to convey water, primarily from the northern part of the state to the drier South to cities like Los Angeles. Arizona uses 4.2 TWh or 5.6% of its electricity for water transport to the southern cities.

5.4 ENERGY USE FOR WATER TREATMENT

Energy use for treating and delivering potable water to our homes is one of the single most important luxuries in our life, although this is under-appreciated by many people in industrialized countries that take such services for granted. The energy used for water treatment varies with the source,

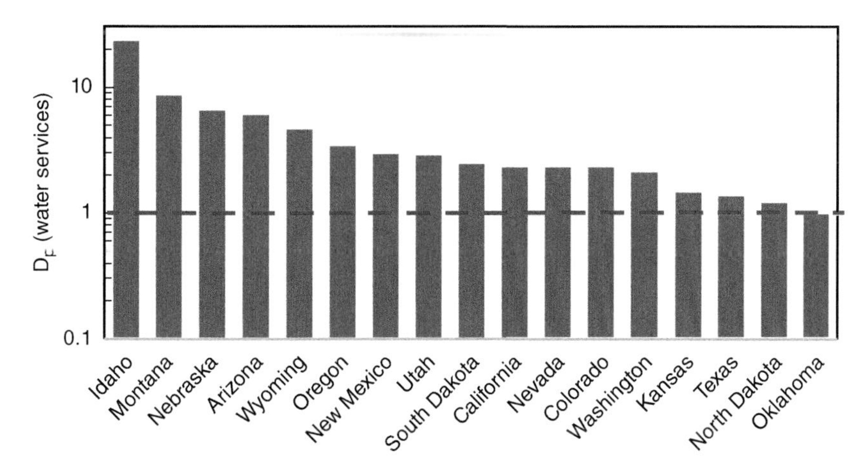

Figure 5.5 Energy use in D units for water services based on primary energy assuming 32.2% efficiency for electricity generation. The red dashed line shows energy use of 1 D. *Source*: Data from Tidwell et al. (2014).

Table 5.2 Electrical energy use for water treatment based on either public water supplies.

Source	Electricity (kWh/m^3)	D
Surface water	0.42	0.12
Groundwater	0.37	0.15
Seawater	3.2	0.88

D units are calculated for electricity (not primary energy) assuming 171 w.
Source: Based on Pabi et al. (2013).

with surface water generally requiring more energy to treat than groundwater, but the need for pumping groundwater can produce a rise in overall energy demands. EPRI and the WRF examined energy treatment for different water technologies and water sources (Pabi et al., 2013) that updated a previous 2002 study (Goldstein and Smith, 2002). This 2013 assessment showed that the public water supply required 0.42 kWh/m^3 when using a surface water and 0.37 kWh/m^3 for a groundwater (Table 5.2). These translate to daily energy units ranging from 0.12 to 0.88 D (not including seawater desalination) assuming a use of 171 w based on overall water use and the population served for the study (258 million). Overall, the annual electricity use was calculated to be 39.2 TWh or 1% of electricity production in 2011. For primary energy, these numbers would increase by approximately 3.1 times assuming a typical energy efficiency of 32.2% for power plants in 2010.

An analysis of typical energy use within a water treatment plant shows that most energy goes into pumping water (Fig. 5.6). Water is usually pumped to a plant where it is then pumped through a series of processes that include water filters, with the finished water sent to distribution systems to customers. The actual treatment process consumes only 14% of the overall energy. This analysis does not include the energy for making chemicals that are used in the process (coagulants such as alum, for example) or disposing and handling of sludges produced from treatment.

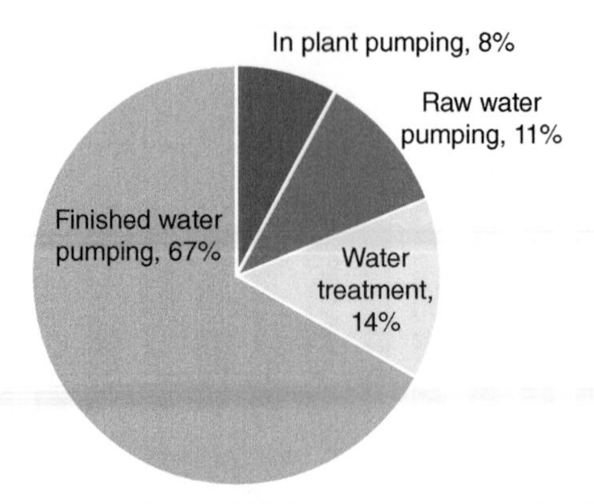

Figure 5.6 Distribution of energy for water treatment through water intake to distribution to customers. *Source*: Pabi et al. (2013).

Table 5.3 Electrical energy use for used water treatment based on either public water supplies.

Source	Electricity (kWh/m^3)	D
Less than secondary	0.20	0.048
Partial	0.22	0.054
Pumping reused water	0.34	0.083
Secondary	0.55	0.13
Greater than secondary	0.71	0.17
No discharge	0.71	0.17

D units are calculated for electricity (not primary energy) assuming 150 w.
Source: Based on Pabi et al. (2013).

5.5 ENERGY FOR USED WATER TREATMENT

The energy use for used water (commonly referred to as wastewater) treatment can be slightly higher than that of water treatment. For domestic wastewater treatment plants, a often cited number is 0.6 kWh/m^3 (McCarty et al., 2011). This energy use assumes a modern facility in the United States that uses full secondary treatment (required by law) and meets typical nutrient (nitrogen and phosphorus) removal standards. An EPRI study in 2013 calculated a slightly higher energy use of 0.71 kWh/m^3 for plants classified as "greater than secondary" based on the use of nutrient removals with conventional secondary treatment using activated sludge or with membrane bioreactors (MBRs) (Table 5.3) (Pabi et al., 2013). However, the amount of energy varies widely in a used water plant with the total energy dependent on plant size and flowrates, location in the country, and specific discharge standards. Some plants have waivers for full secondary treatment due to discharge to the ocean or to accomplish partial treatment prior to sending the water on for additional treatment, and thus, these categories have lower energy demands. Plants with "no discharge" typically have land application of the treated water and do not discharge directly to a navigable water body.

The EPRI study assessed annual energy use for wastewater treatment at 30.2 TWh or 0.8% of electricity generation (Pabi et al., 2013). The energy used in these different processes can be

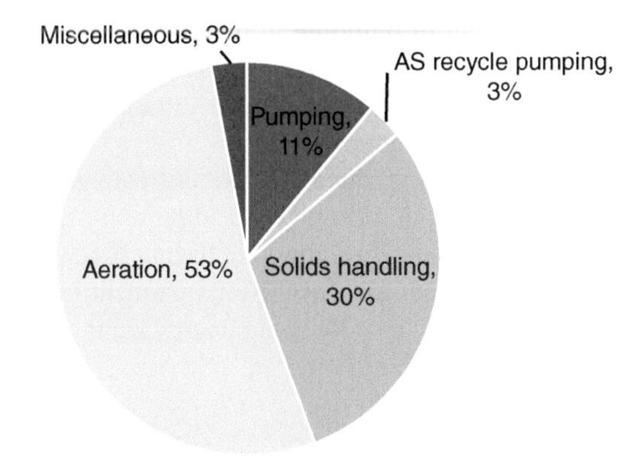

Figure 5.7 Distribution of energy for a used water treatment plant. *Source*: Adapted from Pabi et al. (2013).

normalized to volume of water and then converted to units of D using 1 kWh/d = 1 D. The D values in Table 5.3 are estimated for a reduced amount of used water being treated relative to the population of 150 w, which is less than the water delivered to a home (170 w) due to overall water losses for irrigation or leaks in sewers. In practice, water delivered to a plant can also exceed water flows sent to customers due to infiltration, so it is also possible to have higher used water flowrates than water use rates.

Energy for a used water plant is much less dependent on pumping due to primarily gravity flow to the plant. In-plant and other pumping consumes an estimated 14% of energy (Fig. 5.7) compared to 86% of electricity used in a potable water treatment plant. Used water flowing by gravity to the treatment plant avoids energy for collection, and the treated used water discharge site is usually close to the plant which can eliminate or minimize those pumping costs. In a used water plant, about half of the electricity for conventional treatment is used for aeration of the water in tanks to supply oxygen for bacteria to remove the organic matter. The use of an aerobic process for treatment produces a high amount of waste biomass (sludge) which must be further treated in anerobic digesters. This treatment in digesters produces methane and eventually biosolids that are safe for land disposal or reuse (depending on the plant and quality). At smaller plants, there is only enough methane gas to heat the digestors or produce a small excess that is flared, while in the largest plants, there is sufficient methane to further warrant its recovery and production of electricity.

The amount of energy used to supply treated water and to collect and treat the used water is usually not included in energy estimates for a home. Energy use for supplying water was estimated at 39.2 TWh/y (Pabi et al., 2013). This electricity use is significant as it adds 0.15 D per person based on electricity use or 0.37 D for a home (assuming 2.52 people in an average home). For used water treatment, it was estimated that 30.2 TWh/y was used, which would add an additional 0.11 D per person or 0.29 D per home, for a total of 0.66 D per home. If we factor in the additional energy in the primary fuel, then this would increase by about three times the energy needed to make that electricity. This amount of energy use was estimated to increase by 39% for water and by 74% for used water compared to a 1996 assessment of energy needed for the water infrastructure. These increases could be due to the growth in population, increased percentages of the population being connected to modern treatment systems (i.e. from septic tanks to central treatment facilities), and energy use to meet more stringent effluent standards for nutrients such as nitrogen and phosphorus.

Nitrogen removal increases the amount of pumping in the plant and results in greater electricity used for pumping. Using MBRs for treatment instead of conventional activated sludge units also increases energy use due to a need to pull the treated water through membranes with small pores as well as aeration to avoid membrane clogging.

Different approaches are being used to reduce used water treatment energy consumption and operating costs such as more tailored aeration based on loading to the plant and improving solids handling. Some used water treatment plants are now purchasing renewable energy or setting up their own solar power plants to offset electricity use. New treatment technologies that can avoid aeration are discussed in the section below on producing energy using wastewater.

5.6 DESALINATION

Desalination is examined here as a separate water production technology due to its increasing global importance and more frequent use in the United States than is generally recognized by even water treatment professionals. There are more than 649 desalination plants in the United States with a capacity of exceeding 402 million gallons per day, with >95% of these plants used to treat brackish groundwater (USGS, 2018). Most of this water is not for potable use and instead is used for cooling water for power generation, aquaculture, and the oil and gas industry for drilling, enhanced oil and gas recovery, and hydraulic fracturing. Globally, almost 65% of all desalination plants use reverse osmosis (RO), with 21% using multistage flash (MSF) distillation, and 7% using multi-effect distillation (MED) (Rao et al., 2016). In 2012, 75.2 TWh of electricity was used to desalinate 65.2 million m^3 of water, consuming ~0.4% of global electricity production.

RO is the preferred method for seawater desalination as it has the lowest energy requirement of 1.6–4.8 kWh/m^3 compared to other desalination technologies, with an average of ~3.2 kWh/m^3 based on a US Department of Energy (DOE)-funded study (Rao et al., 2016). This electrical energy use is about 10 times that of conventional water treatment due to the energy consumed to produce the high pressure needed to force freshwater through the RO membrane. However, even conventional treatment technologies can have high costs due to pumping and conveyance over large distances as noted above. The theoretical minimum for seawater desalination under idealized conditions was estimated as 1.06 kWh/m^3 based on minimal water recovery or only a small portion of water passed through the membrane and desalinated relative to the total amount of water used. A more practical minimum of 1.56 kWh/m^3 was estimated based on a 50% water recovery (Elimelech and Phillip, 2011), while another estimate was 1.48 kWh/m^3 based on practical operational conditions (US Department of Energy, 2017). RO plants designed for high energy efficiency can achieve ~2 kWh/m^3, so they operate at close to the theoretical minimum amount of energy for practical operation.

Water recoveries using RO are typically around 80%. Higher water recoveries are possible but high treated water recoveries will produce salt concentrations in the brine that can result in precipitation and excessive membrane fouling. Low recoveries are undesirable as they would increase the relative amount of water requiring pretreatment and pumping, which are costs which can become a substantial portion of the total desalination electrical energy costs. While DOE estimated that 89% of total electrical energy was for desalination with 11% as for other uses, Voutchkov (2018) estimated that 29% of the energy costs were associated with pretreatment and other processes (Table 5.4). These estimates for pretreatment are not comparable, but the two estimates of 0.27–0.39 kWh/m^3 show that associated treatment costs cannot be neglected in the overall costs for desalination.

Table 5.4 Different estimates of electricity use for desalination facilities assuming ocean water with a subsurface intake.

Operation	US DOE (kWh/m^3)	Voutchkov (kWh/m^3)
Intake	0.0038	0.19
Pretreatment	0.27	0.39
Desalination	3.3	2.54
Post-treatment	0.11	—
Discharge	0.0038	—
Other[a]	—	0.27
Product water delivery	—	0.18
Total	3.69	3.57

[a] Includes post-treatment and discharge.
Source: Adapted from US Department of Energy (2017) and Voutchkov (2018).

Example 5.3

How much electricity energy is used to desalinate water, in units of D_e, if a person uses 52 w?

If we assume an average desalination energy of 3.69 kWh/m^3 that includes all the other related energy technologies (Table 5.4), then we have

$$52\text{ w}\frac{\frac{1\text{ gal}}{\text{d cap}}}{\text{w}}\ \frac{3.78\text{ L}}{\text{gal}}\ \frac{1\text{ m}^3}{10^3\text{ L}}\ \frac{3.69\text{ kWh}}{\text{m}^3}\ \frac{1\text{ D}}{\frac{2.32\text{ kWh}}{\text{d cap}}} = 0.31\text{ D}_e$$

Compared to the electricity use for a home of 13 D_e, this is not a lot of energy, but this is only for home use and a lower estimate of per-person use. If the daily water use was 157 w, then we would calculate 0.94 D_e. Furthermore, if we assumed all water use in the country normalized per person, or 1300 w, then we would have 7.8 D_e. However, about half (48%) of this 1300 w water use is for power plant cooling, and thus, it would not need to be desalinated for that purpose.

5.7 ENERGY STORAGE USING WATER

Energy storage has become more critical to the renewable energy infrastructure because of the intermittent nature of electricity generation from solar and wind sources. On the smaller scale of homes or businesses, batteries can be used and these battery systems are becoming more affordable and common. Some new larger-scale solar farms also are being designed with huge battery storage capabilities, for example, the City of Los Angeles has contracted for a solar farm to produce 0.4 GW or 876 GWh of electricity, with 800 MWh of battery backup (Service, 2019).

Large-scale energy storage for electrical power generation utilities predates the rise of wind and solar systems as energy storage has been used for decades by coal and nuclear power plants for load balancing or for storing energy so that it can be used during periods of peak load when demand may exceed that possible by the power plant capacity. In the United States, over 95%

of storage utility energy is by pumped storage hydropower (PSH) based on pumping water using turbines into a higher elevation reservoir when electricity demands are low and then releasing the water through the same turbines to produce electrical power when needed. The overall efficiency is generally in the range of 70–80% based on the electrical power produced and that recovered from the turbines, although some optimized systems have indicated up to 87% energy recovery (Wikipedia, 2020b). The amount of energy that can be stored worldwide was estimated at 184 GW, with 25 GW in the United States in 2017 equal to about 2% of power production. Once constructed, the water systems have long lifespans typical of reservoirs where the main challenge is infill due to sediment accumulating in the reservoir and reducing its storage capacity. PSH water use has a net energy loss estimated to be 5 TWh or about 0.1% of electricity use in 2019 (US Energy Information Administration, 2020b).

Any existing hydropower dam is a potential resource for PSH if it has sufficient water capacity. Often referred to as hybrid hydroelectric systems, water that has traveled through the turbines to generate electrical power can be pumped back using other energy sources (such as wind or solar) and then be available as needed based on electricity demand. Such systems require availability of a downstream water storage reservoir with a size dependent on the amount of time water would need to be stored before being returned to the uphill damned storage reservoir.

Reservoirs used for PSH must have sufficient elevation differences to make the system economical and practical for energy storage. The maximum efficiency for water storage is given as ρgh, where ρ is the density of water, g the gravitational constant, and h the elevation difference. Pumping the water into the reservoir will result in some energy loss, and then generating the power using a turbine results in a second energy loss. These two losses combined contribute to the 70–80% overall energy recovery. The average water height for PSH is 152 m (500 ft) (Hunt et al., 2020). Water tanks could be used for water storage, but these require natural elevation changes larger than those practical based on water tank heights used for storing potable water that is then released to provide sufficient water pressure for community water systems.

Water storage capacity is often reported in terms of power, P [W], calculated as

$$P = \frac{Q\,\rho g h}{\eta} \tag{5.1}$$

where Q (m^3/s) is the flowrate, ρ water density (10^3 kg/m^3), g the gravitational constant (9.81 m/s^2), h the pumping height (m), and η the pump efficiency. For the given units, power is in W, where 1 kg m^2/s^3 = 1 W. If Q is given in units of m^3/h and units of kW are required, then the result on the right-hand side of the equation must be divided by 3.6×10^6 for the unit conversion. This concept of energy for pumping water into a storage reservoir is illustrated in the example below.

Example 5.4

(a) If we want to use a water tower to store energy, how much energy does it take to completely fill a typical water tower 30 m high, holding 1 million gallons of water? (b) Using units of D, how many homes could be powered if each home uses 13 D? (c) How much power would it take to pump that water into the tower in 6 h for a pump efficiency of 90%?

(a) The energy to pump water 1 m in height *theoretically* (i.e. with no energy losses), and for a 30 m high tower, for 1 m^3 of water can be calculated as:

$$\rho gh = (10^3 \frac{kg}{m^3})(9.81 \frac{m}{s^2})(1\ m) = \frac{9810\ \frac{kg\ m^2}{s^2}}{1\ m^3} \frac{1\ J}{\frac{kgm^2}{s^2}} \frac{1\ kWh}{3.6 \times 10^6\ J} = 0.00273 \frac{kWh}{m^3}$$

$$30\ \text{m high tower} = 30 \times 0.00273 \frac{kWh}{m^3} = 0.082 \frac{kWh}{m^3}$$

If the water tower can hold 1 million gallons = 3785 m^3, then the energy to fill that tower is

$$E = (3785\ m^3)(0.082 \frac{kWh}{m^3}) = 310\ kWh$$

(b) Converting this into units of D and then calculating the number of houses (assuming one person per house):

$$\text{Water Tower D} = 310 \frac{kWh}{d\ cap} \frac{1\ D}{2.32 \frac{kWh}{d\ cap}} = 135\ D$$

$$N_h = \frac{(135\ D)}{(13\ D)} = \sim 10\ \text{houses}$$

From this calculation, we can appreciate that water for energy storage requires a large elevation change since 30 m (~100 ft) does not store much energy.

(c) The flowrate is calculated from the given water volume to be pumped into the tower in 6 h, and the power is calculated using Eq. (5-1) as:

$$Q = \frac{V}{t} = \frac{(3785\ m^3)}{(6\ h)} = 631 \frac{m^3}{h}$$

$$P = \frac{Q \rho gh}{\eta} = \frac{(631 \frac{m^3}{h})(1000 \frac{kg}{m^3})(9.81 \frac{m}{s^2})(30\ m)}{(0.90)} \frac{1\ J}{\frac{kg\ m^2}{s^2}} \frac{1\ kWh}{3.6 \times 10^6\ J} = 57\ kW$$

The theoretical power is 52 kW, but at a pump efficiency of 90% the power requirement is 57 kW (76 hp) due to energy losses in pumping.

PSH systems have been used for many years, with most of the PSH facilities built within the period of 1950–1990. Since then, there have been very few new PSH facilities in the United States (US Department of Energy, 2016), although growth of PSH facilities is still increasing globally. One example of a PSH facility in the United States is the Muddy Run reservoir system built by the Philadelphia Electric Company and completed in 1968. At that time, it was the largest PSH in the world with a capacity of 1.1×10^{10} gal of water in the upper reservoir and 10^{12} gal in the lower reservoir. The system had an elevation change of 120 m (410 ft), and a capacity for 1.072 GW, and produced 361 GWh in 2016 (Wikipedia, 2020a). The largest pumped storage facility in the United States now is the Bath County pumped storage station in Virginia, with a capacity of 3.03 TW (Wikipedia, 2020b) and a storage capacity of 24 GWh (Fig. 5.8). A 3.6 GW PSH plant is currently

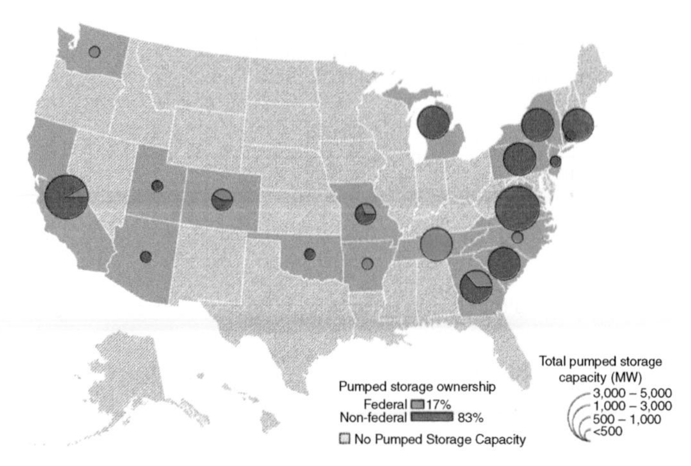

Figure 5.8 Locations of pumped storage facilities in the United States which have a total capacity of 21.6 GW. *Source*: US Department of Energy (2016).

under development in the Hebei Province in China which will have a capacity of 3.6 GW when complete in 2021.

At the end of 2015, the United States had a total of 42 PSH plants with a total capacity of 21.6 GW, nearly all of this storage (97%) for utility-scale electricity storage (US Department of Energy, 2016), with 25 GW estimated by 2017 (Wikipedia, 2020b). The same DOE study reported that there could be an additional 16.2 GW by 2030, with 35.5 GW of new storage possible by 2050, providing nearly 57 GW of total pumped energy storage. Globally, there is sufficient storage capacity for 17.3 PWh (1 PWh = 10^{15} Wh) or enough to store 79% of the electricity produced today (Hunt et al., 2020). The cost of this storage is estimated to be <\$50 MWh^{-1}, with the lowest cost estimates of \$1.8 MWh^{-1}. Some of these sites would be expensive compared to battery storage, but the scale of energy storage might preclude using batteries. For example, the solar farm planned by the City of Los Angeles has an estimated cost of \$13 MWh^{-1} for a storage capacity of 0.8 GWh (Service, 2019) which would be less than the cost for some of these PSH sites but also smaller in energy storage capacity.

5.8 CO_2 EMISSIONS AND PROJECT DRAWDOWN SOLUTIONS

Project Drawdown identified two water-related technologies to reduce CO_2 emissions: low-flow water fixtures, for a reduction of 0.9 Gt; and improved water distribution efficiencies, for a reduction of 0.7 Gt. These combined approaches present an opportunity for 1.6 Gt in reduced CO_2 emissions or 0.02 C averaged over the world population from 2020 to 2050. Given the large amounts of energy use for the water infrastructure, these would appear to be modest goals. For example, if the western US states were to continue to use the same amount of energy, with no change in population, from 2020 to 2050, the total CO_2 emissions would be 80.4 C based on using 2.68 C from Example 5.1. An assessment of Sanders and Weber (2012) concluded that 12.6% of primary energy use in the United States in 2010 was related to our water infrastructure. Based on total CO_2 emissions in 2010 (17.44 Gt) that would equate to 2.18 C for just a single year, or using the same number over 30 y, a total of 65.4 C. All of these estimates imply that much needs to be done to reduce CO_2 emissions related to the water infrastructure over the coming decades.

Drawdown solutions do not "double count" changes, and therefore, these estimates related to water use do not consider that renewable electricity production will greatly impact water consumption and thus water use. For example, in the Sanders et al. (2012) study, 5.3% of the energy was used for water services, while the balance (7.3%) was waste heat. If wind and solar can be used to provide electricity for the water infrastructure, then emissions associated with water use would essentially vanish. The key for avoiding carbon emissions from operation of the water infrastructure is therefore to sever the ties between the pumps and treatment plants and fossil fuel energy sources and rely on renewable energy sources.

The sheer amount of electrical power used by the United States for its water infrastructure cannot be adopted by the world as that would overwhelm any envisioned capacity for electricity generation. Total water flow in the United States was estimated at 1300 w (Fig. 5.1) based on conveyance and the population, which includes power generation, agriculture, industries, and domestic uses. For domestic water use, estimates are still 52–171 w depending on the specific survey and locations of the population. The greatest gains in reducing energy use for water will therefore be to decrease reliance on fossil fuel energy for water pumping, but that could be quite challenging for certain states such as California and Arizona that rely on water pumped over many hundreds of miles, and Idaho which can use nearly half its electricity generation primarily for irrigation. Better integration of water use with renewable energy cycles, coupled to conservation and better farming techniques, will all be needed to produce a more energy-sustainable water infrastructure.

References

BAUER, D., PHILBRICK, M. & VALLARIO, B. 2014. The water-energy nexus: challenges and opportunities.

ELIMELECH, M. & PHILLIP, W. A. 2011. The future of seawater desalination: Energy, technology, and the environment. *Science,* 333, 712–717.

GOLDSTEIN, R. & SMITH, W. A. 2002. *Water & sustainability (Volume 4): U.S. Electricity consumption for water supply & treatment -- The next half century.* Electrical Power Research Institute (EPRI).

HUNT, J. D., BYERS, E., WADA, Y., PARKINSON, S., GERNAAT, D. E. H. J., LANGAN, S., VAN VUUREN, D. P. & RIAHI, K. 2020. Global resource potential of seasonal pumped hydropower storage for energy and water storage. *Nature Communications,* 11, 947.

LAWRENCE LIVERMORE NATIONAL LABORATORY. 2020. *Energy flow charts* [Online]. Lawrence Livermore National Laboratory. Available: https://flowcharts.llnl.gov/commodities/energy [Accessed April 18 2021].

MAYO CLINIC. 2020. *Nutrition and healthy eating* [Online]. Available: https://www.mayoclinic.org/healthy-lifestyle/nutrition-and-healthy-eating/in-depth/water/art-20044256?p=1 [Accessed June 28 2020].

MCCARTY, P. L., BAE, J. & KIM, J. 2011. Domestic wastewater treatment as a net energy producer – can this be achieved? *Environmental Science & Technology,* 45, 7100–7106.

OSSEIRAN, N. 2017. *2.1 billion people lack safe drinking water at home, more than twice as many lack safe sanitation* [Online]. World Health Organization (WHO). Available: https://www.who.int/news/item/12-07-2017-2-1-billion-people-lack-safe-drinking-water-at-home-more-than-twice-as-many-lack-safe-sanitation#:~:text=lack%20safe%20sanitation-,2.1%20billion%20people%20lack%20safe%20drinking%20water%20at%20home%2C%20more,as%20many%20lack%20safe%20sanitation&text=Some%203%20in%2010%20people,report%20by%20WHO%20and%20UNICEF [Accessed February 15 2021].

PABI, S., AMARNATH, A., GOLDSTEIN, R. & REEKIE, L. 2013. *Electricity use and management in the municipal water supply and wastewater industries*. Report No. 3002001433, Palo Alto, CA, USA. Available: https://www.sciencetheearth.com/uploads/2/4/6/5/24658156/electricity_use_and_management_in_the_municipal_water_supply_and_wastewater_industries.pdf [Accessed September 19 2020].

POKHREL, Y. N., HANASAKI, N., YEH, P. J. F., YAMADA, T. J., KANAE, S. & OKI, T. 2012. Model estimates of sea-level change due to anthropogenic impacts on terrestrial water storage. *Nature Geoscience,* 5, 389–392.

RAO, P., AGHAJANZADEH, A., SHEAFFER, P., W. R. MORROW, III, BRUESKE, S., DOLLINGER, C., PRICE, K., SARKER, P., WARD, N. & CRESKO, J. 2016. *Volume 1: Survey of available Information in support of the energy-water bandwidth study of desalination systems*. https://eta.lbl.gov/publications/volume-1-survey-available-information. LBNL-1006424.

SANDERS, K. T. & WEBBER, M. E. 2012. Evaluating the energy consumed for water use in the United States. *Environmental Research Letters,* 7, 034034.

SEDLAK, D. L. 2014. *Water 4.0: The past, present, and future of the world's Most vital resource*, Yale University Press.

SERVICE, R. F. 2019. Solar plus batteries is now cheaper than fossil power. *Science,* 365, 108.

STOKES, J. R. & HORVATH, A. 2009. Energy and air emission effects of water supply. *Environmental Science & Technology,* 43, 2680–2687.

TIDWELL, V. C., MORELAND, B. & ZEMLICK, K. 2014. Geographic footprint of electricity use for water services in the western U.S. *Environmental Science & Technology,* 48, 8897–8904.

US DEPARTMENT OF ENERGY 2016. *Hydropower vision: A new chapter for America's 1st renewable energy source*.

US DEPARTMENT OF ENERGY 2017. Bandwidth study on energy use and potential energy savings opportunities in U.S. seawater desalination systems.

US ENERGY INFORMATION ADMINISTRATION. 2020a. *How is electricity used in U.S. homes?* [Online]. Available: https://www.eia.gov/tools/faqs/faq.php?id=96&t=3 [Accessed May 10 2020].

US ENERGY INFORMATION ADMINISTRATION. 2020b. *What is U.S. electricity generation by energy source?* [Online]. Available: https://www.eia.gov/tools/faqs/faq.php?id=427&t=3 [Accessed July 16 2020].

USGS. 2018. *How is brackish groundwater being used?* [Online]. Available: https://water.usgs.gov/ogw/gwrp/brackishgw/use.html [Accessed July 16 2020].

USGS. 2020. *Water Q&A: How much water do I use at home each day?* [Online]. Available: https://www.usgs.gov/special-topic/water-science-school/science/water-qa-how-much-water-do-i-use-home-each-day?qt-science_center_objects=0#qt-science_center_objects [Accessed July 1 2020].

VIESSMAN, W. J. & HAMMER, M. J. 2005. *Water Supply and Pollution Control*. 7th ed., Upper Saddle River, NJ, Pearson Prentice-Hall.

VOUTCHKOV, N. 2018. Energy use for membrane seawater desalination – current status and trends. *Desalination,* 431, 2–14.

WIKIPEDIA. 2020a. *Muddy Run pumped storage facility* [Online]. Available: https://en.wikipedia.org/wiki/Muddy_Run_Pumped_Storage_Facility [Accessed July 8 2020].

WIKIPEDIA. 2020b. *Pumped-storage hydroelectricity* [Online]. Available: https://en.wikipedia.org/wiki/Pumped-storage_hydroelectricity [Accessed July 8 2020].

WIKIPEDIA. 2020c. *Residential water use in the U.S. and Canada* [Online]. Available: https://en.wikipedia.org/wiki/Residential_water_use_in_the_U.S._and_Canada#:~:text=The%20equivalent%20average%20use%20per,gphd%20(or%20472%20lphd). [Accessed July 1 2020].

CHAPTER 6

RENEWABLE ENERGY

6.1 INTRODUCTION

Wood is the oldest material used by mankind for heating, but using wood might not be the best choice from an environmental perspective due to the energy needed to haul it to your home and the environmental consequences of combustion by-products. Modern wood stoves can operate with clear flue gases and near-zero emissions, but they are relatively rare in homes in the United States (although now required in many European countries). Project Drawdown considers biomass to be a transition fuel: It is better than fossil fuels as there is no net release of CO_2 into the atmosphere from the fuel itself (not including other aspects of growing and delivering the fuel), but ultimately it is not as clean a source of heat or electricity compared to other non-combustion-based renewable options. Therefore, although wood is a viable method of producing heat for your home, it is not further considered here in this chapter as a renewable fuel for electricity generation.

From the perspective of most homeowners, solar energy is the clear choice for renewable energy. Wind energy on the scale of a home is expensive, although it can provide electricity in the absence of sunlight. To what extent do these two technologies make sense for a homeowner? If we use these technologies for our homes, we can develop efficient systems based on clean electricity and, with battery storage, be completely free from grid electricity. Timing electricity use to be coupled to when power is most available (during the middle of the day for solar) can help minimize battery storage. Even if you are connected to the grid, timing electricity use (for charging an electric vehicle) for when it is most abundant (or least expensive) can save money. In this chapter, we consider the potential benefits of solar and wind energy primarily for home applications as these constitute an important part of your daily energy consumption and production. Other renewable technologies, such as using geothermal heat sources, biomass energy, and production of biofuels, are also covered here although these technologies are not currently available at the single-home scale. Residential heating with geothermal systems is based on using the ground as a constant temperature source for a heat pump (and not a deep geothermal heat source to produce power) and so that geothermal heat pump system is covered in Chapter 9.

6.2 SOLAR PHOTOVOLTAICS

Understanding how much electricity your solar panels produce is not as straightforward as you might expect! Solar panels are rated in terms of peak power, or the electrical power produced when

Daily Energy Use and Carbon Emissions: Fundamentals and Applications for Students and Professionals, First Edition. Bruce E. Logan.

sunlight at 1000 watt per square meter (W/m^2) or 1 kW/m^2 shines on a panel. A typical solar panel today produces more than 300 W, so if you purchased 30 panels that were each rated at 330 W, you are buying a 9.9 kW system. The average size home solar installation is about 10 kW, so that means on average people are buying around 30 panels for a house. However, what you can extract from these panels depends on many different factors including placement, location, and the weather.

For a system rated at 9.9 kW of power, if you had peak solar irradiation for 24 hours a day, you would produce 238 kWh of energy. Of course, the sun does not shine all day (well, except in polar areas in the summer like Iceland), and your panels are likely not perfectly aligned with the sun. Most flat panels placed on roofs or land are placed in a stationary position and at a fixed angle, and therefore, they do not move to track optimal sun angles during the day. The angle of the panels is usually the same as your roofline (i.e. they are not installed at an angle different than your roof) if they are placed on the roof, and they may not point at the most optimal angle or direction. If a solar panel system on your roof were to receive 8 h of peak sunlight, that would mean 79 kWh/d of electricity produced per panel. How much energy could you really expect? Companies that sell and install solar panel systems can give you good estimates based on your particular roofline, house orientation, panel performance, and weather in your area, but we can make some approximations based on general locations.

You can roughly estimate the annual production based on general location in the country using Figure 6.1. Each of these regions in the figure shows the sunlight over a day converted to the number of *peak solar hours*. For example, in much of the Northeastern United States that is 4.2 h, while much of coastal California would be 5 h, with a small part of Arizona and the California desert topping the list at 6 h (Fig. 6.1). A more detailed table listing specific locations is in Appendix 6. These tabulated data provide different estimates than Figure 6.1 for the zone marked 4.2 h, for example, 3.91 h in State College and 4.0–4.2 h for Harrisburg, both in PA; 4.08 h for New York City and 3.6–3.9 h for Syracuse, both in NY; and 4.0–4.51 h for Portland ME.

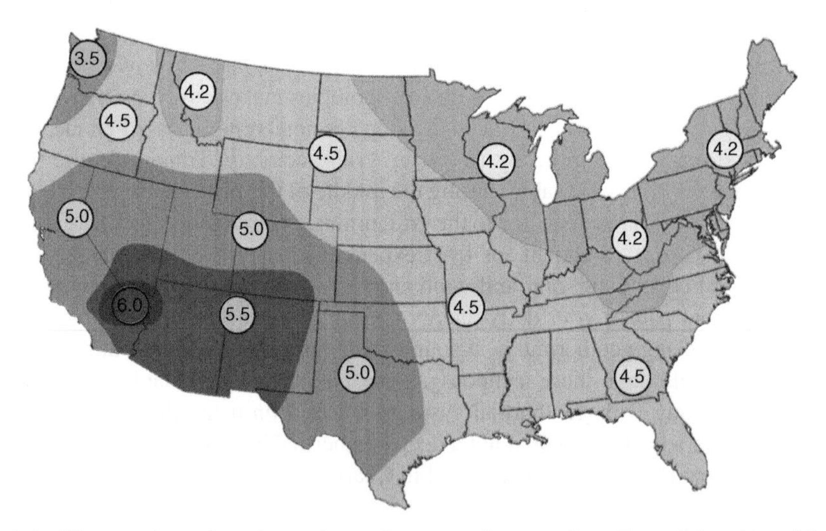

Figure 6.1 The number of peak sun hours by zone. *Source*: Sun Hours Map: How Many Sun Hours Do You Get?. Unboundsolar.com. Link: https://unboundsolar.com/solar-information/sun-hours-us-map.

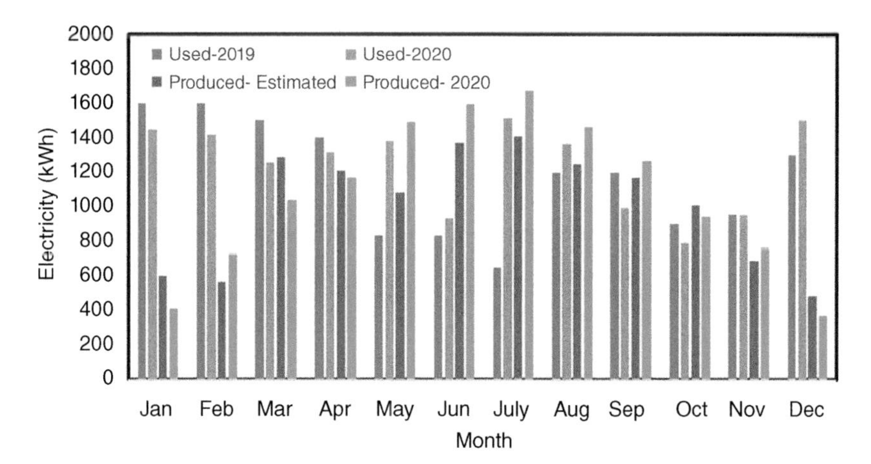

Figure 6.2 Comparison of electricity generation produced and used in 2020 versus that used in 2019 production estimated for 2020, for a home in State College, PA. (Production for a 33-panel array at 320 W per panel, or 10.56 kW at peak power). The summer of 2020 was the hottest year on record for this town.

For our base case of approximately 10 kW, a house in Pennsylvania with 4.2 peak hours could produce on average 42 kWh/d, compared to 45 kWh/d in Florida and 55 kWh/d in Arizona based on zones in Figure 6.1. These values may seem surprisingly close given the relatively large differences in latitude and climate between Florida and Pennsylvania, but specific locations within any of these states are more important than general areas (i.e. direction they point, angle of the roof, and possible obstructions to sunlight over the day).

How well do these predictions compared to actual systems? Energy use in the author's home in 2019 (30 panels, 10.56 kW) is compared to that estimated by a solar company and actual 2020 data in Figure 6.2. For most months, the contractor estimates of performance were in fairly good agreement with the actual energy production, but differences were clearly evident for some other months due to weather conditions. For example, the estimate for electricity produced based on 2019 data was less than that actually produced in 2020 for March, but in the month of May, electricity generation appreciably exceeded that predicted for the solar panel system for that month. Using an estimate of 4.2 h for Zone 5 in Figure 6.1, the amount of electricity predicted for this home would be 19 D (assuming one person), while the contractor estimated 15 D. The number of peak hours for State College using Appendix 6 is 3.91 h, so that is equivalent to 18 D. The actual results, based on the data in Figure 6.2, is that 15.7 D was produced in 2020, which was an overall result closest to the estimate of the contractor.

On average, a home in the United States uses about 13 D (30 kWh/d) averaged over the year, and therefore, a 9.9 kW system could produce 18.1 D (42 kWh/d) or about 40% more energy than your home might need throughout the year, even in the Northeastern United States. However, in any month, there can be a large difference between solar electricity produced and that used, such as January when electricity production is low relative to use or July when production is high. This variability in production over hours, days, and months is why it is so difficult to provide a stable amount of electricity to home-based renewable energy systems.

Example 6.1

Calculate the energy from a single solar panel in units of D for zones producing 4.2 h (for example in Maine) and 5.5 h (for example in Arizona) of peak sunlight. Assume the panel produces 320 W.

For the zone where there is 5.5 h/d for peak sunlight, we can convert this to kWh and then to units of D:

$$320\ \text{W}\ \frac{5.5\ \text{h}}{\text{d}}\ \frac{\text{kWh}}{10^3\ \text{Wh}} = 1.76\frac{\text{kWh}}{\text{d}}\ \frac{1\ \text{D}}{\frac{2.32\ \text{kWh}}{\text{d}}} = 0.76\ \text{D}$$

For 4.2 h, the result is 0.58 D. Therefore, to provide 13 D, you would need 22 panels in Maine compared to 17 panels in Arizona.

Cost of Solar Energy for Homes

Understanding how those panels relate to your energy use will eventually bring you to the question: What does it cost? The short answer is that it costs less than buying all the electricity that you will need over the next 25 years. Solar panels are currently rated to last 25 years, although there is a gradual decrease in performance over time. Panels today might degrade 0.5% per year in performance or <5% over 10 years. That might impact the payback time slightly, but electricity prices are likely to go up, not down, over the same period of time. For example, the national average cost of electricity was 6.81 ¢/kWh in 2000, compared to 9.83 ¢/kWh in 2010 and 10.6 ¢/kWh in 2019. So, there was a 44% increase between 2000 and 2010 and a 7.8% increase between 2010 and 2019.

The payback period for installing a 10 kW system ranges from about 8 to 17 years, depending on your location, the cost of the system, and most importantly the cost of electricity in your region. Several examples are calculated for a range of states in Table 6.1. Some results might be quite surprising, for example, the payback is estimated to be about the same in sunny Arizona as Maine. While Arizona averages more sunlight hours over the year, the cost of electricity is 41% more expensive in Maine than it is in Arizona, thus reducing the number of years needed for recovering your initial costs. Most states fall into the 10- to 13-year payback period as the costs of the solar installation do not vary much, and most states have on average about 4.2–4.5 h of peak equivalent sunlight based on Figure 6.1 and other data in Table 6.1. This analysis does not consider all costs, such as interest and finance charges if you are obtaining a loan for purchasing the solar array or the opportunity cost of not making other investments with the money you would spend on the solar panels.

The tax credit for installing solar panels has been changing recently, and therefore, it is difficult to predict how long federal tax credits will continue into the future. In 2019, the tax credit was 30% but it decreased to 26% in 2020, and it was planned to be further reduced to 22% in 2021 (Rhodes, 2020). However, due to the COVID-19 pandemic, federal tax credits were extended and were not predicted to decrease from 26% to 22% until 2023. It is possible that tax credits could again be extended into future years. However, as shown in Table 6.1, the payback periods for all states were less than the 25-year warranty offered by most manufacturers without any federal tax credits, with many areas having a payback period of 11–16 years. The cost of solar panels may also continue to decrease which could further drive down residential costs.

Table 6.1 Costs for solar panel installations sorted by number of years for payback (flat rate) based on cost of installation for a 33 panel (10.56 kW) system capturing the number of peak hours in Figure 6.1 and electricity costs in January of 2019 (Payless power, 2020). The discounted price assumes a 30% discount for the purchased system cost.

State	Installed ($/W)	Paid ($)	Sunlight (h/d)	Delivered (kWh/d)	D	Worth (¢/kWh)	Payback Full (y)	Disc. (y)
Maine	2.86	30,202	4.2	44	19	17.26	11	8
Arizona	2.76	29,146	5.5	58	25	12.21	11	8
Texas	2.79	29,462	5.5	58	25	11.65	12	8
New York	3.10	32,736	4.2	44	19	14.83	14	10
Florida	2.72	28,723	4.5	48	20	11.99	14	10
Delaware	2.87	30,307	4.5	48	20	12.25	14	10
Pennsylvania	3.06	32,314	4.2	44	19	12.53	16	11
S. Carolina	3.13	33,053	4.5	48	20	11.89	16	11
Colorado	3.18	33,581	4.5	48	20	11.89	16	11
Illinois	3.21	33,898	4.2	44	19	12.19	17	12
Iowa	3.39	35,798	4.5	48	20	11.44	18	13
Washington	2.69	28,406	3.5	37	16	9.31	23	16
Averages	**2.98**	**31,469**	**4.5**	**47**	**20**	**12.45**	**15**	**11**

Source: Adapted from Payless power (2020).

Example 6.2

Calculate the payback period for a 10.56 kW solar panel array (33 panels) that cost $32,138 in 2019 prior to a 30% tax credit in State College, Pennsylvania, assuming an electricity price of 10.53 ¢/kWh.

To compare this example to other numbers in Table 6.1, first we can calculate the cost of the 33-panel system per Watt before any discounts, as

$$\text{Cost/W} = \$32{,}138 \, \frac{1}{10.56\,\text{kW}} \, \frac{1\,\text{kW}}{10^3\,\text{W}} = \$3.04\ \text{W}^{-1}$$

This cost is quite close to the average given in Table 6.1 of $3.06 W^{-1}. The value of the electricity produced each day for a house in this region, with 4.2 h of peak light times 10.56 kW equal to 44 kWh, is therefore,

$$\text{Daily average worth} = \frac{44\,\text{kWh}}{\text{d}} \, \frac{10.53\,¢}{\text{kWh}} \, \frac{\$}{100\,¢} = \$4.63\ \text{d}^{-1}$$

Using the daily income to balance the cost of the system after taxes, we have the number of years, n, for payback for the discounted price is:

$$n = \$32\,(1 - 0.30) \frac{\text{d}}{\$4.63} \, \frac{\text{y}}{365\,\text{d}} = 13.3\ \text{y}$$

That result is slightly more than that shown in Table 6.1 due to the lower cost of the electricity at this location. Given all the uncertainties (weather, cost of electricity over the next 10 or so years, and so forth) that go into this calculation, the payback time could likely be in the range of 10–15 years for this location. This payback does not take into consideration that after this period, the electricity produced by these panels is essentially "free" (although there are likely to be maintenance costs such as inverter replacements).

Cost of Solar Energy for Large-Scale Solar Farms

Large-scale solar farms located on open land are easier to install than those on the roof of a home and benefit from volume discounts on the massive number of panels. For example, in 2019, the City of Los Angeles signed a deal to purchase electricity from a solar farm at 1.997 ¢/kWh, with battery storage at 1.3 ¢/kWh (Service, 2019). When you consider that the national average cost of electricity in 2020 was 10.43 ¢/kWh, this represents amazing savings for the city. The system will be rated at 400 MW and will annually produce 0.876 GWh of electricity which is estimated to provide power equivalent to 65,000 homes. Unfortunately, this amount of electricity will only satisfy 7% of Los Angeles's total electrical power needs, but it is a step in the right direction and quite economical. Unlike fossil fuels which can vary in cost and availability over time, the sun will still be shining brightly for at least the next billion years!

Solar farms now cost about $1/kW, compared to $3/kW for residential solar arrays on your roof, due to their economy of scale (EnergySage, 2020). Thus, a 1 MW solar farm would cost about $1 million. Prices continue to fall, and the number of large solar farms is also on the rise. The land needed for producing 1 MW varies, but typically, it is 2.5 acres of panels with a total of 4 acres with access space and other accessories.

What happens when you have excess solar power generation capacity? During a sunny day in California, the cost of electricity can essentially go negative due to production that outstrips demand with maximum sunshine. Thus, energy storage systems such as batteries have become of great interest to fully utilize the electricity produced from these systems.

6.3 WIND ELECTRICITY

The rise of wind energy over the past several years to become a critical and major component of global electricity production is nothing short of amazing. The "standard" for a large power plant used to be a nuclear power plant, which typically produces 1 GW of electrical power. The net capacity of nuclear power in the United States is 98 GW, produced by 96 commercial reactors. By September 2019, total wind energy *capacity* reached 106 GW, more than that of all US nuclear power plants (US Energy Information Administration, 2019) and more than the capacity of all hydropower sources. However, this is capacity and not the actual electricity production, which was 34.2 GW (1.08 EJ) for wind and 92.3 GW (2.91 EJ) for nuclear in 2019 (Table 2.7). It was estimated that onshore wind capacity would grow by an additional 19.4 GW (before the COVID pandemic), but by mid-2020 wind capacity had only increased by 2.5 GW with estimates lowered to around 13 GW of growth for the year. One important advantage of wind power is the relatively short time required for a wind tower installation. A single nuclear power plant can take 5–10 years to build, while wind farms can be built in only a few months although planning (project assessment, land leases, and other processes) can take 2–3 years. Thus, once planning is completed for many new wind projects, they could be installed in relative short periods of time.

The amount of power produced by a single wind turbine varies, with large onshore wind turbines with blades on the order of 60–70 m long each generating about 3–5 MW, producing a total of 10–15 million kWh in a year. Offshore wind turbines are even larger, with new towers capable of producing 14 MW using blades over 100 m long. The capacity factor for a wind turbine is scaled relative to its maximum generator output, with a historical average of around 35% of the maximum capacity for onshore systems (32.2% in the United States in 2019) and 41% for offshore wind turbines. The power (*P*) that can be produced by capturing the kinetic energy in wind having an average velocity (*v*) is:

$$P = \frac{1}{2} \rho A v^3 \cdot \eta C_p \tag{6.1}$$

where ρ is the density of air (which is a function of temperature), A is the area swept out by a blade of radius R, or $A = \pi R^2$, C_p the power coefficient or the aerodynamic efficiency of the turbine, which has a theoretical upper limit of 0.59 (known as the Betz limit), and η is the efficiency of the rest of the machine (including the generator, gearbox, and other mechanical and electrical components). Thus, when the efficiency of a wind turbine system is stated, it is the product of ηC_p, based on the available energy of 0.5 $\rho A v^3$.

Example 6.3

Calculate the power produced by a wind turbine with three blades of length of 52 m in air at a density of 1.23 kg/m³ and at a wind speed of 10 m/s, assuming an efficiency of 40%.

To use Eq. (6.1), we need to calculate the swept area of the blades. We assume the length of the blade is the radius, so the area is:

$$A = \pi R^2 = \pi \quad (52\ \text{m})^2 = 8500\ \text{m}^2$$

We assume the efficiency indicated is the product of ηC_p, so the power is:

$$P = \frac{1}{2} \eta C_p \rho A v^3 = \frac{1}{2}(0.40)\left(1.23\frac{\text{kg}}{\text{m}^3}\right)\left(8500\ \text{m}^2\right)\left(10\frac{\text{m}}{\text{s}}\right)^3 \frac{1\ \text{W}}{1\frac{\text{kg m}^2}{s^3}} \frac{1\ \text{MW}}{10^6\ \text{W}}$$

$$P = 2.1\ \text{MW}$$

Where Can We Find Good Sites to Capture Wind Energy?

AWS Truepower and the National Renewable Energy Laboratory (NREL) have mapped the country for locations most amenable to wind energy, and many of the "best" locations in the main 48 states are in the middle of the country, ranging from North Dakota to Texas (Fig. 6.3).

An examination of where wind energy is actually captured mostly follows the areas identified as being optimal locations, but Texas stands out as having the largest wind capacity by far compared to other states (Fig. 6.4). As of September 2019, Texas had 26.9 GW of wind energy capacity, followed by Iowa (8.9 GW), Oklahoma (8.1 GW), Kansas (6.2 GW), and California (6.1 GW). Four other states had 3.2–3.9 GW, with the rest of the country making up the balance of 25.1 GW (US Energy Information Administration, 2019).

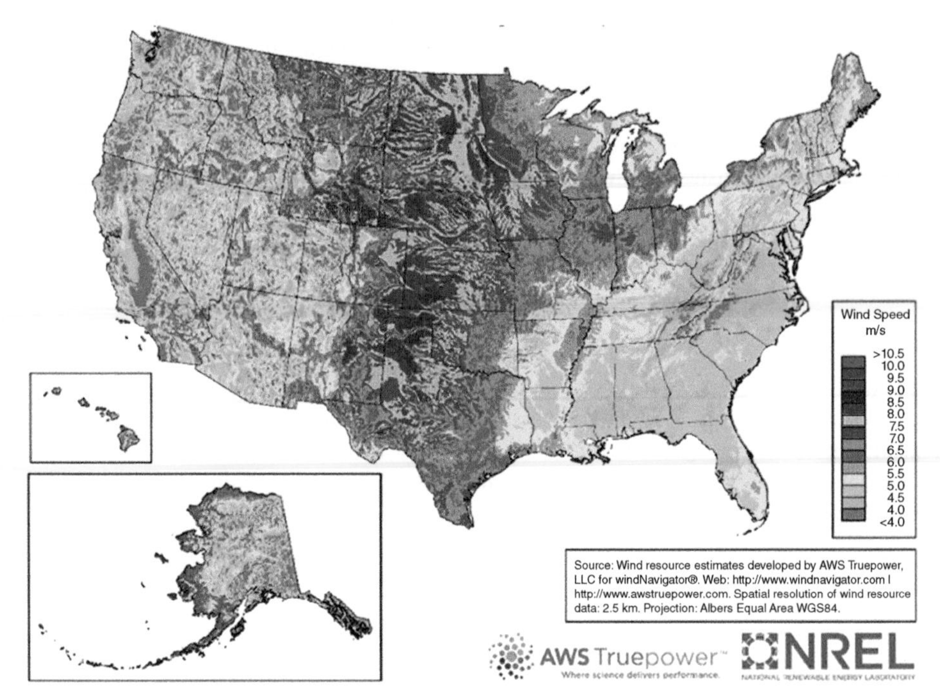

Figure 6.3 Wind speed map for the US states in m/s at a height of 80 m. *Source*: US Energy Information Administration (2020).

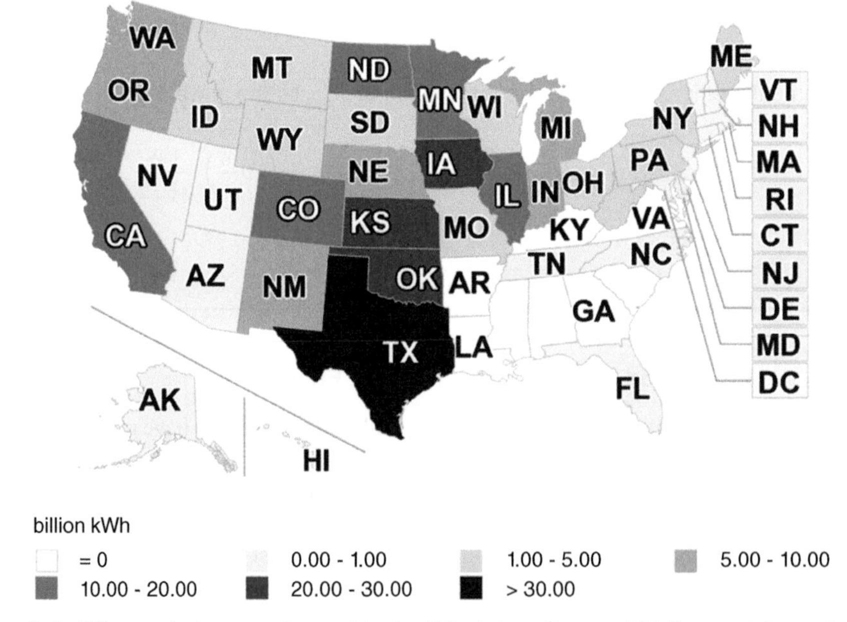

Figure 6.4 Where wind energy is used in the US states. *Source*: US Energy Information Administration (2020).

Offshore Wind Energy

The United States is lagging the rest of the world in developing offshore wind energy. The first US offshore wind farm was installed near Block Island, RI, and was completed in 2016 with a capacity of 30 MW. At the end of 2018, the worldwide offshore capacity was estimated to be 23.1 GW (Wikipedia, 2020d). Most offshore wind farms are located in Europe, which had a capacity of 22.1 GW at the end of 2019 (Wind Europe, 2020). The largest wind farms are in Northern Europe, with the UK and Germany accounting for 2/3 of the total. While construction and maintenance costs for offshore wind farms can be larger than those on land, the wind is more consistent at these offshore sites, and the systems can be larger and capture more energy per turbine than land-based systems.

Should I Get a Wind Turbine for My Home?

Small wind turbines can be installed for homes and are used for on- or off-grid applications that also include telecom towers, offshore platforms, boats, rural schools, and other applications. Wind turbines for homes are typically 2–10 kW, with those for boats being as small as 50 W. In general, your home would need to be in an area with an average annual wind speed of more than about 13 mph (6 m/s) for a potentially economical installation. Reasons for installing wind turbines include energy independence from grid power (off-grid application) or producing net power using a metered system connected with the grid (the meter runs backwards if more power is generated than consumed). These turbines typically rise 50–100 ft into the air and require at least an acre of open land, so they may not be desirable for residences in many suburban neighborhoods. An example of a residential-scale wind turbine is shown in Figure 6.5.

While very large wind turbines are relatively inexpensive per unit of capacity, residential-scale systems do not benefit from an economy of scale and thus are relatively more expensive than larger

Figure 6.5 Photograph of the MorningStar solar home at Penn State showing its wind turbine. This house was built by more than 800 Penn State students for the 2007 Solar Decathlon in Washington, DC. It is meant to exemplify a net-zero energy home. *Source*: © Kevin Sliman/Penn State University.

wind systems or solar panels for your home. The average cost for a home wind turbine system (batteries not included!) is currently around \$8–9 W^{-1} (Direct Energy, 2020) or about three times that of a solar photovoltaic panel system. Wind turbines are most useful in remote applications where energy is desired when the sun is not shining, and a combination of wind and solar systems provides increased reliability for available electricity when linked to an appropriately sized battery storage system. Also, the cost of connecting to the grid in these scenarios may be cost-prohibitive. Otherwise, the economic viability of a residential wind system increases with the cube of the average wind speed.

For now, a residential-scale wind generation system begins to make economic sense when average wind speeds at a residence are quite high (above 13 mph or 6 m/s) relatively close to the ground (50–100 ft) or if competitive costs of electricity are high. Large-scale wind generation is worth supporting as a clean and renewable method of electricity generation wherever it is affordable.

6.4 GEOTHERMAL ELECTRICITY

There are sources of renewable energy other than wind and solar that, while generally not available to a single homeowner, need to be considered for electricity and fuel production relative to total energy use in the United States. Geothermal systems used in homes are based on using the ground as a heat sink or source for heat pumps and thus are not used for electricity production. Geothermal power systems, which produce electricity, are only practical on large scales as they draw heat from deep underground to run a steam turbine, and thus, they require all the components of a complex power plant to produce electricity.

Geothermal power plants that use heat from thermal energy sources deep below the ground can provide sustainable energy due to the essentially unlimited heat that arises from natural radioactive decay that occurs within the earth and the relatively small amount of heat extracted from geothermal power sites relative to the total heat available. Worldwide use of geothermal power was projected to be 15.4 GW in 2020 (IGA, 2020) and is expected to reach 28 GW in the next 15–20 years, with an overall estimated potential of 2000 GW (Wikipedia, 2020b). Geothermal heating is already used in more than 70 countries, with geothermal electricity generation in 29 countries. About 2.6 GW of geothermal electricity is presently produced in the United States, with an estimated 60 GW that could be produced in the United States by 2050 with sufficient research investments (Hamm, 2019). Most heat sources available in the United States are in the western states (Fig. 6.6).

Electricity generation using geothermal sources requires high-temperature sources that are usually available only by drilling to depths of several kilometers. Most hydrothermal wells, drawing up hot fluids, are currently less than 3 km (1.9 mi) deep, with the most optimal heat sources at 10 km deep or more. Dry hot rock at 4–8 km deep is much more common than hot water or steam underground. Plans are underway to utilize this energy source by sending cold water through an injection well, where the water gets heated by the rock. The hot water or steam then exits through a production well. These systems are called Engineered Geothermal Systems or Enhanced Geothermal Systems (EGS). The sustainable and essentially constant availability of heat makes geothermal sources very desirable sources of renewable energy in comparison with more intermittent solar and wind energy sources.

There are three main types of geothermal power stations: dry steam (rare), flash steam (most common), and binary cycle (more recently developed and required for EGS). Flash steam systems require temperatures above 182°C. However, binary cycle systems can operate at 110–180°C and in some cases down to temperatures as low as 57°C, although the lower the temperature, the less

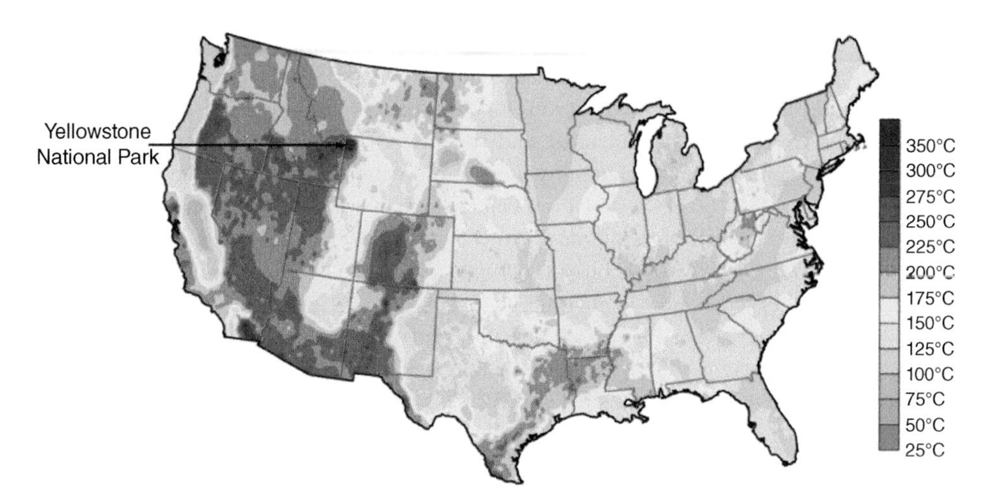

Figure 6.6 Map of temperatures at a drilling depth of 7 km (~4 mi). *Source*: Hamm (2019).

efficient the energy extraction (Wikipedia, 2020b). Binary cycle systems are therefore used in most new applications, with energy efficiencies of about 10–13%.

Geothermal power plants are not without potential environmental concerns. Fluids drawn from deep locations can contain CO_2, hydrogen sulfides, methane, ammonia, and radon, as well as toxic metals and salts, and these chemicals are all pollutants if released into the environment. Proper mitigation efforts are therefore required, not unlike unconventional oil and natural gas extraction (i.e. fracking) to avoid release of these chemicals. Binary plants that are the most common type of geothermal power system use heat exchangers and pump the fluids back into the deep subsurface to avoid releasing these chemicals into the surrounding environment.

6.5 BIOMASS ENERGY

Biomass can be used for heat and electricity generation in many different forms, with the most common one being as a fuel for electrical power generation. In the United States, biomass provided 5.26 EJ of electricity in 2019, with 49% of that used for industrial processes (primarily ethanol production), 28% for transportation, and 9% for electricity production (Lawrence Livermore National Laboratory, 2019). Biomass categorized as wood solids (~33% of biomass energy) is often included in the same category as waste fuels which include black liquor, a by-product of the Kraft mill pulping process (27%), other paper-making wastes, municipal solid waste from landfills (20%) landfill gas, and other waste-derived sources (US Energy Information Administration, 2016).

Combustion of Cellulosic Biomass

Biomass can be harvested as a side product when growing food crops, and in that case, it is classified as waste biomass, or it can be specifically grown for the purpose of generating heat or electrical power. The efficiency of heat production using dried biomass is around 75–80%. Here, we will consider using purpose-grown biomass for combustion to make electricity. The efficiency of conversion of biomass to electricity is around 20–25%, with the EIA reports for 2019 indicating an overall average of 21.9% efficiency based on primary energy consumed and electricity generated. Steam power plants generally have higher efficiencies of around 33–43%.

As in all agriculture systems, growing crops for electricity production requires careful attention to planting, water amounts and frequencies, and soil nutrients. The biomass must be harvested and transported to a central site for electricity generation, which can increase the cost and reduce the overall efficiency of net power production. One advantage of growing biomass crops, such as switchgrass, over food crops is that biomass crops often require relatively little direct attention during the growing season. However, biomass crops such as grasses or wood have lower energy densities than fossil fuels such as coal and oil. The most important advantage of biomass is that the carbon in the biomass has been captured very recently from the atmosphere, so that final combustion of the fuel does not release any net CO_2 into the atmosphere aside from the fuels used to plant, harvest, and transport the fuels to the site used for electricity generation. Certain biomass crops can even be net carbon negative, since they "bury" CO_2 into root systems that are not harvested.

The amount of biomass that can be grown annually per acre limits overall energy production relative to land use. One crop that has gained considerable attention for biomass production is Miscanthus, which can be grown to produce 27 tonne per hectare (t/ha, equal to 10 ton/acre) (Jacobson et al., 2013). The energy value of this crop can reach ~17 MJ/kg, and it requires very few nutrients compared to other crops such as corn. While biomass can be used for electricity production, the overall efficiency of conversion of sunlight to biomass is low. Therefore, the conversion of sunlight to electricity using biomass can be quite low compared to solar energy using photovoltaics, as examined in Example 6.3. However, plants "build" their solar energy-capturing bodies for free, so the cost of solar capture by plants can be very low.

Example 6.4

It is estimated that 27 tonnes per year of Miscanthus can be harvested per hectare. (a) If the dried biomass can be produced 17 MJ/kg, what would be the power produced assuming an energy efficiency of 22%? (b) How much power could be produced from solar panels covering the same area assuming a 1.7 m² panel produces 1 kWh/d?

(a) With the given information and with some conversion factors, we have

$$17\,\frac{\text{MJ}}{\text{kg}}\,\frac{27\,\text{tonne}}{\text{ha y}}\,\frac{10^3\,\text{kg}}{\text{tonne}}\,\frac{\text{GJ}}{10^3\text{MJ}}\,\frac{277.8\,\text{kWh}}{\text{GJ}}\,\frac{1\,\text{y}}{365\,\text{d}}\,\frac{1\,\text{d}}{24\,\text{h}}\,0.22 = 3.2\,\text{kW}$$

(b) For a solar panel, with 1 ha = 10,000 m², we have

$$\frac{1\,\frac{\text{kWh}}{\text{d}}}{1.7\,\text{m}^2}\,\frac{10{,}000\,\text{m}^2}{\text{ha}}\,\frac{1\,\text{d}}{24\,\text{h}} = 245\,\text{kW}$$

This comparison shows that a solar panel farm is about 77 times more efficient than biomass for capturing energy that will be turned into electricity.

The integration of solar panels and biomass growth in a single field are not mutually exclusive activities since solar panels do not completely shade fields in most configurations and biomass typically only grows for part of the year. The combination of photovoltaics and farming is a new area of research and development called agrivoltaics. If solar panels are placed on dark, gravel surfaces they can result in increased temperatures relative to surrounding areas. The growth of certain crops, such as chiltepin pepper, jalapeño, and cherry tomatoes, was found to be improved in terms of food production or water savings in their appropriate environments (Barron-Gafford et al., 2019). Thus, more careful integration of solar panels and farming, like wind turbines and farming, can yield greater advantages than either process alone.

Ethanol from Biomass

Bioethanol is currently mostly generated from corn and sugarcane, although many other feedstocks can be used as well. Glucose, which can be generated by the breakdown of either cellulose or starch from plants, can be fermented to ethanol according to

$$C_6H_{12}O_6 \rightarrow 2C_2H_5OH + 2CO_2 \tag{6.2}$$

Production of ethanol from corn starch is generally viewed to be non-sustainable due to the energy use for production, nutrient consumption, impacts on water quality, and competition of fuel production with corn as a food staple (Keeney, 2009). Some studies suggest that an equal volume of gasoline is consumed to produce the same volume of ethanol from corn (Patzek et al., 2005), which would mean less energy can be produced by a mixture of gas and ethanol since gasoline has a 1.5 times greater energy content than ethanol.

Ethanol production from the non-edible parts of plants, including cellulose-containing waste materials or crops grown specifically for biofuel production, such as Miscanthus, shows much better prospects as a sustainable and carbon-neutral/negative resource for bioethanol production especially when connected to engineered carbon sequestration systems (Field et al., 2020). It was estimated that 1198 gal of ethanol could be produced from Miscanthus based on a harvestable biomass of 13.2 tons/acre, compared to 456 gal/acre for corn (4.5 ton/acre) and switchgrass at 421 gal/acre (4.6 tons/acre) (Heaton et al., 2019). Unfortunately, cellulosic biofuel plants have failed in recent years to demonstrate economic profitability due to lower prices for oil and other fuels than in preceding years. A review of cellulosic ethanol production suggested that delays in its deployment on a large scale were primarily due to technical difficulties, high costs, and competition with other fuels (Padella et al., 2019). In order to make cellulosic ethanol production economically viable, it has been suggested that other high-value products will need to be produced along with the ethanol (Rosales-Calderon and Arantes, 2019). These findings suggest that bioethanol derived from corn starch may not be a viable renewable fuel for near-term solutions to climate change.

The long-term prospects for biomass energy are much better for biomass that is used to produce biofuels for transportation applications, rather than electricity generation, especially if the overall process is linked with carbon capture and only land that is not useful for food production is used for biomass production. Biofuels are critical for carbon-neutral transportation fuels in the United States, and they are already extensively used for this purpose throughout the world. To avoid fossil fuel consumption in the production of biofuels, the vehicles that are used for the biomass system to plant, harvest, and transport biomass should be fueled by those same biofuels, avoiding carbon emissions from additional fossil fuels. How the land is chosen for conversion to biofuel production is important to overall carbon emissions as well. Converting existing forests to switchgrass production, for example, does not appear to be viable for sustainable biofuel production (Field et al., 2020). However, the advantages of biofuels are more evident when consideration is given to converting land that is transitioning from pasture lands or cropland into land for cellulosic biofuel production. Linking biofuel production with carbon capture and sequestration into the deep subsurface has the potential to accomplish 4 times the net carbon capture of forest restoration and 15 times the net capture of grassland restoration (Field et al., 2020). The topic of producing biofuels as a method for net greenhouse gas reduction if further addressed in Chapter 14.

Biodiesel from Biomass

The fuel used in diesel combustion engines that is derived from biomass sources is called biodiesel or green diesel. It consists of long-chain fatty acid esters which are usually produced from the reaction of soybean oil, animal fat (tallow), or other oils such as vegetable oils (canola or rapeseed) that are reacted with an alcohol to produce methyl, ethyl, or propyl esters (Wikipedia, 2020a). This fuel is compatible with diesel engines used in cars and other vehicles, and thus, it can be used alone or blended with diesel fuels produced from fossil fuels. Recycled oils often require more specific conditions to be used in combustion engines. When biodiesel is sold, it is labeled with a B prefix to indicate the percentage of biodiesel relative to the total content. For example, B100 is only biodiesel fuel, while B20 indicates 20% is biodiesel and 80% is derived from fossil fuels.

For any biofuel produced from cultivated biomass, there are always concerns about the energy used to produce that fuel as well as possible CO_2 emissions due to land use changes. If forests are displaced by crops used to produce biodiesel, there can be net positive CO_2 emissions due to the release of stored biomass in both the aboveground biomass and the soils. Assuming that land is only converted from other crops for biofuels production, a life cycle analysis found that there could be a 6–25% reduction in total energy use for soybean-based biofuels for most production methods (transesterification and proprietary hydrogenation by a process called SuperCetane) (Huo et al., 2008). However, when biodiesel was produced using another hydrogenation process (UOP, called diesel II), it was calculated to lead to a 13–31% increase in energy use. Most importantly, all soybean-based approaches were found to result in a 52–107% reduction in fossil energy use and a 64–174% reduction in overall GHG emissions. A study by the US EPA determined an average of 50% reduction in GHG emissions compared to fossil fuels. These values can be compared to an average reduction of 20% of GHG emissions for corn ethanol plant and 110% reduction using cellulosic biomass (switchgrass) in the same report.

Biodiesel production in Europe and Canada typically is based on canola (rapeseed) oil. One study based on procedures used in Canada found that the use of canola oil to produce biodiesel could reduce energy dependence on fossil fuels by 85% and decrease GHG emissions by 85–100% (Ainslie et al., 2006). In Europe, the conclusions regarding possible benefits of biodiesel are less optimistic with the European Commission finding that biodiesel reduced GHGs by only 38%, compared to even less (30%) in an independent report (Gilbert, 2012, Pehnelt and Vietze, 2012). The advantages of biodiesel for reducing GHG are therefore likely only to be useful under certain optimum conditions (see Chapter 14).

Biomethane from Biomass

The biological production and capture of methane from the degradation of organic matter in a controlled environment could provide a valuable and renewable energy resource that could be compatible with any technology developed today to use natural gas (90% methane, with other gases such as propane and butane). Anaerobic digestion (AD) is a process where organic matter is first broken down in an environment with no oxygen and then eventually converted to methane, CO_2, and a small amount of residual solids. AD is commonly used at wastewater treatment plants around the world to take the large volumes of water with relatively high concentrations of solids (~0.5–3% by weight) produced by aerobic treatment systems and reduce the amount of solids requiring final treatment and disposal. AD is a relatively mature technology with Europe leading in its incorporation into industrial and agricultural processes. In Germany alone, there are over 4000 digesters installed that produce about 4 GW of electrical power. Wastewater applications of AD are further discussed in Chapter 7.

Nearly any source of biodegradable organic matter can be converted in an AD into methane, called biomethane or biogas. The overall process of biomethane production results in decomposition to primarily acetate and hydrogen gases, which are then converted to methane according to the reactions:

$$\text{Hydrogenotrophic}: 4H_2 + CO_2 \rightarrow CH_4 + H_2O \tag{6.3}$$

$$\text{Acetoclastic}: CH_3COOH \rightarrow CH_4 + CO_2 \tag{6.4}$$

When wastewater solids are used to make methane in an AD, the gas produced is around 50–70% CH_4 with CO_2 as the other main gas. The source of the CO_2 coming from the AD is from food and therefore atmospheric CO_2, so that this process does not release net CO_2 to the atmosphere.

Organic matter dissolved in solution can produce the highest rates of methane. However, when solids are used such as wood pellets, crop residues, and food waste, long retention times are needed to convert these solids into soluble organics such as acetate and hydrogen. These long detention times require large tanks that can make the process expensive to build and operate. At many sites where AD is used, adding easily degraded waste materials (such as used cooking or vegetable oils) will greatly increase methane production rates and make the AD process more economical. Crops can be grown specifically for methane production either on land or in water (algae). For land-based crops, the same challenges encountered for bioethanol sources are relevant for methane production.

One advantage of producing methane, compared to ethanol, is that methane can be easily recovered from the solution due to its very low solubility in water, for example, 0.024 g/L at 25°C. In contrast, ethanol is highly soluble in water, and thus, it takes a large amount of energy to separate the ethanol from the water–ethanol solution. An additional advantage of methane is being able to transport the fuel in gas lines. Biomethane can be cleaned up to remove CO_2 and contaminants (such as sulfides) and injected into natural gas lines, providing a "gas grid" that is a stable source of energy just like an electricity grid is used to transfer electricity over long distances.

It is estimated that 0.7% of global electricity generation (100 TWh) could be generated using biomethane by 2050. Avoiding the use of fossil fuels to produce this electricity could eliminate 9.8 Gt of CO_2 equivalent greenhouse gas emissions (Wilkinson, 2020). The uncontrolled release of methane, however, is a great concern for climate change as methane is 84 times more potent a greenhouse gas than CO_2 (20 year period). The control of unintentional gas emissions from AD locations is therefore very important for limiting GHG emissions.

Project Drawdown Assessment of Biofuels

Biofuels in general are not suggested by Project Drawdown as a solution for CO_2 emissions due to the need for fuels that are consumed while producing the biofuels, reliance on non-renewable nutrients such as phosphate, and the environmental effects related to all combustion-based processes (Wilkinson, 2020). However, several of these disadvantages could be overcome with more attention paid to the biomass source and production methods. It has been estimated that biomass energy provided by perennial feeds stocks (not forests, annuals, or waste) could reduce 2.5 Gt

(Hawken, 2017) to 7.5 Gt (Wilkinson, 2020) of CO_2 emissions from 2020 to 2050 if they replace coal and natural gas in electricity production.

6.6 HYDROGEN GAS PRODUCTION USING RENEWABLE ENERGY

Hydrogen gas (H_2) is in many ways the ultimate fuel in terms of avoiding environmental impacts: Its use in fuel cells or combustion engines results in only the emission of water and therefore no residues (although if used in a combustion process there could be by-products formed due to nitrogen in the air and high temperatures). H_2 can be compressed into a high-density gas or liquids in tanks or transported in gas lines. Commercially available cars can now run on H_2. Forklifts powered by H_2 are particularly useful when they need to run 24/7 as they can be more quickly refueled using H_2 than forklifts that are recharged using electricity. The main challenge with H_2 is that it does not exist in appreciable amounts in the environment so has to be produced using another energy source. The most common method to produce H_2 is from fossil fuels but using H_2 produced in this way would not help to decrease CO_2 emissions. The use H_2 gas is discussed in several other sections, for example, to produce ammonia for fertilizers (Chapter 8), as a fuel for cars (Chapter 11), and for heavy-duty vehicles, maritime shipping, and airplanes (Chapters 12 and 14).

Different Renewable Approaches for Producing Hydrogen Gas

There are many different ways to produce H_2 gas, and each of these methods is often assigned a color. The three most common colors are as follows: green, for using renewable electricity and water electrolysis, blue, when using natural gas and carbon sequestration (subsurface injection), and grey, for production using natural gas without carbon capture. At the present time, the least expensive method for H_2 gas production is steam reforming using natural gas, without any carbon capture (Chen et al., 2018) (Table 6.2). Coal can also be used with steam reforming, but natural gas has a much lower cost based on energy content. Biogas produced from anaerobic digestion can be used to make the H_2, and it is classified as carbon-neutral since the origin of the carbon in the methane was atmospheric. Capture and sequestration of the carbon from biogas could make it the only process that could be carbon negative. However, this classification neglects the source of the energy that went into food production. Biogas is mostly likely to be used as a biofuel, however,

Table 6.2 Fuels and methods of H_2 gas production with typical colors assigned to the various approaches.

Color		Carbon	Fuel and method
H_2	Green	Neutral	Renewable electricity for water splitting
H_2	Green	Neutral	Biomass (fermentation or microbial electrolysis cells)
H_2	Blue	Negative	Biogas (renewable methane) and sequestration
H_2	Blue	Neutral	Biogas (renewable methane)
H_2	Blue	Neutral	Natural gas with sequestration with CO_2 injection
H_2	Turquoise	Neutral	Natural gas with solid carbon capture
H_2	Pink	Neutral	Nuclear power for water splitting
H_2	Gray	Positive	Natural gas using steam reforming
H_2	Brown	Positive	Coal using steam reforming
H_2	White	Positive	Various, with H_2 as a by-product of other processes

rather than as a source for H_2 gas production. Biomass can be used to make H_2 gas directly via fermentation or indirectly as the fuel and source of electrons for microbial electrolysis cells (MECs) (along with renewable electricity). Biomass can also be used for producing H_2 via gasification (for example, using waste biomass), but to date, these biomass-based processes have not considered to be as economical or viable as green hydrogen gas production. Electricity from nuclear power plants can also be used for water electrolysis, but so far, this does not appear to be economical given the much lower costs for utility-scale solar power compared to new nuclear power. Among these carbon-neutral methods, the greatest commercial interest is in green H_2 gas production using renewable electricity, so green hydrogen production via water electrolysis if examined further below. Biological routes to H_2 gas production are also briefly reviewed.

Green Hydrogen Gas Produced by Water Electrolysis

The conversion of water into H_2 gas requires ~1.2 V in theory, with 1.8–3.5 V used in practice. The fundamental process of water splitting is quite simple: H_2O is split into O_2 gas with the release of protons (H^+) into water at one electrode (for acidic processes), and H_2 is evolved at the counter electrode from protons and the electrons delivered in the circuit. Implementation of the process is complicated by the need to reduce energy losses and that can require expensive catalysts such as platinum, and the gas products of the two electrodes (oxygen from the anode and water from the cathode) should not recombine so that requires a gas separator or more expensive ion exchange membrane. In addition, salt and organic matter impurities such as chloride ions need to be removed from the water as chloride ions can react with the anode and form toxic by-products such as chlorine gas (Cl_2) that can also damage the membrane or react with organic matter to form additional toxic compounds (Jasper et al., 2017).

The two most common types of water electrolyzers are alkaline water electrolyzers and proton exchange membrane (PEM) (acidic) water electrolyzers (Figure 6.7). In an alkaline WE, the electrolyte is a very high pH solution of potassium hydroxide where water splitting produces hydroxide ions and oxygen at the anode, and H_2 gas is produced at the cathode from water dissociation by the two half-reactions (Carmo et al., 2013):

$$\text{Anode}: 2OH^- \rightarrow \frac{1}{2}O_2 + H_2O + 2e^- \tag{6.5}$$

$$\text{Cathode}: 2H_2O + 2e^- \rightarrow H_2 + 2OH^- \tag{6.6}$$

$$\text{Overall}: H_2O \rightarrow H_2 + \frac{1}{2}O_2 \tag{6.7}$$

Alkaline water electrolyzers systems are usually operated at around 40–90°C to improve reaction rates compared to room temperature. The highly caustic electrolyte requires the use of very stable materials under these harsh conditions. A battery-type (non-ion selective) inexpensive separator is generally used to keep the solutions isolated from each other, but the cell cannot be highly pressurized as too much gas crossover could occur through the separator resulting in either gas losses or possible generation of explosive gas mixtures. Anion exchange membranes (AEMs) would be preferred as separators, but commercially produced membranes are currently not sufficiently stable in highly alkaline solutions. The choice of catalysts is larger than those possible for PEM-based systems (which must use precious metals) but also constrained by the highly alkaline conditions, with Ni/Co/Fe mixtures often used on the anode and Ni on the cathode of the alkaline water electrolyzers.

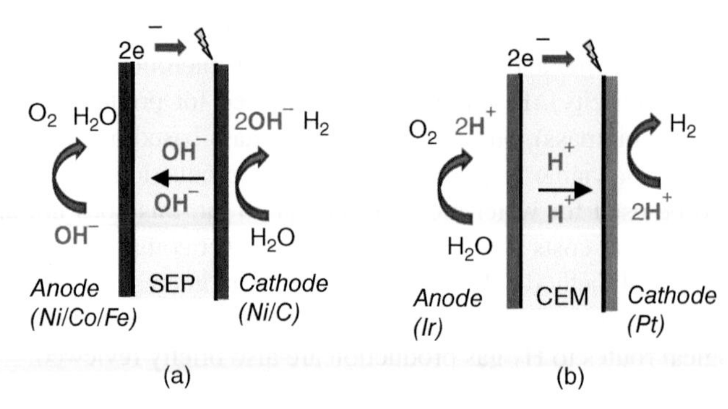

Figure 6.7 Schematic representation of the reactions and ions for two types of common water electrolyzers: (a) alkaline and (b) PEM-based systems, with typical catalysts shown in parentheses (SEP = separator).

PEM water electrolyzers use highly purified water and operate under conditions similar to that of a hydrogen fuel cell, except in reverse as hydrogen and oxygen gases are generated rather than being consumed, and power is input into the system rather than being produced (Fig. 6.7b). The anode is usually catalyzed by a precious metal such as iridium to drive oxygen evolution on the anode, and the cathode contains a catalyst such as platinum for the reactions:

$$\text{Anode}: \text{H}_2\text{O} \rightarrow 2\text{H}^+ + \frac{1}{2}\text{O}_2 + 2\text{e}^- \tag{6.8}$$

$$\text{Cathode}: 2\text{H}^+ + 2\,\text{e}^- \rightarrow \text{H}_2 \tag{6.9}$$

where the overall reaction is the same as that for alkaline water electrolyzers. These systems can be operated over a wide range of temperatures (20–100°C). Water splitting releases protons which produce a highly acidic conditions in the PEM, but this ion transfer is needed to provide protons at the cathode. The use of a PEM also allows for gases to be pressurized in the vessel which helps reduce costs for hydrogen gas compression and storage. The main challenge for PEM systems is the high costs of the catalysts and membranes as these can account for up to half the capital costs (Mayyas et al., 2019). Precious metals are usually used because of their stability compared to alternatives that can degrade under the acidic conditions produced in the PEM membranes.

Water electrolyzers are already highly energy-efficient (70–80%), with this efficiency chosen based on minimizing costs due to balancing capital and operation expenses, so these energy efficiencies could likely be improved if other costs could be lowered. If powered by renewable energy such as solar and wind sources, H_2 by water electrolysis provides a path for renewable H_2 production. There is already ~0.32 GW of energy used to produce H_2 by water electrolysis, although most H_2 is produced by steam reforming of natural gas with a similarly high thermal efficiency of 70–85% (Wikipedia, 2020c). The main challenges for renewable H_2 production by water electrolyzers are the costs as water electrolysis is currently more expensive than conventional steam reforming methods using fossil fuels. The DOE set a 2020 goal for H_2 gas production that required an electricity cost of \$0.037 kWh (Energy.gov, 2020, Wikipedia, 2020c) which is now achievable using large-scale solar or wind arrays. Therefore, the current barrier for more affordable H_2 production is that the cost of the catalysts and membranes (~50% of capital costs) and the highly corrosion-resistant materials used for the electrolyzer structure itself remain high (Mayyas et al.,

2019). Thus, if the cost of the catalysts and membranes could be reduced, renewable hydrogen production could have a large impact on the generation of a renewable and transportable fuel.

Biohydrogen from Biomass Fermentation

Renewable H_2 can be produced by bacterial fermentation of glucose, with substrates like those suitable for production of ethanol from glucose and other cellulosic substrate. The complete oxidation of glucose to H_2 and CO_2 suggests the possibility of 12 mol of H_2 per mole of glucose, via:

$$C_6H_{12}O_6 + 6H_2O \rightarrow 12H_2 + 6CO_2 \tag{6.10}$$

However, due to thermodynamic constraints and biological pathways, the highest practical yields via fermentation (in appreciable quantities) are 4 mol of H_2 and 2 mol of acetic acid (Logan, 2004), as shown by:

$$C_6H_{12}O_6 + 2H_2O \rightarrow 4H_2 + 2C_2H_4O_2 + 2CO_2 \tag{6.11}$$

Once formed, the further conversion of acetate to useable levels of H_2 is thermodynamically unfavorable, and therefore, there is no biological route for further production of H_2 from fermentation end products. However, due to a lack of sufficient conversion of the substrate to the desired product fermentation alone is not likely to ever provide a suitable route for large-scale H_2 production unless the remaining products can be used in other ways.

Biohydrogen Using Photobiological or Microbial Electrochemical Technologies

Fermentation end products can be transformed into useful end products such as H_2 using an additional source of energy input into the system, either using sunlight in a photoelectrochemical process or by using microbial electrochemical technologies (METs). Two different photobiological routes of H_2 production are possible: using algae to directly produce H_2 gas using energy in sunlight or using photosynthetic bacteria that use organic matter in water and produce H_2 using sunlight as the needed energy source (photofermentation). However, these two photobiological approaches have so far not been economical compared to other H_2 production routes. In addition, photobiological processes using sunlight do not appear to be viable as engineered processes linked to organic matter in water due to the complexity of the process that requires engineered systems, sunlight, and large amounts of water and the long periods of inactivity outside of daylight hours.

MECs can be used to produce H_2 from organic matter if supplied by an additional source of power (electricity). In an MEC, certain types of bacteria can oxidize the substrates such as acetate on an electrode, and hydrogen can be evolved electrochemically from water at the other electrode (Logan et al., 2008, Logan, 2008). The voltage that needs to be added in theory is ~0.14 V, although in practice, higher voltages of 0.6–0.9 V are used. These voltages are less than those needed for water splitting to produce hydrogen gas. While high conversion yields of acetate to H_2 (~11.6 mol per mole of cellulose) using MECs have been shown using cellulose fermentation end products (Lalaurette et al., 2009), the current densities are low and costs of the materials are relatively expensive given the rate of product production for an engineered process to convert biomass to H_2. Thus, this approach has not yet found commercial applications for large-scale H_2 production. It may be possible that MECs could find commercial applications in wastewater treatment where the main costs for treatment can be offset based on the production of H_2 gas. MECs to produce H_2 or methane and microbial fuel cells (MFCs) that can produce electricity from biomass are discussed further in Chapter 7.

Hydrogen Storage

One of the major challenges for using H_2 as a transportable fuel is the cost and energy losses for storage. When H_2 is produced by water electrolysis, it can be partly pressurized to 15 bar (1 bar = 0.968 atm) using a PEM electrolyzer, 7 bar using alkaline electrolyzers, or at 1 bar with solid oxide or membraneless water electrolyzers (Van Hoecke et al., 2021). The most economical method of storage is through compression to 100–700 bar in tanks. Other processes include conversion of H_2 to ammonia and storage as a liquid, or infusing the H_2 into other materials, such as liquid organic hydrogen carriers (LOHCs) that include toluene or methylcyclohexane (MCH), or in solid form with metal hydrides. A typical energy cost for producing 1 kg of H_2 containing 33 kWh is 55 kWh using alkaline water electrolysis (energy efficiency of 60%) (Van Hoecke et al., 2021). The energy needed for compression of 1 kg of H_2 to 70 MPa, which is typical for a hydrogen fuel cell vehicle (HFCV), is an additional 3.1 kWh or about 5.3% of the total energy of 58.1 kWh.

In 2020, the costs of hydrogen as produced (without additional compression) by water electrolysis ranged from \$2.50 kg^{-1}, or about the same price of a gallon of gasoline (US \$2.90) with the same energy content, to \$6.80 kg^{-1}. H_2 produced from natural gas costs much less (\$1.00–1.80 kg^{-1}). Refueling stations in California sell pressurized H_2 at United States \$16.50 kg^{-1} (using from fossil fuels), reflecting the additional costs for compression, transportation, and delivery. HFCVs can go ~100 km (69 mi) on 1 kg of H_2 or about 2.76 times as far compared to a car using gasoline that gets 25 mpg. Thus, the comparable cost based on vehicle range and efficiency is closer to ~\$6 for H_2 compared to \$2.90 for the gasoline. It is expected that the costs of H_2 gas produced by water electrolysis and renewable electricity will continue to be reduced in the coming years.

6.7 COSTS OF RENEWABLE VERSUS CONVENTIONAL ENERGY SOURCES

The costs of electricity generation using renewable energy sources are one of many important considerations, but energy costs for some renewable energy technologies can be less than those of conventional sources. The great advantage of renewable energy is that there is no "fuel" needed except that freely provided by nature. In the analysis by *Lazard* (2021), utility-scale wind energy is currently the least expensive energy source of \$26–\$54 MWh^{-1}, but utility-scale solar is close at the lower end of the estimates, but it has a narrower range of \$29–\$38 MWh^{-1} (Fig. 6.8). Rooftop solar (on homes) can be one of the most expensive methods of renewable energy sources due to the small scale and relatively complex installation environments compared with other situations for solar electricity such as commercial rooftop and community-scale sites. Offshore wind is estimated to be above that utility-scale solar or land-based wind, but it has the advantage of being a much more reliable and continuous source than those available for land-based systems.

Natural gas using combined cycle (CC) gas turbines is the least expensive conventional fuel technology at \$44–\$73 MWh^{-1}. Conventional power plants using gas, coal, and nuclear fuels run more efficiently under continuous rather than intermittent operation. However, electricity demands are not constant. To supply electricity during the times of highest electricity demands, some plants only operate when the demand is high, and thus, electricity costs are high. These plants are known as peaking plants, and they use natural gas due to an ability to quickly ramp up power production when needed. The intermittent operation of peaking plants, however, means costs are high to produce that electricity (\$151–\$198 MWh^{-1}). Therefore, peaking power plants electricity costs are higher than natural gas power plants that operate under more continuous conditions. Nuclear power is also estimated to have a high cost that is like peaking plants and residential solar installations.

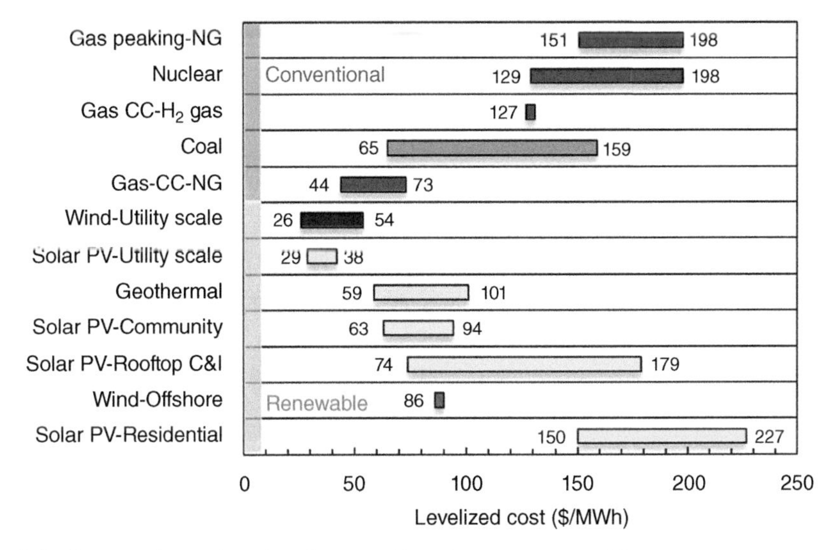

Figure 6.8 Comparison of costs of conventional versus renewable energy sources. (NG = natural gas, CC = combined cycle, C&I = commercial and industrial). The costs for Gas CC using H_2 and offshore wind are future estimates. *Source*: Adapted from Lazard (2021).

The use of green hydrogen, produced by water electrolysis, is estimated to have a cost 1.7–2.9 times higher than electricity production using natural gas. However, this cost is still less than that of nuclear or natural gas peaking. This comparison indicates that hydrogen using existing technologies could replace nuclear power as a cost-effective method of electricity production, but incentives other than electricity costs would be needed to replace natural gas plants with H_2-fueled gas turbines.

6.8 ENERGY STORAGE IN BATTERIES

Rechargeable large-scale batteries are widely used, but they are not currently the main way that energy is stored for electricity generation. Pumped hydropower is the single most common method to store energy generated by power plants. Water turbines are used to pump water to an uphill reservoir at a high efficiency, and then, the water is fed back through those turbines to generate electricity. Energy storage using water is further discussed in Chapter 5. Electricity storage using batteries has historically been widely used for energy storage for starting cars and smaller devices, such as toys, phones, and other rechargeable devices. With the increase in solar electricity generation and the use of electricity for transportation, batteries are being more frequently used to partly or completely power vehicles and to provide electricity storage in homes and businesses. Non-rechargeable batteries, for example, zinc–carbon batteries, produce electricity from chemical reactions and thus are not considered here to be a battery-based method of electrical energy storage.

The two desirable attributes of rechargeable batteries are energy density, or how much kWh you can pack into the mass of the battery (kWh/kg), and power density, or how fast the batteries can be discharged (and recharged) normalized to batter mass. These two properties of batteries are often compared in Ragone plots, which show the range of battery types for energy density and discharge rates (Fig. 6.9). Lithium–ion batteries are currently the best for energy storage based on

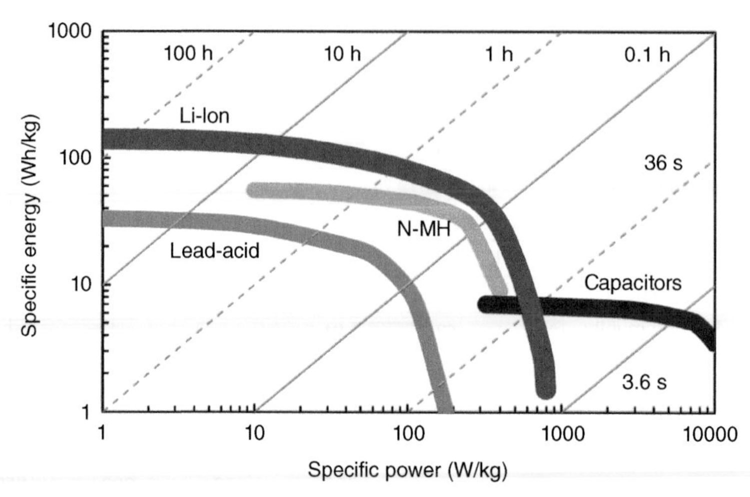

Figure 6.9 Ragone plots show how the different types of batteries and capacitors can hold energy compared to how fast they can deliver it. While Li–ion batteries are efficient in terms of energy storage based on weight, capacitors have the unique capability to rapidly discharge energy. The most desirable energy storage materials would fall into the upper right of the plot. *Source*: Adapted with permission from McCloskey (2015) American Chemical Society.

costs and these two metrics, although research is progressing on newer types of lithium batteries with improved performance and on other types of batteries. Improvements are being made in energy densities of batteries, but they are very much lower than that of gasoline that has an energy density of 12,900 Wh/kg. For example, the Tesla home battery currently stores 14 kWh and weighs 114 kg, for an energy density of 120 Wh/kg. Thus, gasoline has an energy density 100 times larger. A 85 kW battery for a Tesla model S that weighs 540 kg would have an energy density 83 times less than gasoline. A high energy density is important as the work to move the vehicle increases with the weight of the vehicle. Therefore, if the battery gets too heavy, the vehicle, like a truck or airplane, cannot haul anything other than the battery.

A life cycle analysis determined that 42 kg of CO_2 are emitted per 1 kWh of storage capacity (Ciez and Whitacre, 2019). If we consider that a single solar panel produces 0.58 D or 1.3 kWh, that is 55 C for total CO_2 emissions in just one day or 0.015 C if that were spread out over 10 years. Thus, the carbon footprint for batteries is relatively low for the expected lifetimes of the batteries.

Another alternative to energy storage is to use hydrogen gas (H_2). Currently, the round-trip efficiency (energy to produce H_2, store it, and then produce electricity) can be lower than those based on water storage, and costs for this approach using H_2 needs to be better defined. Water electrolyzers can function at about 80% efficiency, with conventional PEM (PEM fuel cells [the type currently used in cars]) having about 60% energy efficiency. Thus, an overall round-trip efficiency would be 48%. However, high-temperature solid oxide hydrogen fuel cells can reach 90% energy efficiency (Fuel Cells Bulletin, 2019) which could translate to 72% efficiency, neglecting any energy losses for hydrogen gas energy storage. This range using the high-temperature fuel cells is therefore at the lower range of hydropower storage efficiencies.

6.9 IMPACT OF RENEWABLE ENERGY ON REDUCING CARBON EMISSIONS

Project Drawdown has estimated the potential reduction of CO_2 emissions due to enhanced renewable energy activities, ranging from increased deployment of utility- and distributed-scale solar photovoltaics to landfill methane capture (Table 6.3) (Wilkinson, 2020). The total emissions calculated for a 30-year period from 2020 to 2050 for these categories are 215 Gt of CGH equivalents. Note that Project Drawdown also includes electric cars (EVs) in this category of renewable energy technologies, but here that topic is grouped into the topic of transportation in Chapter 12. By assuming a 30-year period for these Drawdown solutions and a population of 9.4 billion people, the daily C values (as a negative C as these are reductions in emissions) are calculated to put this into the perspective of other energy-consuming and CO_2 emitting activities. These reductions, that sum to a total of −2.3 C, can be compared to production of 48.9 C (Chapter 4) based on fossil fuel use in the United States in 2019. It is clear from the sum of all the C values in Table 6.3, which are only 7% of the current total use, that all these suggested changes are needed to reduce CO_2 emissions into the environment at levels needed to minimize increases in temperature and effects of climate change.

Table 6.3 Possible reductions in CO_2 emissions (Gt) over a period of 30 years (from 2020 to 2050) due to increased activities in these renewable energy areas (selected from 76 activities).

No.	Solution	CO_2 (10^9 tons)	−C
6	Onshore wind turbines	47.2	0.50
8	Utility-scale solar photovoltaics	42.3	0.45
9	Improved clean cookstoves	31.3	0.33
10	Distributed solar photovoltaics	28.0	0.30
15	Concentrated solar power	18.6	0.20
26	Offshore wind turbines	10.4	0.11
28	Methane digesters	9.8	0.10
40	Geothermal power	6.2	0.07
43	Biogas for cooking	4.6	0.05
47	Perennial biomass production	4.0	0.04
51	Nuclear power	2.7	0.03
53	Biomass power	2.5	0.03
56	Landfill methane gas capture	2.2	0.02
58	Waste-to-energy	2.0	0.02
59	Small hydropower	1.7	0.02
61	Ocean power	1.4	0.01
76	Micro-wind turbines	0.1	0.00
	Total Energy	**215.0**	**2.30**

The value of C is shown as a negative sign as these are carbon reductions.
Source: Adapted from Wilkinson (2020).

References

AINSLIE, B., DOWLATABADI, H., ELLIS, N., RIES, F., ROUHANY, M. & SCHREIER, H. 2006. *A review of environmental assessments of biodiesel displacing fossil diesel*. https://www.canolacouncil.org/download/215/pages/5281/review_of_environmental_assessments. ©2006 Paya Solutions Inc.

BARRON-GAFFORD, G. A., PAVAO-ZUCKERMAN, M. A., MINOR, R. L., SUTTER, L. F., BARNETT-MORENO, I., BLACKETT, D. T., THOMPSON, M., DIMOND, K., GERLAK, A. K., NABHAN, G. P. & MACKNICK, J. E. 2019. Agrivoltaics provide mutual benefits across the food–energy–water nexus in drylands. *Nature Sustainability,* 2, 848–855.

CARMO, M., FRITZ, D. L., MERGEL, J. & STOLTEN, D. 2013. A comprehensive review on PEM water electrolysis. *International Journal of Hydrogen Energy,* 38, 4901–4934.

CHEN, J. G., CROOKS, R. M., SEEFELDT, L. C., BREN, K. L., BULLOCK, R. M., DARENSBOURG, M. Y., HOLLAND, P. L., HOFFMAN, B., JANIK, M. J., JONES, A. K., KANATZIDIS, M. G., KING, P., LANCASTER, K. M., LYMAR, S. V., PFROMM, P., SCHNEIDER, W. F. & SCHROCK, R. R. 2018. Beyond fossil fuel–driven nitrogen transformations. *Science*, 360, eaar6611.

CIEZ, R. E. & WHITACRE, J. F. 2019. Examining different recycling processes for lithium-ion batteries. *Nature Sustainability,* 2, 148–156.

DIRECT ENERGY. 2020. *Residential solar and wind systems: What are the energy costs?* [Online]. Available: https://www.directenergy.com/learning-center/green-living/residential-solar-and-wind-systems#:~:text=Wind%20turbine%20systems%20can%20run,could%20cost%20upwards%20of%20%2440%2C000 [Accessed June 3 2020].

ENERGY.GOV. 2020. *DOE technical targes for hydrogen production from electrolysis* [Online]. Available: https://www.energy.gov/eere/fuelcells/doe-technical-targets-hydrogen-production-electrolysis [Accessed June 10 2020].

ENERGYSAGE. 2020. *Solar farms: what are they, and how do you start one?* [Online]. Available: https://news.energysage.com/solar-farms-start-one/ [Accessed June 3 2020].

FIELD, J. L., RICHARD, T. L., SMITHWICK, E. A. H., CAI, H., LASER, M. S., LEBAUER, D. S., LONG, S. P., PAUSTIAN, K., QIN, Z., SHEEHAN, J. J., SMITH, P., WANG, M. Q. & LYND, L. R. 2020. Robust paths to net greenhouse gas mitigation and negative emissions via advanced biofuels. *Proceedings of the National Academy of Sciences*, 117, 21968–21977.

FUEL CELLS BULLETIN. 2019. VTT's reversible fuel cell for highly efficient hydrogen production. *Fuel Cells Bulletin.*2019, 14–15.

GILBERT, N. 2012. *Rapeseed biodiesel fails sustainability test* [Online]. Available: https://www.nature.com/news/rapeseed-biodiesel-fails-sustainability-test-1.11145 [Accessed December 4 2020].

HAMM, S. G. 2019. *Geo vision: Harnessing the heat beneath our feet*, Department of Energy. https://www.energy.gov/sites/prod/files/2019/06/f63/GeoVision-full-report-opt.pdf.

HAWKEN, P. 2017. *Drawdown: The most comprehensive plan ever proposed to reverse global warming*, Penguin Books.

HEATON, E. A., BOERSMA, N., CAVENY, J. D., VOIGT, T. B. & DOHLEMAN, F. G. 2019. *Miscanthus (Miscanthus x giganteus) for biofuel production* [Online]. USDA Cooperative Extension System. Available: https://farm-energy.extension.org/miscanthus-miscanthus-x-giganteus-for-biofuel-production/#Potential_Yields_.28Tonnage_and_Energy_Content.29 [Accessed December 29 2020].

HUO, H., WANG, M., BLOYD, C. & PUTSCHE, V. 2008. Life-cycle assessment of energy and greenhouse gas effects of soybean-derived biodiesel and renewable fuels. ANL/ESD/08-2.

IGA. 2020. *2020 To become a milestone year for the global geothermal energy sector* [Online]. Available: https://www.geothermal-energy.org/2020-to-become-a-milestone-year-for-the-global-geothermal-energy-sector/ [Accessed October 28 2020].

JACOBSON, M., MARRISON, D., HELSEL, Z., RAK, D., FORGENG, B. & HIEL, N. 2013. *Miscanthus budget for biomass production* [Online]. Penn State Extension. Available: https://extension.psu.edu/miscanthus-budget-for-biomass-production [Accessed October 16 2020].

JASPER, J. T., YANG, Y. & HOFFMANN, M. R. 2017. Toxic byproduct formation during electrochemical treatment of latrine wastewater. *Environmental Science & Technology,* 51, 7111–7119

KEENEY, D. 2009. Ethanol USA. *Environmental Science & Technology,* 43, 8–11.

LALAURETTE, E., THAMMANNAGOWDA, S., MOHAGHEGHI, A., MANESS, P.-C. & LOGAN, B. E. 2009. Hydrogen production from cellulose in a two-stage process combining fermentation and electrohydrogenesis. *International Journal of Hydrogen Energy,* 34, 6201–6210.

LAWRENCE LIVERMORE NATIONAL LABORATORY. 2019. *Estimated U.S. energy consumption in 2019: 100.2 quads* [Online]. Available: https://flowcharts.llnl.gov/content/assets/docs/2019_United-States_Energy.pdf [Accessed May 9 2020].

LAZARD 2021. Lazard's levelized cost of energy analysis, Version 14.0.

LOGAN, B. E. 2004. Extracting hydrogen and electricity from renewable resources. *Environmental Science & Technology,* 38, 160A–167A.

LOGAN, B. E. 2008. *Microbial fuel cells,* Hoboken, NJ, John Wiley & Sons, Inc.

LOGAN, B. E., CALL, D., CHENG, S., HAMELERS, H. V. M., SLEUTELS, T. H. J. A., JEREMIASSE, A. W. & ROZENDAL, R. A. 2008. Microbial electrolysis cells for high yield hydrogen gas production from organic matter. *Environmental Science & Technology,* 42, 8630–8640.

MAYYAS, A. T., RUTH, M. F., PIVOVAR, B. S., BENDER, G. & WIPKE, K. B. 2019. Manufacturing cost analysis for proton exchange membrane water electrolyzers. National Renewable Energy Lab. (NREL), Golden, CO (United States).

MCCLOSKEY, B. D. 2015. Expanding the Ragone plot: Pushing the limits of energy storage. *The Journal of Physical Chemistry Letters,* 6, 3592–3593.

PADELLA, M., O'CONNELL, A. & PRUSSI, M. 2019. What is still limiting the deployment of cellulosic ethanol? Analysis of the current status of the sector. *Applied Sciences,* 9, 4523.

PATZEK, T. W., ANTI, S. M., CAMPOS, R., HA, K. W., LEE, J., LI, B., PADNICK, J. & YEE, S. A. 2005. Ethanol from corn: Clean renewable fuel for the future, or drain on our resources and pockets? *Environment, Development and Sustainability,* 7, 319–336.

PAYLESS POWER. 2020. *2020 Electricity rates by state (updated April 2020)* [Online]. Available: https://paylesspower.com/blog/electric-rates-by-state/ [Accessed June 3 2020].

PEHNELT, G. & VIETZE, C. 2012. Uncertainties about the GHG emissions saving of Rapeseed biodiesel. *Jenna Economic Research Papers*, No. 2012, 039, Friedrich Schiller University Jena and Max Planck Institute of Economics, Jena.

RHODES, J. 2020. *The future of US solar is bright* [Online]. Available: https://www.forbes.com/sites/joshuarhodes/2020/02/03/the-us-solar-industry-in-2020/#6348f5065ed3 [Accessed June 6 2020].

ROSALES-CALDERON, O. & ARANTES, V. 2019. A review on commercial-scale high-value products that can be produced alongside cellulosic ethanol. *Biotechnology for Biofuels,* 12, 240.

SERVICE, R. F. 2019. Solar plus batteries is now cheaper than fossil power. *Science,* 365, 108.

US ENERGY INFORMATION ADMINISTRATION. 2016. *Biomass and waste fuels made up 2% of total U.S. electricity generation in 2016* [Online]. Available: https://www.eia.gov/todayinenergy/detail.php?id=33872#:~:text=Biomass%20and%20waste%20fuels%20generated,based%20(biogenic)%20energy%20sources [Accessed June 9 2020].

US ENERGY INFORMATION ADMINISTRATION. 2019. *U.S. onshore wind capacity exceeds 100 gigawatts* [Online]. Available: https://www.eia.gov/todayinenergy/detail.php?id=42235 [Accessed June 3 2020].

US ENERGY INFORMATION ADMINISTRATION. 2020. *Wind explained: Where wind power is harnessed* [Online]. Available: https://www.eia.gov/energyexplained/wind/where-wind-power-is-harnessed.php [Accessed June 2 2020].

VAN HOECKE, L., LAFFINEUR, L., CAMPE, R., PERREAULT, P., VERBRUGGEN, S. W. & LENAERTS, S. 2021. Challenges in the use of hydrogen for maritime applications. *Energy & Environmental Science,* 14, 815–843.

WIKIPEDIA. 2020a. *Biodiesel* [Online]. Available: https://en.wikipedia.org/wiki/Biodiesel [Accessed December 4 2020].

WIKIPEDIA. 2020b. *Geothermal power* [Online]. Available: https://en.wikipedia.org/wiki/Geothermal_power [Accessed June 9 2020].

WIKIPEDIA. 2020c. *Hydrogen production* [Online]. Available: https://en.wikipedia.org/wiki/Hydrogen_production [Accessed June 10 2020].

WIKIPEDIA. 2020d. *Offshore wind power* [Online]. Available: https://en.wikipedia.org/wiki/Offshore_wind_power [Accessed June 3 2020].

WILKINSON, K. 2020. *The drawdown review: Climate solutions for a new decade*. Project Drawdown.

WIND EUROPE 2020. Offshore wind in Europe.

CHAPTER 7

WATER – AN ENERGY SOURCE

7.1 EXTRACTING ENERGY FROM WATER

The water infrastructure consumes a large amount of electricity, but our natural and engineered water infrastructure also provides opportunities for electrical power generation. The most obvious example for energy from water is hydropower where electricity is generated by flowing water through turbines. There are other opportunities for capturing energy from water due to specific water composition, either due to its use or its properties (such as salinities) in nature. Water used for some purpose that is returned to the environment will have been modified. The two largest water uses, water for cooling power plants and water used for irrigation, do not provide sufficient changes to the water properties to make them useful for electricity generation. However, when water is used for other purposes it can be possible to create processes that can generate electricity by extracting energy in the water if it contains sufficiently high concentrations of organic matter. When water is used for domestic or many industrial processes it can become enriched in organic matter. We now call this "used water," rather than wastewater, as the water is not a waste because after treatment it is still a resource. Thus, wastewater treatment plants are now called used water or resource recovery plants, so that we can recovery the water for other applications and use the things in water for specific purposes. Alternatively, it may be possible to generate electricity from salinity differences between water that is going to be discharged from an engineered process, such as a used water treatment plant and seawater, or between naturally occurring salinity gradients between river water and seawater.

In this chapter we look at energy from water based on hydropower, but we also examine at methods for extracting energy out of the water during or after the used water treatment processes. There are two general ways that energy is extracted: using the organic matter to create electricity or a fuel such as hydrogen or methane; or using the treated water as the fresh water or less-salty water in a salinity gradient energy technology. There are many proposed technologies for energy extraction of the organic matter, but here we focus on three basic approaches: using anaerobic digestion-based technologies of solids which can be used in conjunction with traditional aerobic treatment technologies; using fully anaerobic treatment technologies; and using electrode-based technologies such as microbial fuel cells (MFCs) and microbial electrolysis cells (MECs).

The salinity differences between a relatively non-salty water, such as a river water or treated effluent from a resource recovery plant, and a relatively salty water such as seawater have a thermodynamic potential that can be extracted as electricity. Although the energy content due to the entropic energy differences between river water and seawater are relatively low, the large amounts

Daily Energy Use and Carbon Emissions: Fundamentals and Applications for Students and Professionals, First Edition. Bruce E. Logan.

of water flowing into the sea could provide opportunities for harvesting those chemical potential differences in the form of electrical power. If properly engineered, these many different uses of water could significantly contribute electrical power to the water energy infrastructure, perhaps even one day making the water infrastructure completely energy sustainable based only on water energy.

7.2 HYDROPOWER

Electricity production using hydropower has been the single most important renewable energy source for decades. In 1950, hydropower provided 101 TWh of electricity, or about 30% of electrical power in the United States at that time. By 2019, the amount of hydropower electricity had grown to 274 TWh. However, due to the growth of other methods for electricity generation hydropower provided only 6.6% of electricity in 2019 but it still accounted for 38% of renewable electricity generation (US Energy Information Administration, 2020a) (Fig. 7.1).

In 2015 there were a total of 2198 hydropower sites in the United States, providing a total capacity of 79.6 GW. Worldwide, hydropower is estimated to provide 16.6% of all electrical power (Wikipedia, 2020b). The largest hydropower dam in the United States, located on the Columbia River at the Washington and Oregon borders, has a total of 7.07 GW (61.9 TWh) of total generation capacity. The largest plant in the world is the Three Gorges Dam in China, which has a capacity of 22.5 GW. Globally, hydropower could increase by more than 3% a year for the next 25 years (Wikipedia, 2020b).

There are more than 80,000 dams in the United States, although many are not suitable for electrical power production. The number of locations and the extent for additional power generation has been reported in several studies, providing a number of different assessments on the capacity for additional electricity generation. A 2012 report based on a survey of 54,000 sites concluded that there were 597 sites with capacities of >1 MW, with 100 sites that could contribute 8 GW of additional power (Reid, 2016). A total of 81 of those sites were operated by the US Army Corps

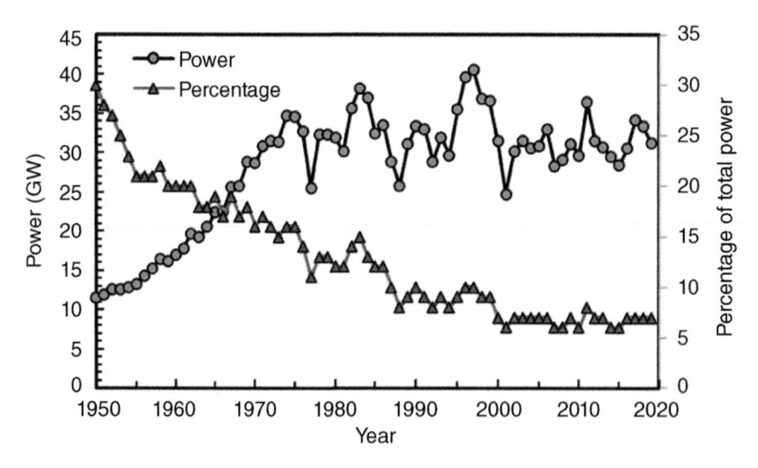

Figure 7.1 Hydropower production in the United States from 1950 to 1990, and the percentage of total power produced. In past and present, for a total capacity of 79.6 GW, and possible future additional power through improvements and added sites. *Source*: Data from US Energy Information Administration (2020a).

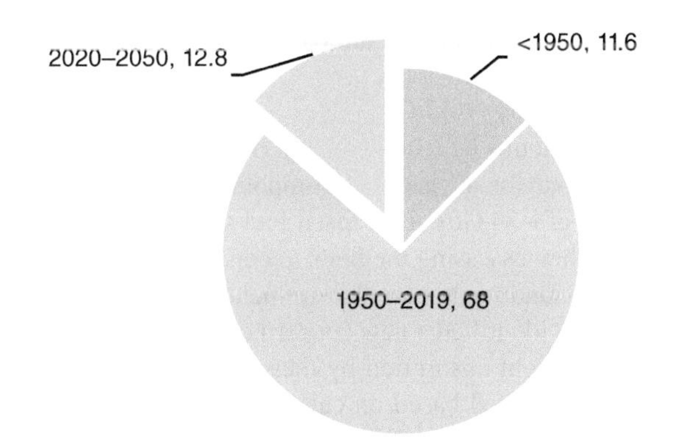

Figure 7.2 Hydropower past and present, for a total capacity of 79.6 GW, and possible future additional power through improvements and added sites. *Source*: Data from US Department of Energy (2016).

of Engineers, thus requiring federal action to convert them from only providing flood control to also producing electrical power. A 2013 report by the Corps of Engineers suggested that 6.2 GW of power was possible, but only 2.8 GW of this total was considered to economically feasible at that time.

A more extensive study by the Department of Engineering in 2016 concluded that 7.6 GW of electricity generation would be possible and that it could be increased to a total of 12.8 GW of new power by 2050 if there was a more dedicated effort, eventually producing a total of 92.4 GW from hydropower in the United States (US Department of Energy, 2016) (Fig. 7.2). Of this 12.8 GW only 4.8 GW would be power from currently non-powered dams. Thus, a large part of further hydropower development rests on upgrades, technology innovation, improved market conditions, environmental assessments, and public acceptance. Even maintaining the existing systems for hydropower generation is a challenge. Most of the existing hydropower systems were built in the 1970s. While hydropower systems are designed to last about 50 years, 40% of the Core of Engineer's sites are now 50 or more years older (Reid, 2016).

Environmental Considerations of Water Impoundments for Hydroelectric Power Generation and Other Uses

Growing hydropower is good for addressing climate change by providing carbon-neutral electricity, but only if it can be done within acceptable limits on impact on the environment. No method of energy capture and electricity production is without some environmental consequences and impounded water, for reservoirs or for hydropower generation, is no exception. Dams can greatly disrupt river and land ecosystems, altering sediment transport, water temperatures, fish populations and migration routes, and flooding areas of the land that are either critical for other ecosystems or covering land that may have previously been occupied by people. For several of these different reasons, some dams are being removed while others continue to be built around the world for hydropower generation (sometimes despite recognized large impacts on the environment).

A further consideration when erecting a dam, or more generally for impounding water, is secondary impacts on the environment due to increased greenhouse gas (GHG) emissions of CO_2, CH_4, and N_2O gases. GHG emissions from reservoirs created for hydroelectric power generation were estimated to contribute 0.31–0.94 Gt/y of CO_2 equivalents (CO_2e) due to CO_2, CH_4, and

N_2O gases and their estimated lifetimes in the atmosphere (Deemer et al., 2016). These accuracy of estimates of GHG emissions, however, are still being debated. For example, reservoirs in general were calculated to be responsible for ~0.77 Gt/y of CO_2e, with a range of 0.5–1.2 Gt/y, while other reports determined the range should be 0.14–3.4 Gt/y (Deemer et al., 2016). Even higher rates of 8.4 Gt/y have been calculated for all lakes and impoundments around the world (DelSontro et al., 2018). Given emissions of ~34 Gt/y from fossil fuel sources, these estimates require further analysis and considerations when assessing methods to control GHG emissions.

The impact of water impoundments can also influence climate change in ways other than GHG emissions. A recent study identified a new concern of impounded reservoirs relative to electricity generation based on the light absorption by these reservoirs (Wohlfahrt et al., 2021). The break-even time (BET) was compared based on calculating solar warming of the reservoir relative to the amount of warming that would offset CO_2 emissions due to the use of the carbon-free electricity from the hydropower plants. These researchers found that for about half (46%) of the hydropower plants the BET was about 4 years, or a total of 5% of the lifetime of the reservoir and hydropower plant (80 years). However, for 19% of the plants the BET was 40 years, and for 13% it was the complete lifetime of the system (80 years). This suggests additional considerations are needed to curtail such warming events through better designs of these systems or other methods to mitigate the extent of this warming.

Hydropower for Your Home?

If you have a stream running through your property it may be possible to produce your own electricity using a microhydropower system that can produce <100 kW, or a pico-hydro that produces <5 kW (Energy.gov, 2020). There are commercially available turbines that can be anchored to the bottom of a river, a bridge support, or to a structure on the side of a river, and used for electricity generation. Typically, these small systems produce around 5 kW, and have methods to prevent debris from entering the turbine (Smart hydro power, 2020). In the United Kingdom, it is estimated that a 5-kW system could last 40–50 years and would cost about 25,000 £ ($31,500 US). A typical 10 kW photovoltaic system costs about $30,000–$35,000, although it only produces power during the day while the hydropower system would produce power nearly continuously (although subject to weather and river flow changes throughout the year).

Tidal Power

The moon and the sun combine to produce gravitational effects on water that can result in large changes in water heights due to the tides, resulting in large flows of seawater upriver and higher water levels, followed by the outflow when the tide shifts and lower water levels. Some tides are relatively small but in certain locations the change in water height due to tides can reach 40 ft. One approach to capture this tidal energy is to direct the water into a holding area, called a tidal barrage, and then during the receding tide water through a turbine to generate electricity from the difference in water height similar to methods used for dams on rivers. Another method is called a tidal stream generator where water turbines, based on the same approach to capture energy from air flow using wind turbines, can be installed in the river to capture kinetic energy from the water flow (Wikipedia, 2020c). Energy capture using barrages has occurred in select locations around the world, such as a 254 MW plant in South Korea and a 240 MW plant in La Rance, France (US Energy Information Administration, 2020b). There are several other installations of tidal power generation systems around the world, but they have much lower generating capacities. The United

States does not currently have any tidal power plants, and the US EIA reports that there are only a few sites in the United States where such energy systems could prove to be economical. Concerns over tidal structures are that they can have adverse effects on the ecosystem by injuring fish and plants or interfering with fish migration patterns, or adversely impacting water quality through increased water turbidity.

Wave Power

Energy can be captured from the motion of waves, with estimates of 25 kW/m^2 for the coast of San Francisco (Wikipedia, 2020d). The energy is captured from the swells of the waves and thus large wave heights can produce more power. While there have been experimental tests of such technologies, there are currently no commercial applications.

7.3 HOW MUCH ENERGY IS IN USED WATER (WASTEWATER)?

The energy in used water predominantly produce from people in towns and businesses, also called domestic wastewater, has been estimated to be 13.9 kJ/g-COD (Owen, 1982, McCarty et al., 2011), 14.7 kJ/g-COD (Shizas and Bagley, 2004), and 17.8 kJ/-g-COD (Heidrich et al., 2011). Using an estimate of the COD produced by a person of 90 g COD/d (range of 60–120 g COD/d) (Heidrich et al., 2011), the lower and upper limits of these different numbers suggest a possible range of 10–25 W and therefore 0.10–0.26 D. Wastewater reaching a treatment plant also contains material from other sources, and therefore these estimates would likely be lower than that based on the actual COD of a wastewater, as explored in the example below.

Example 7.1

The total energy contained in waters from domestic and commercial buildings sent to used water treatment plants can be estimated in a variety of ways. (a) Estimate the power (kW) and D values for just human waste in the water assuming 17.8 kJ/g-COD and 100 g-COD/d per person. (b) If the calculation is now made using the wastewater COD, what are the power and D values for a used water containing 400 mg/L of COD?

(a) The calculation relies on a unit conversion of kJ to kWh, producing

$$17.8\frac{\text{kJ}}{\text{g COD}}\;100\frac{\text{g COD}}{\text{d cap}}\;\frac{1\ \text{kWh}}{3600\ \text{kJ}}\;\frac{1\ \text{d}}{24\ \text{h}} = 0.021\frac{\text{kW}}{\text{cap}}$$

For the range of 60–120 g-COD/d, the power would be between 0.012 and 0.025 W. With the use of 2.32 kWh/d cap = 1 D, we have

$$0.021\ \text{kW}\;\frac{24\ \text{h}}{\text{d cap}}\;\frac{1\ \text{D}}{2.32\ \frac{\text{kWh}}{\text{d cap}}} = 0.22\ \text{D}$$

This result suggests that food consumption of 1 D results in human waste and other organic matter added into the used water which is equivalent to about 22% of that 1 D.

(b) Using 400 mg/L of COD in the used water we have for 150 gal/d

$$17.8\frac{\text{kJ}}{\text{g COD}}\,400\,\frac{\text{mg COD}}{\text{L}}\,\frac{1\text{ g}}{1000\text{ mg}}\,150\,\frac{\text{gal}}{\text{d cap}}\,\frac{3.78\text{ L}}{\text{gal}}\,\frac{1\text{ kWh}}{3600\text{ kJ}}\,\frac{1\text{ d}}{24\text{ h}}=0.047\,\frac{\text{kW}}{\text{cap}}$$

$$0.047\text{ kW}\,\frac{24\text{ h}}{\text{d cap}}\,\frac{1\text{ D}}{2.32\,\frac{\text{kWh}}{\text{d cap}}}=0.49\text{ D}$$

The result of the calculations in Example 7.1 show that there is quite a large range in the estimated energy content of used waters. The amount of organic matter in a used water is highly variable, with the average COD of a used water in Europe, for example, about twice that typical of the same type of used water (domestic sources) in the United States. The upper limit for energy based on 60–120 kJ/g COD is 0.025 kW, and the lower limit for the COD of a wastewater based on 200 mg/L of COD is 0.024 kW. Therefore, 0.025 kW will be assumed for the average power based on a person, **equivalent to 0.26 D based on energy**. Thus, around one quarter of the 1 D of energy in our food translates to about 0.26 D of energy due to human waste and other organic matter added into the used water.

The total amount of energy in all used waters in the United States is not a well-researched value. One approach to estimate this energy is to consider three sources: domestic, animal, and food processing. In the United States, in 2019, the number of animals that were a part of the modern lifestyle or food chain, were (millions): 93.3 cows, 75 pigs, 9.2 horses, 5.2 lambs, and 505 chickens. The food requirements of these animals are quite different, but in general the amount of food needed is related to body weight. Food processing used waters vary in both quantity and organic matter content, with very large water flows and organic matter from vegetable washing operations to lower flows but very high organic matter concentrations from meat and dairy operations. One estimate was that the industry released 0.1 quad of energy annually in the used waters (Logan, 2004), equivalent to 0.035 kW per person or 0.36 D (assuming 328.2 million people in 2019). Another way to estimate the impact of food processing in terms of waste energy is to recognize that ~40% of food produced is wasted, which based on ~0.1 kW per person is 0.4 D.

Example 7.2

Assuming 0.025 kW per person, what is the total power (GW) in used waters in the United States originating from domestic sources, animal wastewaters, and food processing wastewaters? (a) Assume animals have a contribution equal to that of people and add an additional 40% to account for food waste. (b) The same assumption for people but assume twice the wastewater for animals and additional 0.1 quads/y from the food processing industry.

(a) If we use 0.025 kW per person, and we have 328.2 million people, the total for power in domestic used water sources is:

$$0.025\,\frac{\text{kW}}{\text{cap}}\,328.2\times10^{6}\text{ cap}\,\frac{\text{GW}}{10^{6}\text{ kW}}=8.2\text{ GW}$$

For the other sources, we have

$$\text{Total} = 8.2\text{ GW}(1 + 1 + 0.4) = 8.2\text{ GW} + 8.2\text{ GW} + 3.28\text{ GW}$$
$$= 20\text{ GW}$$

(b) To add in the food processing industry, we need to covert quad to kWh, as

$$\frac{0.1\text{ quad}}{\text{y}}\ \frac{293{,}071\text{ GWh}}{\text{quad}}\ \frac{1\text{ d}}{24\text{ h}}\ \frac{\text{y}}{365\text{ d}} = 3.34\text{ GW}$$
$$\text{Total} = 8.2\text{ GW}(1 + 2) + 3.34 = 8.2\text{ GW} + 16.4\text{ GW} + 3.34\text{ GW}$$
$$= 28\text{ GW}$$

Note that the 0.1 quad estimated for food processing energy is about the same as that estimated for 40% based on food waste.

From the above examples, we have 20–28 GW of energy in all used waters, and 8.2 GW for domestic used water. These can be compared to a use across the United States of 3.3 GW for treatment, suggesting that treatment requires about 40% of the energy in the used water. This estimate is further examined in the example below using wastewater-specific assumptions.

Example 7.3

Calculate the percentage of energy in a used water compared to that needed to treat it, assuming 0.6 kWh/m³ for treatment and an energy content of 17.8 kJ/g COD with 400 mg/L of COD.

Using the given information on the energy and COD contents, we have

$$17.8\frac{\text{kJ}}{\text{g COD}}\ 400\ \frac{\text{mg COD}}{\text{L}}\ \frac{1\text{ g}}{1000\text{ mg}}\ \frac{1\text{ kWh}}{3600\text{ kJ}}\ \frac{1000\text{ L}}{\text{m}^3} = 2.0\frac{\text{kWh}}{\text{m}^3}$$

Based on the given energy requirement, the percentage of energy is

$$\%\text{energy} = \frac{0.6\ \frac{\text{kWh}}{\text{m}^3}}{2\ \frac{\text{kWh}}{\text{m}^3}} \times 100\% = 30\%$$

This result indicates that there is 3.3 times more energy in the used water than that needed to treat it, or that we would need to extract 30% of the available energy to have an energy-neutral used water treatment plant considering only the energy in the COD. Recovering the ammonia could further offset energy costs due to the high energy demands for producing ammonia using the Haber-Bosch process (McCarty et al., 2011).

7.4 METHANE PRODUCTION FROM BIOMASS IN WASTEWATERS

Anaerobic Digestion (AD)

AD systems are used at most larger-scale modern used water treatment plants to reduce the amounts of solids that require disposal by converting those solids primarily to methane gas, with only a relatively small amount of biomass and solids remaining (Viessman and Hammer, 2005). There are four basic steps in anaerobic digestion using wastewaters with solids:

- Solubilization of particulate solids.
- Breakdown of soluble organics by fermentation into simple substrates (volatile fatty acids and hydrogen gas).
- Conversion to these small compounds primarily to acetate.
- Production of methane from acetate and hydrogen gas.

In the last step, methane is produced by two reactions via hydrogenotrophic methanogenesis (using H_2 gas), and acetoclastic methanogenesis (using acetate), according to the two balanced reactions

$$4\,H_2 + CO_2 \rightarrow CH_4 + 2\,H_2O \tag{7.1}$$

$$CH_3COOH \rightarrow CH_4 + CO_2 \tag{7.2}$$

Since the first reaction consumes CO_2, and the second step produces CO_2, the overall balance typically is that the final gas has typically has a 2:1 mixture of CH_4:CO_2. Rates of methane production are slow primarily due to solubilization of solids so that AD reactors usually require hydraulic retention times (HRTs) of around one month. Higher rates of methane production with greatly reduced HRTs are possible if the wastewater has a low solids content.

Energy recovery in the form of methane can be maximized by transferring as much influent organic matter into the AD reactor as possible. Some wastewater treatment plants, such as the Strass plant in Austria, produce sufficient methane in the AD reactors to fully power the treatment plant (Wett et al., 2007). Achieving energy neutrality using AD can be difficult if there are low concentrations of organic matter in the water (<3 g/L of COD) fed to the AD reactors as they need to be heated to ~38°C, which can consume a large part of the methane that is produced. At large treatment plants once the AD reactors are heated the remaining methane can be used to make electricity, but such electricity generation is not economical at smaller treatment plants. Thus, some treatment plants take in additional liquid wastewater that is highly concentrated directly into the AD reactors to increase methane production rates and produce electricity.

One key source of organic matter for methane production that is often neglected, particularly in the United States (compared to Europe), is animal wastewaters. The energy in animal wastewaters is potentially equal to that of all human wastewaters, with about 0.0094 EJ/y in wastewaters from pigs, cows, and other penned animals, an amount that is similar to that generated by the human population in the United States (Logan, 2004). Animal used waters in the United States are usually conveyed to open lagoons where the organic matter is degraded, and methane and CO_2 are released directly into the atmosphere. The release of methane without capture or treatment is a particular concern as methane is a potent GHG as discussed below. Treating animal wastewaters with AD provides opportunities to both avoid uncontrolled methane emissions to the atmosphere from waste lagoons as well as to generate renewable electricity. On small scales, or where biomass concentrations are too low, methane cannot be economically produced in digesters due to a need to keep the solutions heated to ~38°C. However, for large agricultural operations AD can be both economical and a benefit to renewable power generation.

Fully Anaerobic Biological Treatment Process Trains Producing Methane

There are several fully anaerobic technologies being developed to replace aerated activated sludge systems (Li and Yu, 2016). Instead of recycling biomass and aerating a tank to drive aerobic degradation of the organic matter, these systems operate anaerobically with the end products being methane and CO_2. Several of these systems use membranes for biomass retention rather than settling of biomass using clarifiers. Two systems are particularly promising for applications: anaerobic membrane bioreactors (AnMBR), and anaerobic fluidized bed reactors (AFBR) coupled with an anaerobic fluidized membrane bioreactor (AFMBR) (Fig. 7.3). In AnMBRs, biomass is retained in the system either using an external membrane system with high velocities of the fluid across the membrane to reduce biofouling, or recirculation of the biogas across the membranes to use the gas bubbles to scour the membrane (Wu and Kim, 2020, Shin and Bae, 2018).

In the AFBR granular activated carbon (GAC) particles are used to provide a high surface area support for anaerobic biofilms (McCarty et al., 2011, Kim et al., 2011) (Fig. 7.3). The GAC particles are kept suspended by recirculation of the fluid in the reactor providing an upflow velocity that helps to mix the reactor and prevent biomass from agglomerating the particles and clogging the reactor. The effluent from this fluidized bed reactor does not meet effluent quality requirements, and therefore the effluent flows into an AFMBR reactor which also contains GAC particles and

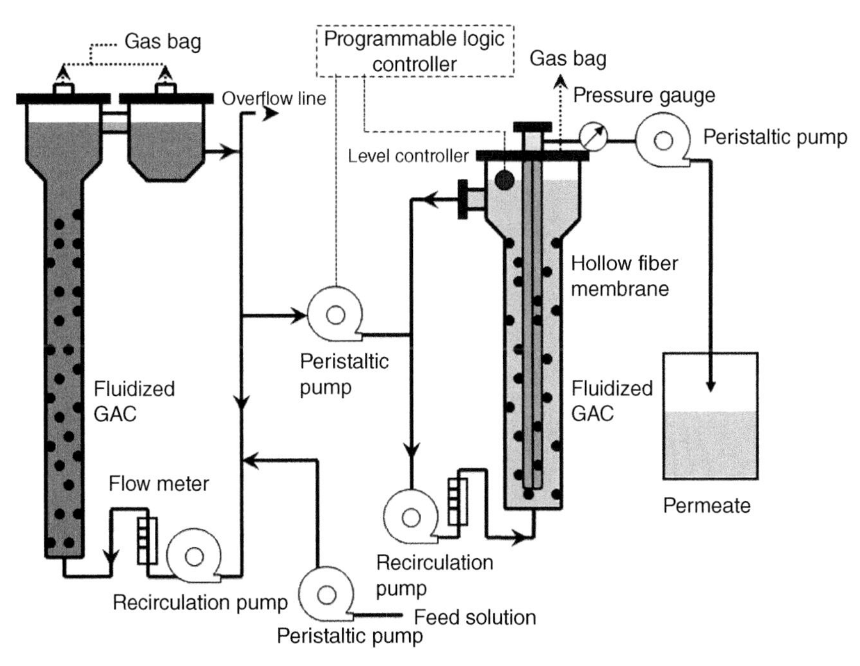

Figure 7.3 Methane generation from organic matter treating relatively dilute concentrations of organic matter using a two-stage process, using a fluidized bed reactor (AFBR) followed by an anaerobic fluidized bed membrane bioreactor. *Source*: Kim et al. (2011).

additionally an array of hollow fiber membranes (Kim et al., 2011). The fluid in the AFMBR is pulled through the membrane, producing a very high-quality effluent. Research has shown that this system can operated for years without requiring membrane cleaning, and that both processes are stable down to sufficiently low temperatures to make them useful for domestic wastewater treatment.

Methane is a Potent Greenhouse Gas (GHG)

The main concern with application of these AnMBR is the potential for release of methane into the atmosphere. Over a 20-year period, methane is 84 times more potent than CO_2 as a GHG (Wikipedia, 2020a). This decreases to a factor of 28 times when considering a longer time frame of 100 years which is the half-life of CO_2 in the atmosphere. If an anaerobic process is used to treat wastewater and the organic matter is converted to methane and CO_2, then the methane must be removed from the treated water to avoid making the wastewater treatment plant a source of GHGs. If the carbon in organic matter is released all as CO_2 then the wastewater treatment plant would be carbon neutral as the origin of the CO_2 would have been atmospheric carbon that was originally fixed by plants and used as food. However, converting that carbon into methane can make the plant a net GHG emitter. Therefore, it is critical that the methane be removed from the wastewater and not released into the environment.

Example 7.4

A wastewater has 400 mg/L of $C_6H_{12}O_6$ (glucose), equivalent to a chemical oxygen demand (COD) of 427 mg/L. (a) How much CO_2 (g/L) is released if it is fully oxidized in an aerobic treatment process (i.e. using oxygen)? (b) If oxidation of the glucose produces a 2:1 molar ratio of CH_4:CO_2, what is the mass of CO_2 and CH_4 that would be produced? (c) What is the total GHG emissions equivalent to CO_2 if CH_4 has a GHG potential 28 times that of CO_2 by mass (100-year time average)?

(a) To calculate the amount of CO_2 released, we must balance the oxidation equation for glucose, which produces

$$C_6H_{12}O_6 + 6O_2 \rightarrow 6CO_2 + 6H_2O$$

Thus, 6 mol of CO_2 are produced for each mole of glucose. The mass concentration of CO_2 for 0.4 g/L of glucose is therefore

$$0.4\frac{\text{g glu}}{\text{L}}\,\frac{\text{mol glu}}{180\text{ g glu}}\,\frac{6\text{ mol CO}_2}{1\text{ mol glu}}\,\frac{44\text{ g CO}_2}{\text{mol CO}_2} = 0.59\,\frac{\text{g CO}_2}{\text{L}}$$

(b) If the ratio of CH_4 to CO_2 is 2:1, then for the 6 carbons in glucose we would form $4CH_4$ and $2CO_2$. The concentrations are:

$$0.4\frac{\text{g glu}}{\text{L}}\,\frac{\text{mol glu}}{180\text{ g glu}}\,\frac{4\text{ mol CH}_4}{1\text{ mol glu}}\,\frac{18\text{ g CH}_4}{\text{mol CH}_4} = 0.16\frac{\text{g CH}_4}{\text{L}}$$

$$0.4\frac{\text{g glu}}{\text{L}}\,\frac{\text{mol glu}}{180\text{ g glu}}\,\frac{2\text{ mol CO}_2}{1\text{ mol glu}}\,\frac{44\text{ g CO}_2}{\text{mol CO}_2} = 0.20\frac{\text{g CO}_2}{\text{L}}$$

(c) The equivalent mass concentration of CO_2 for CH_4 would be:

$$0.16\frac{\text{g CH}_4}{\text{L}}28\frac{\text{g CO}_2}{\text{g CH}_4}=4.48\frac{\text{g CO}_2}{\text{L}}$$

The total for the anaerobic treatment process is therefore

$$0.20\frac{\text{g CO}_2}{\text{L}}+4.48\frac{\text{g CO}_2}{\text{L}}=4.68\frac{\text{g CO}_2}{\text{L}}$$

Based on this calculation we see that if CH_4 is made and released into the atmosphere that the net result is a 7.9 times increase in CHG emissions from the anaerobic process compared to the aerobic process.

Methane is very insoluble in water, for example 23 mg/L for pure methane gas at 20°C. Therefore, with high methane generation rates the methane will not be able to remain dissolved in the water and it will rapidly volatilize out producing methane gas. While this low solubility helps to facilitate the separation of the gas from water in an anaerobic process, even a small amount of methane left as a residual in the water is a concern if it is released to the air. Using the values from Example 7.4, if 23 mg/L of methane were dissolved in the water then that would be equivalent to 0.64 g/L of CO_2. This methane, combined with the 0.16 g/L released would total 0.8 g/L of CO_2, or an additional 35% increase in CO_2 relative to that in the sample with aerobic oxidation (0.59 g/L). If the wastewater is colder, for example at 15°C, then methane is even more soluble (32 mg/L). Therefore, from the perspective of GHG emissions, removing methane from treated water is very important. Currently, wastewater treatment is only responsible for a small amount of GHG emissions but that could change if more methane were emitted from wastewater treatment plants.

7.5 ELECTRICITY GENERATION USING MICROBIAL FUEL CELLS (MFCs)

Energy is directly captured from organic matter as electricity in an MFC by using bacteria that oxidize organic matter and release the electrons to the anode of the MFC (Logan, 2008). These bacteria, called exoelectrogens, have evolved to use electron acceptors outside the cell such as solid metal oxides. Thus, exoelectrogens "breathe" by reducing metal oxides such as iron oxides in the environment. In an MFC, these bacteria release the electrons to the anode where the electrons flow to the cathode, creating voltage and current, with oxygen typically reduced at the cathode to water (Fig. 7.4). The overall process accomplishes organic matter degradation and removal from the water, while at the same time generating electricity. The maximum voltage is theoretically 1.101 V based on acetate oxidation at the anode and oxygen reduction at the cathode under typical conditions in the MFC ($E_{Cat} - E_{An} = 0.805\text{ V} + 0.296\text{ V}$) (Logan et al., 2006). However, the cell voltage produced in an MFC is typically ~0.4–0.5 V.

The main challenges for using MFCs for used water treatment are the cost of the electrodes, low power densities, and capturing and effectively using the power that is produced. Early MFC designs used expensive materials consisting of carbon cloth developed for hydrogen fuel cells,

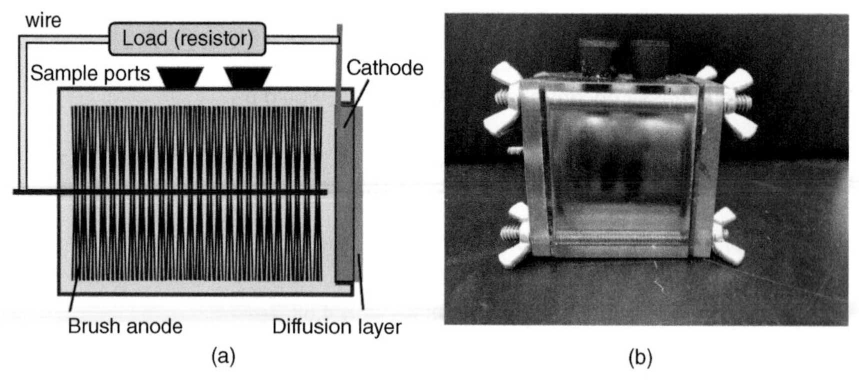

Figure 7.4 (a) Schematic, and (b) photograph of a microbial fuel cell used to produce electricity. Bacteria grow on the graphite fiber brush anode and produce an electrical current that can be harvested as electricity. The cathode (flat slab in schematic, not visible in photograph) catalyzes oxygen reduction to water using an inexpensive catalyst such as activated carbon. The cathode is covered with a diffusion layer so that it does not leak water.

cation exchange membranes made of Nafion™, and platinum catalysts for the cathodes. Since then, research has shown that this expensive cation exchange membrane is not essential for the operation of the MFC, anodes can be made using less-expensive carbon felt or graphite fiber brushes and activated carbon can be used as an oxygen reduction catalyst. The combination of these less expensive materials has reduced the cost of the materials into a range for practical applications, but the manufacturing costs and long-term durability of the materials has so far prohibited commercial applications.

Typical power densities using relatively well buffered laboratory medium with acetate, carbon brushes, and activated carbon cathodes are around $1.36 \pm 0.20\,\mathrm{W/m^2}$ or $34\,\mathrm{W/m^3}$ (Yang et al., 2017) (Table 7.1). When actual used water is used (from domestic sources), the lower buffer capacity of the wastewater combined with a lower solution conductivity, and slow degradation of complex organic matter that must be degraded to simple volatile fatty acids like acetate to produce electricity have resulted in power densities of $\sim 0.4\,\mathrm{W/m^2}$. Treatment of the activated carbon used for the cathode as a catalyst can improve performance, with $4.7\,\mathrm{W/m^2}$ produced using acetate, and $0.8\,\mathrm{W/m^2}$ using domestic wastewater (Yang and Logan, 2016), although the performance decayed over time using wastewater due to cathode fouling (Rossi et al., 2017). Recently it was shown that power densities as high as $7.1 \pm 0.4\,\mathrm{W/m^2}$ could be produced using acetate in a well-buffered medium, but this performance required the use of closely-spaced electrodes, a relatively expensive anion exchange membrane (AEM), and flow of the solution through a carbon felt electrode which would lead to rapid biofouling and clogging if used for wastewater treatment (Rossi et al., 2020). To date there have been no large-scale applications of this new compact design that can produce these higher power densities.

The amount of energy recovered, and the power produced by an MFC is of interest in terms of the magnitude of that energy recovered versus the energy consumed in a conventional treatment plant. The maximum energy content of wastewater was calculated in Example 7.3 as $E_w = 2.0\,\mathrm{kWh/m^3}$-water. The amount of energy that can be extracted or recovered, E_r, will be reduced based on the overall efficiency, where $E_r = E_w\, \eta_e$. For example, if $\eta_e = 10\%$ of the energy was recovered then that would be $0.2\,\mathrm{kWh/m^3}$.

Table 7.1 Examples of power densities in microbial fuel cells normalized to the area of one electrode or the volume of the reactor (anode and cathode chambers) using acetate as a substrate or domestic wastewater.

Substrate	A_{Cat} (m^2/m^3)	V (L)	P_A (W/m^2)	P_v (W/m^3)	References
Acetate	780	0.0089	7.1	560	(Rossi et al., 2020)
	680	0.030	4.3	2900	(Fan et al., 2012)
	25	0.025	4.7	120	(Yang and Logan, 2016)
	25	0.025	1.36±0.20	34	(Yang et al., 2017)
Used water	25	0.025	0.8	20	(Yang and Logan, 2016)
	25	6	0.25	6.3	(He et al., 2016)
	7.3	85	0.10	0.73	(Rossi et al., 2019)

The electrode packing density (A_{Cat}) shows how packing more electrode area per volume of reactor (V) can increase the volumetric power density.

Power production in MFCs (P) is usually reported as a power density based on dividing it by the cross-sectional projected area of the electrodes A_e (m^2), or P_A (W/m^2) = P/A_e. Maximizing the power produced per volume of the reactor is also important, and so power is often reported after dividing by the total volume of the reactor, V_r, or P_V (W/m^3) = P/V_r. The power per volume is related to the area of electrodes by $P_V = P_A/A_s$, where A_s (m^2/m^3) is the area of one of the electrodes (usually the cathode) per volume of reactor. For example, if the area of the cathode per volume of the reactor is $A_{Cat} = 25\,m^2/m^3$, and the power per area is $P_A = 1.36\,W/m^2$, then the volumetric power density per reactor volume is $P_V = 34\,W/m^3$ (Table 7.1).

When a wastewater reactor is operated the power going into the treatment system should be minimized. To calculate the power consumption, since power is the energy used per time, $P_R = E_w/\theta$, where θ is the HRT of the water in the reactor. The power can be a negative number, for example when a conventional treatment system is used, or a positive number when power is produced by an MFC.

Example 7.5

(a) Calculate continuous power consumption for a traditional aerated bioreactor that uses half the overall power to treat the wastewater for aeration, or –0.3 kWh/m³, and a HRT of 6 h. (b) Calculate energy extracted from a used water (kWh/m³) and the efficiency by an MFC assuming that power is 0.8 W/m², with an electrode packing density of 25 m²/m³, a HRT of 6 h, and an energy content of 2 kWh/m³.

(a) Using the given information and the relationship between energy and power, we have

$$P_R = \frac{E_w}{\theta} = \frac{\left(-0.3\,\frac{\text{kWh}}{\text{m}^3}\right)}{\left(6\,\frac{\text{h}}{\text{m}^3}\right)}\,\frac{1000\text{ W}}{\text{kW}} = -50\text{ W}$$

(b) The power per volume of reactor is

$$P_V = P_A A_s = \left(0.8\frac{\text{W}}{\text{m}^2}\right)\left(25\frac{\text{m}^2}{\text{m}^3}\right) = 20\frac{\text{W}}{\text{m}^3}$$

To calculate the energy extracted per volume of the wastewater, we rearrange the above equation, to obtain

$$E_w = P_v\,\theta = \left(20\frac{\mathrm{W}}{\mathrm{m}^3}\right)(6\,\mathrm{h})\,\frac{1\,\mathrm{kW}}{1000\,\mathrm{W}} = 0.12\,\frac{\mathrm{kWh}}{\mathrm{m}^3}$$

This is an energy efficiency of 6% based on the given energy content of the wastewater.

7.6 HYDROGEN PRODUCTION USING MICROBIAL ELECTROLYSIS CELLS (MECs)

In a typical MEC oxidation of organic matter occurs on an anode but no oxygen or other chemical electron acceptor is used at the cathode other than water, with the goal being to produce H_2 gas. The evolution of H_2 occurs either through protons in water combining with electrons to produce H_2 at acidic pHs, or water dissociation with consumption of protons and release of OH^- ions at neutral to alkaline pHs. The overall reaction is not spontaneous as the total voltage that is required is ~0.414 V to evolve the H_2, but the anode supplies only about −0.3 V, and so >0.114 V needs to be added in theory, with ~0.6–1 V added in practice for MECs. These voltages are still lower than those needed for water splitting electrolyzers where 1.23 V is theoretically needed under standard conditions, but 1.6 V or more are applied to the electrolyzer in practice.

Although energy must be input into an MEC to treat the wastewater, the H_2 gas can be captured and either stored or used elsewhere in the plant, for example to heat the anaerobic digesters or for use in boilers for other uses in the plant. If the applied voltage (U_{app}) is less than 1.23 V then the overall process can be energy neutral in terms of the energy recovery compared to the input electrical energy, which can be calculated as:

$$\eta_E = \frac{1.23}{U_{app}} \tag{7.3}$$

At a commonly applied voltage of $U_{app} = 0.8$ V, the maximum energy recovery is therefore 154% if all the current is converted to H_2 gas, with complete gas recovery. This efficiency can be greater than 100% because the energy in the organic matter is not included in the calculation.

The power used by the MEC is calculated as $P_A = I\,U_{App}$, where I is a measured current density per area of electrode. The energy consumed to treat the wastewater can then be calculated as done above for the power produced by MFCs in Example 7.5. The energy put into treating the wastewater by using the electrodes can be less than that needed for conventional aeration of 0.3 kWh/m^3, which is about half the total energy used for treating the wastewater in the plant (0.6 kWh/m^3).

Example 7.6

Calculate the energy needed to treat wastewater (kWh/m^3) by an MEC assuming a measured current density of 0.5 A/m^2, an electrode packing density of 24 m^2/m^3, and a HRT of 6 h. Data from Kim et al. (2017).

The current per volume of reactor as

$$I_v = I_A\,A_s = \left(2.5\,\frac{\mathrm{A}}{\mathrm{m}^2}\right)\left(24\frac{\mathrm{m}^2}{\mathrm{m}^3}\right) = 60\frac{\mathrm{A}}{\mathrm{m}^3}$$

$$P_v = I_v\, U_{app} = \left(60 \frac{A}{m^3}\right)(0.9\,V) = 5.44 \frac{W}{m^3}$$

To calculate the energy used per volume of wastewater, we use the equation relating power and time to obtain

$$E_w = P_v\, \theta = \left(5.44 \frac{W}{m^3}\right)(6\,h)\frac{1\,kW}{1000\,W} = 0.033\,\frac{kWh}{m^3}$$

This result is only 11% of that typically used for wastewater aeration (0.3 kWh/m^3), and it has the additional benefit of producing H_2 gas.

Although the applied voltages for MECs are less than those needed for water electrolyzers, the rates of H_2 generation in MECs are orders of magnitude lower. Therefore, the use of MECs will probably be limited to applications converting waste organics, such as those in used waters, into H_2 gas. Another challenge is recovering the H_2 gas. If a single-chamber MEC design is used, where both electrodes are in the same chamber, the H_2 will get converted to CH_4 by methanogens. For example, in a pilot test using a 1000 L reactor, no H_2 was recovered from the system and the gas produced contained only a mixture of CH_4 and CO_2 (Cusick et al., 2011) (Table 7.2). If a membrane or gas separator is inserted between the electrodes, then it is possible to recover H_2 gas more effectively although some gas is lost due to diffusion through the membrane (Kim et al., 2018, Kim et al., 2017, Cotterill et al., 2017).

Some typical values for H_2 production in MECs are shown in Table 7.2 along with estimated upper limits of electrode packing and area-based current densities. The estimated maximum cathode packing density is set at 50 m^2/m^3 for wastewater to avoid electrode clogging, although higher electrode packing densities have been used with (particle-free) laboratory solutions of acetate or fermentation effluents. The rates of possible H_2 production for a 1 m^3 reactor approach those of smaller H_2 generation units that split water. If an MEC was used on a large scale, for example for used water treatment, these rates of H_2 or CH_4 production could help to offset energy costs at the plant or possibly provide a profit if purified, compressed, and sold.

Table 7.2 Examples of rates of hydrogen or methane production in MECs and estimated maximum rates possible for wastewater applications.

Gas	Technology (L)	Fuel	A_{Cat} (m^2/m^3)	I_A (A/m^2)	I_V (A/m^3)	Rate (kg/d-m^3)	References
H_2	1 Ch (0.028)	Acetate	22[a]	8.5	188	0.14	(Call et al., 2009)
H_2	1 Ch (2.7)	Acetate	23[b]	3.2	74	0.044	(Rader and Logan, 2010)
H_2	2 Ch (0.168)	FE	28	3.9	110	0.11	(Kim et al., 2017)
H_2	2 Ch	WW	13	0.29	3.8	0.0003	(Cotterill et al., 2017)
H_2	Est. Max.	UW	50	50	2500	2.25	—
CH_4	2 Ch (0.2)	Current	5	0.022	0.11	0.0016	(Siegert et al., 2014)
CH_4	1 Ch (1000)	UW	7.2	1.28	9.2	0.15	(Cusick et al., 2011)
CH_4	Est. Max.	UW	50	50	2500	4.5	—

1 Ch = 1 chamber, 2 Ch = 2 chamber with a membrane or separator to recover the H_2 gas; acetate = well buffered laboratory solution using acetate as a substrate, FE = synthetic fermentation effluent, UW = used water containing complex organic matter, current generated by water splitting at the anode.

[a] 810 m^2/m^3 based on bristles on the brush cathode.

[b] Actual area of 64 m^2/m^3. Based on data after 3 days; by the end of the tests all the gas was methane.

7.7 ELECTRICITY GENERATION USING SALINITY GRADIENTS

When river water flows into seawater energy is released that is equivalent to water flowing over a dam 250 m (800 ft) high, or about the height of the Hoover dam in the United States (221 m, 726 ft). This difference in energy is due to the different salt concentrations. To understand why there is such a big energy difference consider that when seawater is purified by reverse osmosis (RO) very high pressures are needed to push water out through a membrane, while excluding salts, so that the permeate water is now fresh water. Energy cycles are reversible, so when run in one direction a device might produce energy, while the other direction it will consume energy (although never with an equal amount of energy in each direction). For example, running up a hill takes energy, but when you go down a hill you take advantage of the energy difference due to gravity and you can you down the hill without energy input by you. Similarly, for RO desalination it takes a lot of pressure to push water through a membrane due to the energy differences in the salt and fresh water. If fresh water and seawater are on either side of that membrane, then there will be an osmotic pressure (energy) difference between the two solutions, and this energy difference creates an opportunity to capture that energy and make electricity.

Several technologies have been developed to capture the energy difference between freshwater and seawater, including pressure-retarded osmosis (PRO), reverse electrodialysis (RED), and capacitive mixing (CapMix) (Logan and Elimelech, 2012). PRO makes use of the osmotic pressure differences as fresh water pushes through the membrane and into the seawater chamber, which pressurizes the chamber. This pressure can then be released through a turbine to generate electricity. In RED, stacks of membranes are used to capture energy based on the salt motion through the membranes (with minimal water transfer), while CapMix produces net electrical power based on capacitive discharge processes (similar to how a capacitor in an electronic device discharges current). PRO is the best known of these approaches and is considered to the most feasible approach for capturing salinity gradient energy.

The energy content of the water, ΔG (kWh/m^3) can be approximated for ideal solutions (Yip and Elimelech, 2012) as:

$$\Delta G = \nu RT\left(\frac{c_m}{\varphi}\ln c_m - c_f \ln c_f - \frac{(1-\varphi)}{\varphi}c_s \ln c_s\right) \tag{7.4}$$

where ν = number of species that the salt dissolves into, R is the gas constant, T the absolute temperature, φ = fraction of freshwater (assumed to be limiting compared to seawater), c_f the molar concentration of salt in the freshwater and c_s the molar concentration in the seawater, and c_m the molar concentration of a mixture of the solutions calculated as $c_m = \varphi c_f + (1-\varphi)c_s$. Based on this equation the energy that could be extracted is a function of the relative volumes of water flowing through a device.

The power produced by an SGE technology such as PRO or RED is usually normalized to the total surface area of the membrane in the reactor. The maximum power density with freshwater and seawater using PRO was estimated to be ~38 W/m^2 of membrane area, which is much higher than those of RED or CapMix (Yip and Elimelech, 2014, Logan and Elimelech, 2012).

Example 7.7

(a) Calculate the maximum energy that could be captured for 10% river water with seawater at 298 K. (b) How many gallons of water would you need to capture 1 D if the efficiency of the process was 50%.

(a) For these conditions, we will assume a river water concentration of 1.5 mM and seawater with 600 mM, and using molar concentrations we have for 10% river water or $\varphi = 0.10$,

$$c_m = \varphi c_f + (1 - \varphi)c_s = (0.1)(0.0015\,\text{M}) + (1 - 0.1)(0.6\,\text{M}) = 0.54\,\text{M}$$

$$\Delta G = \nu RT\left(\frac{c_m}{\varphi}\ln c_m - c_f \ln c_f - \frac{(1-\varphi)}{\varphi}c_s \ln c_s\right)$$

$$\Delta G = 2\left(8.314\frac{\text{J}}{\text{mol K}}\right)(298\,\text{K})\frac{10^3\,\text{L}}{\text{m}^3}\frac{1\,\text{kWh}}{3.6\times10^6\,\text{J}}$$

$$\times\left(\frac{(0.54\,\text{M})}{(0.1)}\ln(0.54\,\text{M}) - (0.0015\,\text{M})\ln(0.0015\,\text{M}) - \frac{(1-0.1)}{(0.1)}(0.6\,\text{M})\ln(0.6\,\text{M})\right)$$

$$\Delta G = 0.77\,\frac{\text{kWh}}{\text{m}^3}$$

(b) The energy is expressed in terms of energy based on freshwater, and therefore for every 1 m^3 of freshwater you would need 9× as much seawater. Therefore, the energy captured for 10 m^3 at 50% efficiency would be:

$$0.77\frac{\text{kWh}}{\text{m}^3\ \text{freshwater}}\ \frac{1\ \text{m}^3\ \text{freshwater}}{10\ \text{m}^3\ \text{total}}(0.50) = 0.038\frac{\text{kWh}}{\text{m}^3}$$

To date, no salinity gradient energy technology has been used commercially to produce power. The reasons include the high cost of membranes, energy used to treat the water before it goes into the devices to produce energy, and the energy losses to pump the water to the device and return it to the source. For example, energy consumption for pretreatment of seawater for RO plants, which would be similar to that needed here for pretreatment, are estimated to be 0.2 – 0.3 kWh/m^3 including pumping costs for only the seawater (Voutchkov, 2018, Elimelech and Phillip, 2011). This amount of energy for pretreatment is much greater than that which can be produced by a 1 m^3 of water as shown in Example 7.7. If the fraction of the freshwater is $\varphi = 1$, as in the example, that means that for every 1 m^3 of freshwater and additional 9 m^3 of seawater would be needed, or a total of 10 m^3. The energy extracted based on all the water is shown in Figure 7.5 using the calculation approach in Example 7.7. Now the maximum power expressed as a total of the volume of water used reaches only ~3 kWh/m^3, or 1.5 kWh/m^3 if 50% of the energy was extracted. This is smaller than the energy that was estimated to be needed for water pretreatment.

One way to reduce pretreatment would be to use a very clean water source, such as fully treated used water prior to its discharge into the sea. Even if pretreatment costs could be reduced, however, the membranes used in these processes remain prohibitively expensive for power production compared to other processes. Therefore, while there have been pilot tests for PRO and RED, these tests have not advanced forward into commercial applications. If pretreatment energy consumption and costs could be reduced, and membranes could be produced that were very inexpensive, it might become economical to harvest salinity gradient energy and produce electrical power using one of these technologies.

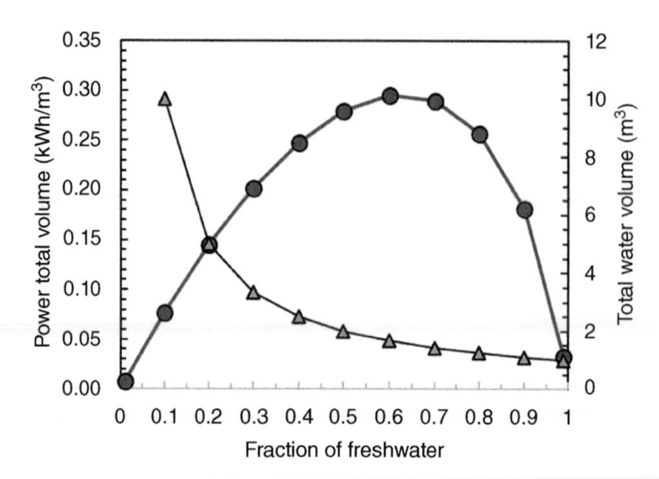

Figure 7.5 The maximum theoretical power per volume of *total water* that can be extracted from river water mixing with seawater. The maximum is a trade off in power extraction and the volume of water needed, with a maximum ~3 kWh/m^3 at $\varphi = 0.6$. Calculations and values of concentrations are the same as those used in Example 7.7. *Source*: Based on Yip and Elimelech (2012).

References

CALL, D., MERRILL, M. D. & LOGAN, B. E. 2009. High surface area stainless steel brushes as cathodes in microbial electrolysis cells (MECs). *Environmental Science & Technology,* 43, 2179–2183.

COTTERILL, S. E., DOLFING, J., JONES, C., CURTIS, T. P. & HEIDRICH, E. S. 2017. Low temperature domestic wastewater treatment in a microbial electrolysis cell with 1 m^2 anodes: Towards system scale-Up. *Fuel Cells,* 17, 584–592.

CUSICK, R. D., BRYAN, B., PARKER, D. S., MERRILL, M., MEHANNA, M., KIELY, P. D., LIU, G. & LOGAN, B. E. 2011. Performance of a pilot-scale continuous flow microbial electrolysis cell fed winery wastewater. *Applied Microbiology and Biotechnology,* 89, 2053–2063.

DEEMER, B. R., HARRISON, J. A., LI, S., BEAULIEU, J. J., DELSONTRO, T., BARROS, N., BEZERRA-NETO, J. F., POWERS, S. M., DOS SANTOS, M. A. & VONK, J. A. 2016. Greenhouse gas emissions from reservoir water surfaces: A new global synthesis. *BioScience,* 66, 949–964.

DELSONTRO, T., BEAULIEU, J. J. & DOWNING, J. A. 2018. Greenhouse gas emissions from lakes and impoundments: Upscaling in the face of global change. *Limnology and Oceanography Letters,* 3, 64–75.

ELIMELECH, M. & PHILLIP, W. A. 2011. The future of seawater desalination: Energy, technology, and the environment. *Science,* 333, 712–717.

ENERGY.GOV. 2020. *Microhydropower systems* [Online]. Available: https://www.energy.gov/energysaver/buying-and-making-electricity/microhydropower-systems [Accessed July 15 2020].

FAN, Y., HAN, S.-K. & LIU, H. 2012. Improved performance of CEA microbial fuel cells with increased reactor size. *Energy & Environmental Science,* 5, 8273–8280.

HE, W., ZHANG, X., LIU, J., ZHU, X., FENG, Y. & LOGAN, B. E. 2016. Microbial fuel cells with an integrated spacer and separate anode and cathode modules. *Environmental Science: Water Research & Technology,* 2, 186–195.

HEIDRICH, E. S., CURTIS, T. P. & DOLFING, J. 2011. Determination of the internal chemical energy of wastewater. *Environmental Science & Technology,* 45, 827–832.

KIM, J., KIM, K., YE, H., LEE, E., SHIN, C., McCARTY, P. L. & BAE, J. 2011. Anaerobic fluidized bed membrane bioreactor for wastewater treatment. *Environmental Science & Technology,* 45, 576–581.

KIM, K.-Y., YANG, W. & LOGAN, B. E. 2018. Regenerable nickel-functionalized activated carbon cathodes enhanced by metal adsorption to improve hydrogen production in microbial electrolysis cells. *Environmental Science & Technology,* 52, 7131–7137.

KIM, K.-Y., ZIKMUND, E. & LOGAN, B. E. 2017. Impact of catholyte recirculation on different 3-dimensional stainless steel cathodes in microbial electrolysis cells. *International Journal of Hydrogen Energy,* 42, 29708–29715.

LI, W.-W. & YU, H.-Q. 2016. Advances in energy-producing anaerobic biotechnologies for municipal wastewater treatment. *Engineering,* 2, 438–446.

LOGAN, B. E. 2004. Extracting hydrogen and electricity from renewable resources. *Environmental Science & Technology,* 38, 160A–167A.

LOGAN, B. E. 2008. *Microbial fuel cells,* Hoboken, NJ, John Wiley & Sons, Inc.

LOGAN, B. E., AELTERMAN, P., HAMELERS, B., ROZENDAL, R., SCHRÖDER, U., KELLER, J., FREGUIAC, S., VERSTRAETE, W. & RABAEY, K. 2006. Microbial fuel cells: Methodology and technology. *Environmental Science & Technology,* 40, 5181–5192.

LOGAN, B. E. & ELIMELECH, M. 2012. Membrane-based processes for sustainable power generation using water and wastewater. *Nature,* 288, 313–319.

McCARTY, P. L., BAE, J. & KIM, J. 2011. Domestic wastewater treatment as a net energy producer – Can this be achieved? *Environmental Science & Technology,* 45, 7100–7106.

OWEN, W. F. 1982. *Energy in wastewater treatment* Englewood Cliffs, Prentice-Hall, Inc., p. 373.

RADER, G. K. & LOGAN, B. E. 2010. Multi-electrode continuous flow microbial electrolysis cell for biogas production from acetate. *International Journal of Hydrogen Energy,* 35, 8848–8854.

REID, R. L. 2016. Blue power. *Civil Engineering*, May, 58–67.

ROSSI, R., JONES, D., MYUNG, J., ZIKMUND, E., YANG, W., GALLEGO, Y. A., PANT, D., EVANS, P. J., PAGE, M. A., CROPEK, D. M. & LOGAN, B. E. 2019. Evaluating a multi-panel air cathode through electrochemical and biotic tests. *Water Research,* 148, 51–59.

ROSSI, R., WANG, X. & LOGAN, B. E. 2020. High performance flow through microbial fuel cells with anion exchange membrane. *Journal of Power Sources,* 475, 228633.

ROSSI, R., YANG, W., SETTI, L. & LOGAN, B. E. 2017. Assessment of a metal–organic framework catalyst in air cathode microbial fuel cells over time with different buffers and solutions. *Bioresource Technology,* 233, 399–405.

SHIN, C. & BAE, J. 2018. Current status of the pilot-scale anaerobic membrane bioreactor treatments of domestic wastewaters: A critical review. *Bioresource Technology,* 247, 1038–1046.

SHIZAS, I. & BAGLEY, D. M. 2004. Experimental determination of energy content of unknown organics in municipal wastewater streams. *Journal of Energy Engineering,* 130, 45–53.

SIEGERT, M., YATES, M. D., CALL, D. F., ZHU, X., SPORMANN, A. & LOGAN, B. E. 2014. Comparison of nonprecious metal cathode materials for methane production by electromethanogenesis. *ACS Sustainable Chemistry & Engineering,* 2, 910–917.

SMART HYDRO POWER. 2020. *Smart turbines* [Online]. Available: https://www.smart-hydro.de/renewable-energy-systems/hydrokinetic-turbines-river-canal/ [Accessed July 15 2020].

US DEPARTMENT OF ENERGY. 2016. *Hydropower vision: A new chapter for America's 1st renewable energy source* http://www.osti.gov/scitech.

US ENERGY INFORMATION ADMINISTRATION. 2020a. *Hydropower explained* [Online]. Available: https://www.eia.gov/energyexplained/hydropower/ [Accessed July 15 2020].

US ENERGY INFORMATION ADMINISTRATION. 2020b. *Tidal power* [Online]. Available: https://www.eia.gov/energyexplained/hydropower/tidal-power.php [Accessed July 15 2020].

VIESSMAN, W. J. & HAMMER, M. J. 2005. *Water Supply and Pollution Control, 7th edition,* Upper Saddle River, NJ, Pearson Prentice-Hall.

VOUTCHKOV, N. 2018. Energy use for membrane seawater desalination – Current status and trends. *Desalination,* 431, 2–14.

WETT, B., BUCHAUER, K. & FIMML, C. 2007. *Energy self-sufficiency as a feasible concept for wastewater treatment systems.* IWA Leading Edge Technology Conference in Singapore.

WIKIPEDIA. 2020a. *Greenhouse gas* [Online]. Available: https://en.wikipedia.org/wiki/Greenhouse_gas [Accessed February 16 2021].

WIKIPEDIA. 2020b. *Hydroelectricity* [Online]. Available: https://en.wikipedia.org/wiki/Hydroelectricity [Accessed July 15 2020].

WIKIPEDIA. 2020c. *Tidal power* [Online]. Available: https://en.wikipedia.org/wiki/Tidal_power [Accessed July 25 2020].

WIKIPEDIA. 2020d. *Wave power* [Online]. Available: https://en.wikipedia.org/wiki/Wave_power [Accessed December 15 2020].

WOHLFAHRT, G., TOMELLERI, E. & HAMMERLE, A. 2021. The albedo–climate penalty of hydropower reservoirs. *Nature Energy.*

WU, B. & KIM, J. 2020. Anaerobic membrane bioreactors for nonpotable water reuse and energy recovery. *Journal of Environmental Engineering,* 146, 03119002.

YANG, W., KIM, K.-Y., SAIKALY, P. E. & LOGAN, B. E. 2017. The impact of new cathode materials relative to baseline performance of microbial fuel cells all with the same architecture and solution chemistry. *Energy & Environmental Science,* 10, 1025–1033.

YANG, W. & LOGAN, B. E. 2016. Immobilization of a metal–nitrogen–carbon catalyst on activated carbon with enhanced cathode performance in microbial fuel cells. *ChemSusChem,* 9, 2226–2232.

YIP, N. Y. & ELIMELECH, M. 2012. Thermodynamic and energy efficiency analysis of power generation from natural salinity gradients by pressure retarded osmosis. *Environmental Science & Technology,* 46, 5230–5239.

YIP, N. Y. & ELIMELECH, M. 2014. Comparison of energy efficiency and power density in pressure retarded osmosis and reverse electrodialysis. *Environmental Science & Technology,* 48, 11002–11012.

CHAPTER 8

FOOD

8.1 THE ENERGY BURDEN OF FOOD

In the industrialized nations, finding food is no longer the burden it once was for mankind as most of us no longer grow or produce our own food: we simply buy it at a store. That convenience comes at a price, both in terms of the money we spend on food and the large amount of fossil fuel energy consumed for the complete food system, consisting of fertilizing and growing crops, feeding animals, processing the food, and then getting it to retail stores, with us finally transporting it to our homes. The 1 D of energy we consume in our food every day, or about 2000 Calories, is carbon neutral as the food was produced through CO_2 capture from the atmosphere. However, from the moment that seeds and fertilizers are added to the soil, and as crops are grown and maintained using fossil fuel-powered machinery, the energy and carbon emissions increase in each additional step that it takes to move that food through farming, transportation, and processing along the way to its final destination.

The energy needed for the typical food system, or the chain of activities to bring food to our table (or restaurant), are presented below along with a few other related topics, such as the energy used to make fertilizer and how food choices can impact overall carbon emissions. As the number of people on the planet increases, our overall CO_2 emissions will increase unless we find a way to make the whole food chain carbon neutral, which is an enormous task given all the different factors that combine to make up the food chain.

8.2 ENERGY NEEDED TO PUT FOOD IN YOUR HOME

One of the most comprehensive studies of the energy used for our food system and carbon emissions was a USDA study by Canning et al. (2017). For the period of 1993–2012, the average energy used for the US food supply according to this study was 13.2 ± 0.4 EJ (range from 11.9 to 13.5 EJ), while total energy used in the United States over that same period averaged 101.6 ± 3.9 EJ (from 92.3 to 107.0 EJ, not corrected for primary energy for renewables). Canning et al. (2017) extensively analyzed energy use for the food supply chain in 2007 and found that in that year the food related energy use was 12.5% of total of 107.0 EJ. Others have also estimated energy use for the food system. For example, in 1974, Hirst determined that the food system energy requirements were about 12% of energy consumption, an earlier study by Canning et al. estimated 15.7% for 2007,

Daily Energy Use and Carbon Emissions: Fundamentals and Applications for Students and Professionals, First Edition. Bruce E. Logan.

while others have concluded energy higher energy use for the food system of around 19% (Heller and Keoleian, 2000, Pimentel et al., 2008).

Energy Sources for the US Food System

Here we will assume that the amount of energy used for the fuel supply is 13.2 EJ in 2019 based on this amount of energy being reasonably constant for about two decades (1993–2012) based on the analysis by Canning (2017). If we simply divide this energy for the food system by that total of 101.2 EJ in 2019, that calculation would suggest that we used 13.0% of our energy for food production or 13.2 D. However, this calculation does not consider that nearly 58% of the energy used for the food system in 2007 was from electricity, with the remainder split between petroleum and natural gas (Fig. 8.1). This electricity was produced using different fossil fuels, renewables, and carbon-neutral sources, and therefore a more careful analysis of energy use needs to consider these different energy sources that went into the production of electricity along with the other fuels such as petroleum and natural gas that were directly used in the process.

To more fully assess the actual amount of actual energy going into the food system, the amount of primary energy used to make that electricity needs to be considered in this energy analysis. Based on the above distribution of fuel sources, we can calculate the D into each category as shown in Table 8.1. The primary energy used to produce electricity is calculated using the ratio of 2.46 (see Table 3.6). This is separated into fossil fuels using a ratio of 1.64, and for carbon-neutral sources using a ratio of 0.83. Using these ratios, a total of 18.1 D_{ff} was used to produce the 24.5 D of energy for the food system. Therefore, if only the energy in the electricity and fuels are included then the total energy was 13.2 D, but when the fuels needed to produce this electricity are included in the analysis the total energy used was 85% higher. The specific fuels that go into electricity production are shown in Figure 8.2 along with a summary of the distribution based on energy source. While coal is listed for the energy sources based on type of non-electrical energy sources used (Fig. 8.2), we can see that coal is a large part of the energy used for electricity production, and thus this fuel needs to be included in the total energy used for the food system.

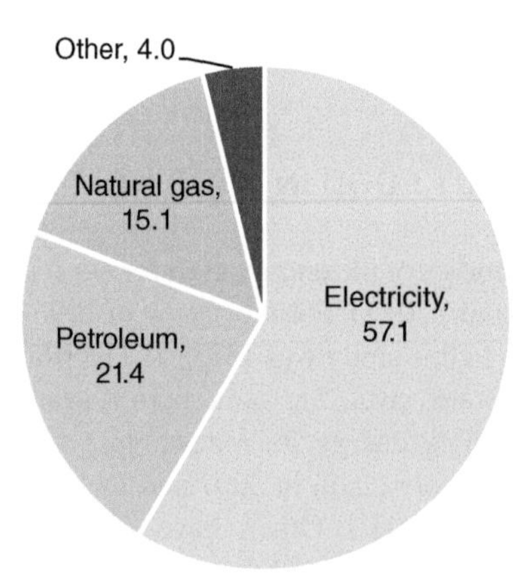

Figure 8.1 Distribution of energy sources by percent for the food system in 2007. *Source*: Data from Canning et al. (2017).

Table 8.1 Calculation of the primary energy sources, in terms of energy used (D), the primary energy used (D_p), the fossil fuel energy (D_{ff}) and the energy that is carbon neutral (D_{nc}). (Nuclear primary energy is used here.)

Source	EJ	D	D_p	D_{ff}	D_{nc}
Electricity	7.70	7.70	18.94	12.62	6.39
Petrol	2.85	2.85	2.85	2.85	—
Natural gas	2.11	2.11	2.11	2.11	—
Other	0.53	0.53	0.53	0.53	—
Total	13.2	13.2	24.5	18.1	6.4

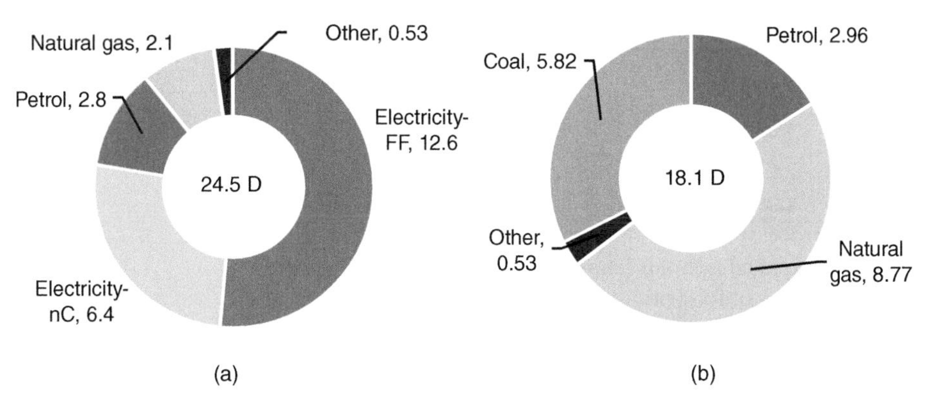

Figure 8.2 Energy sources for food production: (a) Energy sources, with the primary energy for electricity production shown for fossil fuels (FF) or renewable and carbon neutral (nC) sources; (b) only the fossil fuels used to produce the electricity. All numbers in units of D.

Energy for the Food Supply Separated into Different Categories

The energy used for the food system is broken down into eight different categories, with household food services consuming the largest share of the energy use of 31.7% (Fig. 8.3). The categories in are organized in the figure by the size of the consumption category. If organized by the succession of events, then these topics and their meanings, would be:

1. Farm production and agribusiness (agriculture)
2. Food processing and brand marketing (processing)
3. Food and ingredient packaging (packaging)
4. Freight services (transportation needed for wholesale/retail)
5. Wholesale and retail trade and marketing services (wholesale/retail)
6. Away-from-home food and marketing services (food service)
7. Household transportation (transportation, households)
8. Household food services (households)

Food processing is the second largest category for food systems with 17.5% of all energy consumption. Overall, processed food has been estimated to account for 82–92% of all food sales

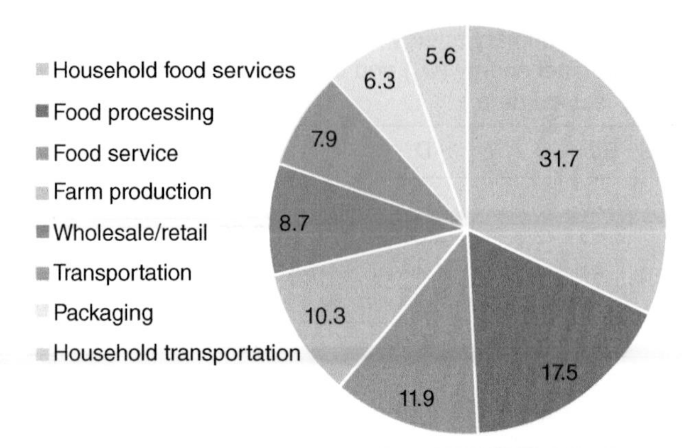

Figure 8.3 Distribution of energy consumption (percent) for eight different categories based on energy use in the United States in 2007. *Source*: Data from Canning et al. (2017).

(Pimentel et al., 2008). In addition, there has been a large change in energy use over time in food processing. For example, between 1997 and 2002 there was an annual average increase in energy used for food processing of 8.3% (Canning et al., 2010). This rise was due to food preparation that was occurring in households being outsourced to food processors (i.e. pre-prepared meals and additional processing of food prior to purchase). The increase in food processing energy is reflected in the time spent in making meals in US homes. For example, the average time spent preparing meals by adults of age 18–64 in the United States decreased from 65 min in 1965 to 31 min in 1995 (Canning et al., 2010).

An increase in food processing energy requirements of 8.3% per year by itself does not seem to be that important to our daily energy consumption. However, Canning et al. (2010) indicated that this amount of energy was equivalent to 2.7 million Btu per person per year. This energy for a single person translates to about 22 gal of gasoline per year, or 0.93 D. Thus, the additional energy expended over 5 years in *food processing alone was equal to adding nearly 1 D of energy consumption just for the food chain*. The other notable increase is in the food service industry, which contributed and additional 0.62 D. Thus, these increases contribute to the energy burden of our food system in ways unrelated to Calorie consumption.

Energy Use for the Food System in Other Countries

While it is not possible to review in detail energy use in other countries it is a safe assumption that the percentage of energy use is high in most countries. While the energy use between countries varies widely, industrialized nations use a high percentage of their energy for growing, processing, and transporting food. For example, in the European Union (EU) 17% of gross energy consumption, and 26% of the final energy, was estimated to be used for the complete food system (from source to table) (Monforti-Ferrario et al., 2015). These percentages are fairly similar to those for the United States, with about 13% of the energy consumption and 25% of gross energy consumption. In Europe, industrial processing consumed 28% of the food system energy. Food waste disposal was only 5% of the total, but that translates for the EU into more than 100 million tonnes of food waste. Renewables in the EU contributed to 7% of the energy used. Food imports (and exports) can also be a large form of energy imported into the country with the EU importing about 20% of its food. In the Middle East, in countries like Saudi Arabia, around 13–16% of annual food consumption is from imported sources. Therefore, even an accounting of the energy going into the food system in one country is difficult to accurately quantify to due to variable of energy used for food production and transport from other countries.

What Should We Eat to Minimize Energy Use for the Food System?

Changes in the food we consume can help to reduce emissions, but a large reduction in food energy does not necessarily occur if you change to a healthy diet. Making a dietary shift from the average distribution of food groups (Fig. 8.4) to healthier foods was concluded to produce only a 3% reduction in energy use (Canning et al., 2010). Adding a carbon tax was predicted to decrease CO_2 emissions even less, or by 1.7% (0.2–5.4% range). However, if there were large-scale changes to the whole energy food system, it might be possible to reduce energy use by up to 74%. Such changes might not be desirable from a dietary perspective, however, as these changes would entail greater consumption of fats and oils, and much more food in the "empty calories" category of sugar and sweets than would be considered wise for a healthy diet (Fig. 8.4). Pimentel et al. (2008) estimated that energy for the food system could be reduced by about 50% through changes in food processing, production, and many other aspects of the food system. In particular, the authors suggested reducing "junk food" by 80%, for example reducing the average annual consumption of 600 cans of soda by a person to 100 cans/y. Eating locally produced, in-season, and non-intensely processed foods can help to lower CO_2 emissions.

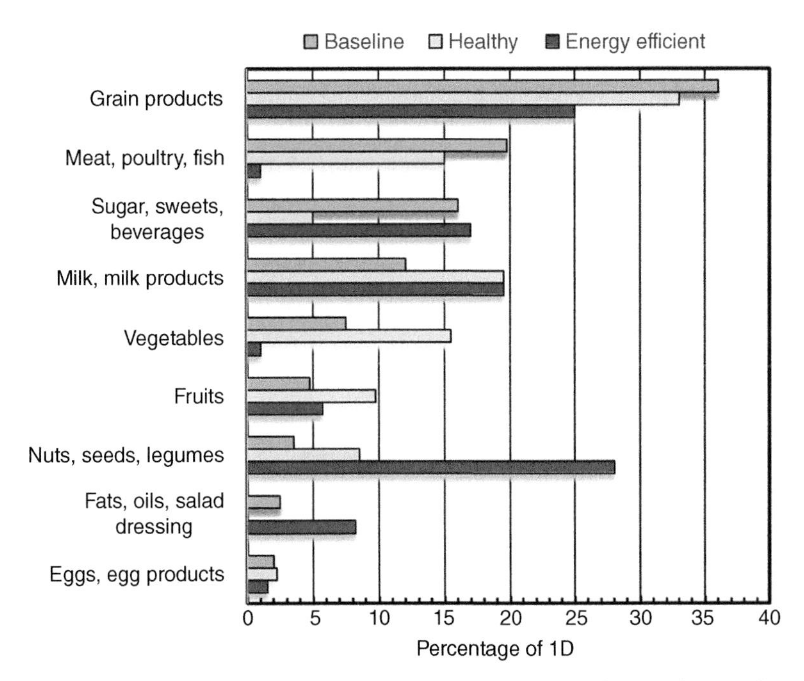

Figure 8.4 Comparison of energy as a percentage of 1 D (2000 Calories) in the food by category for a healthy diet and an energy-efficient diet, relative to a baseline diet. *Source*: Adapted from *Canning et al.* (2010).

8.3 CO_2 EMISSIONS AND OUR CARBON "FOOD PRINT"

The amount of fossil energy that is consumed for the food supply chain creates a very large carbon footprint, or we refer to here as a "carbon food print." The energy use for our food supply was calculated by Canning et al. (2017) to have accounted for 13.6% of all fossil fuel CO_2 emissions in the country in 2007. Based on their estimate of 817 million metric tons of CO_2, this is equivalent to 8.2 C for the population in 2007 (301.2 million) and 7.5 C per person for the food system carbon emissions. A study in 2018 found that food production released the equivalent of 13.7 Gt of CO_2 into the atmosphere, which was more than one quarter of the greenhouse gases released that were caused by human activities (Woolston, 2020).

Using the fossil fuel makeup in Figure 8.2, and the D:C values in Table 4.3 (and assuming D:C = 1.63 for "other"), for a US population in 2020 of 328.2 million it is estimated that CO_2 emissions would be 10.9 C, or 1200 million tons of CO_2 emissions per year (Fig. 8.5), which is 33% larger than that calculated by Canning et al. (2017). These numbers produce the following ratios for food system energy:

- D_p:D ratio (D_p = 24.5, D = 13.2) = 1.86
- D_p:C ratio (D_p = 24.5, C = 10.9) = 2.24
- D:C ratio (D = 13.2, C = 10.9) = 1.21

CO_2 Emissions from Burgers

Americans love to heat hamburgers and hot dogs, especially during the summer months and around the 4th July holidays. Overall, it is estimated that Americans consume 50 billion burgers a year, or about 156 burgers per person in 1 year (Andrews, 2019). A typical ¼ lb burger (just the meat) has 306 Calories (although many restaurants sell burgers that contain ½ lb of meat) which translates to 0.31 D without the bun, toppings, fries or drink. There have been vegetarian options for burgers for many years, but recently there have been several new variations on the burger such as the "Impossible burger," which is claimed to be more authentic in texture, appearance, and taste than previous vegetarian options. Some of these new burger options have only slightly less energy content than those made with meat. For example, the Impossible Burger has 0.12 D (240 Calories) and the Beyond Burger 0.125 D (250 Calories) (Peachman, 2019). Thus, the D values for these burgers

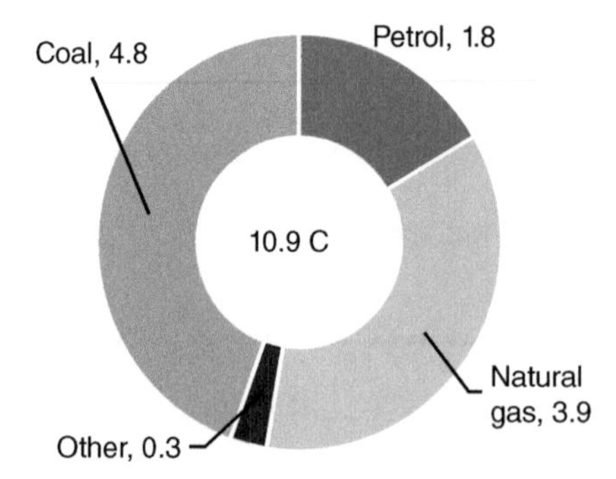

Figure 8.5 Carbon emissions estimated using the fossil fuel data in Figure 8.2, D:C values in Table 4.3, and assuming D:C = 1.63 for the "other" category.

Table 8.2 Carbon emissions and water use for different types of burgers.

Burger type (1/4 lb)	CO_2 emissions (lb)	C	Water (gal)
Industrial beef	12.4	6.2	153
Grass fed beef	0.8–9.1	0.4–4.6	59
Morningstar original burger	2.3	1.2	1.2
Impossible burger	0.8	0.4	3.2
Beyond burger	0.8	0.4	0.8

Source: Data from Peachman (2019).

are not all that different with 0.15 D for the beef burger compared to ~0.12 for these two vegetarian options.

There can be very large differences in carbon emissions from beef versus vegetarian burgers. Beef from cows has the highest C emissions due to the food processing system, with 12.4 lb of CO_2 estimated for just a single ¼ lb burger (Table 8.2). Carbon emissions from a burger could be greatly reduced by using these vegetarian burgers, with C values as low as C = 0.4 for the Impossible Burger and Beyond Burger (Peachman, 2019). Water use is also an issue with producing foods, and the Beyond Burger is clearly a winner in the category of water use, with only 0.8 gal of water used per burger. The clear messages here are: counting Calories or D will not make much difference in your diet for a ¼ lb burger, but in terms of energy and water use, the Beyond Burger is the way to go amongst these contestants!

8.4 WATER FOR FOOD THAT YOU EAT EVERY DAY

The average water use per person in the United States per was calculated to be 1300 w (1300 gal/d) as summarized in Table 5.1. Agriculture accounts for 440 w, or about 34% of the total. Water needed for irrigation can vary widely across the United States depending on the crop and where it is grown. Farmers tend to think about water use in units of "acre inches," or "acre ft," which are the amounts of water applied per acre either 1 in. or 1 ft deep, where 1-acre inch equals 27,000 gal. It is possible to estimate water use for irrigation based on the water applied and the typical yield of the crop. For example, consider the water use for growing corn and soybeans, the two most common crops in the United States. Corn typically requires 12–20 in. of water and yields around 190 bushels or 6.7 m^3 (1 bushel = 35.2 L) per acre (Agpro, 2011). If 1 acre-ft of water was used to produce corn, then that would mean that 1700 gal of water was used to produce a single bushel. Soybeans require less water, but the beans produced are smaller, so they require about 3700 gal per bushel (assuming 9 acre-inches of water and 65 bushels per acre). While such an analysis provides a useful view of water needed to grow crops it does not include processing and handling (for example, cleaning the produce) and other factors that can impact overall water use. Also, these numbers do not provide useful insight into our own personal daily water use based on the food we eat.

Water Needed to Produce the Things that We Eat

The Water Footprint Network is an organization that is dedicated to using the concept of a water footprint to "promote the transition toward sustainable, fair and efficient use of freshwater resources worldwide." They have calculated the water footprint of items or materials that we eat or use (such as clothing). This footprint is calculated as the total amount of water that was consumed for all

Table 8.3 Water used to produce the different items for three different meals, and a few alternative food or snack items.

Meal	Item	Water used (L)	Water used (w)
Breakfast	Cup of coffee (125 mL)	132	35
	Orange juice (200 mL)	200	53
	Two eggs	392	104
	Two slices of bread (0.126 kg)	203	54
Lunch	Pizza Margherita (725 g)	1259	333
	Apple	125	33
	Tea (250 mL)	27	7
Dinner	Beef (0.2 kg)	3080	814
	Potatoes 1 kg (180 g)	52	14
	French baguette (1/4)	39	10
	Glass of milk (250 mL)	255	67
Total meals		**6534**	**1523**
Others	Pasta (dry, 180 g)	333	88
	Beer (250 mL)	74	20
	Glass of wine, 125 mL	125	33
	Chocolate bar (100 g)	1700	449

Source: Data from Fair & smart use of the world's fresh water, https://waterfootprint.org/en/.

processing stages, in units of liters per kilogram, gallons per pound, or some useful volume or weight of the product as sold (such as water used to produce a gallon of milk). For example, consider these crop and animal products in terms of liters per kilogram (L/kg): vegetables, 197; milk, 1020; cereals, 1644; eggs, 3265; pig meat, 4325; and bovine meat, 15,415. This give us a sense of how these different things compare, but it could take a lot of work to figure out how much water was consumed to produce everything you eat in a meal.

Using the "product calendar" developed by the Water Footprint Network it is possible to evaluate water that was used to produce many of the food products we consume at our kitchen table. For example, a breakfast of coffee, orange juice, two eggs and two slices of bread would translate to 245 w to get it to your table (Table 8.3). The other two meals used as an example in this table indicate that lunch would require 373 w, and dinner 905 w, or a total of 1523 w. These are all large meals, and the total exceeds the water use estimated per person of 1300 w, with agriculture estimated at 440 w. However, there are many food items not listed in the table that would have lower water requirements than those listed in Table 8.3.

8.5 ENERGY FOR AMMONIA PRODUCTION (AND H_2) FOR FERTILIZERS

Fertilizers are essential for modern food production. Ammonia in fertilizers is made from hydrogen primarily produced using fossil fuels and nitrogen gas. Global estimates of hydrogen and ammonia production vary. For example, it was recently estimated by the International Energy Agency that 0.83 gigatons (Gt) of CO_2 were released into the atmosphere from global H_2 production (Birol, 2019), which would be equivalent to about 2.2% of global CO_2 emissions 36.6 Gt/y. This is higher than previous estimates of ~1% of all CO_2 emissions due to H_2 gas production. Another recent

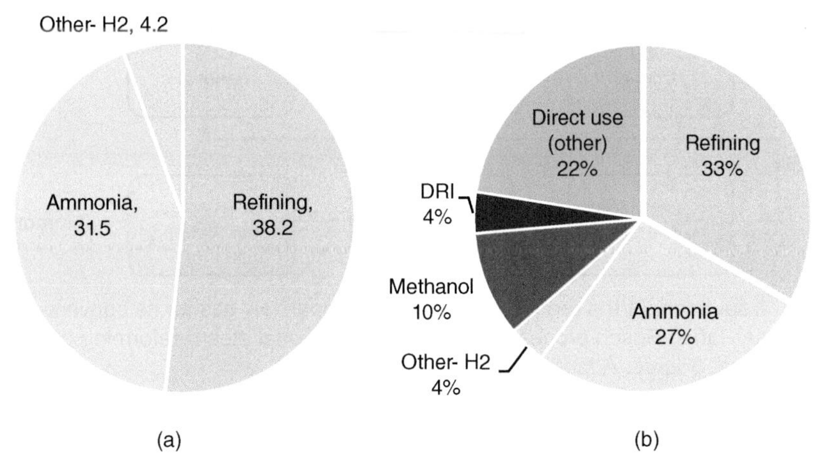

Figure 8.6 Hydrogen production (a) as a pure gas in for a total of 73.9 Mt/y, and (b) as a pure gas and for other purposes as a percentage of the total amount of 115.9 Mt/y. *Source*: Data from Birol (2019).

estimate of global H_2 production, in the form of a pure gas, was 73.9 million tonnes per year (Mt/y $= 10^9$ kg/y), with 42.6% of this (31.5 Mt/y) used for ammonia production (Fig. 8.6). This total for pure H_2 production did not include an additional 45 Mt/y of H_2 produced and then directly used as a part of the production of other products, for example steel and methanol. This total of 31.5 Mt/y of H_2 would be equivalent to 179 Mt/y of ammonia (assuming $3H_2$ are used to produce $2NH_3$, or a ratio of 17 g/6 g = 5.67), consistent with another report of 175 Mt/y (Wikipedia, 2020a). The United States alone produces a substantial amount of ammonia, with estimates ranging from 12.4 Mt/y (7.1% of 175 Mt/y) (Wikipedia, 2020a) to 14 Mt/y (Statistica, 2020).

Ammonia is made primarily using the Haber–Bosch process, consuming about 36.6 GJ/t (GJ/1000 kg) on average at a modern industrial plants that primarily use natural gas (Industrial Efficiency Technology Database, 2020). The most energy efficient ammonia production plants use 28 GJ/t based on steam reforming with natural gas, compared to 42 GJ/t using coal with the latter fuel mostly used for ammonia production in China. Based on this average energy use and 14 Mt of ammonia produced in the United States each year that means that in 2019 this would be equivalent to 0.51 D (0.51 EJ/y). This D calculation suggests that about half of the energy in the food we eat every day is used just to produce ammonia in the United States.

In the conventional Haber–Bosch process, much of the overall process consists mostly of cleaning up the fuel so that the syngas can be produced and used for ammonia production (Fig. 8.7). In the process, the feed (most commonly natural gas) is cleaned up to remove sulfur, and then passed into the steam reformer to produce H_2 gas, followed by CO_2 gas removal, additional CO_2 and CO gas removal through its conversion to methane (shift conversion and methanation), to finally produce the clean H_2 gas source for ammonia production (Wikipedia, 2020a).

Producing Ammonia from H_2 Gas Generated Using Renewable Energy or Carbon-Neutral Approaches

H_2 can be produced from N_2 gas according to the simple stoichiometric formula

$$3H_2 + N_2 \rightarrow 2NH_3$$

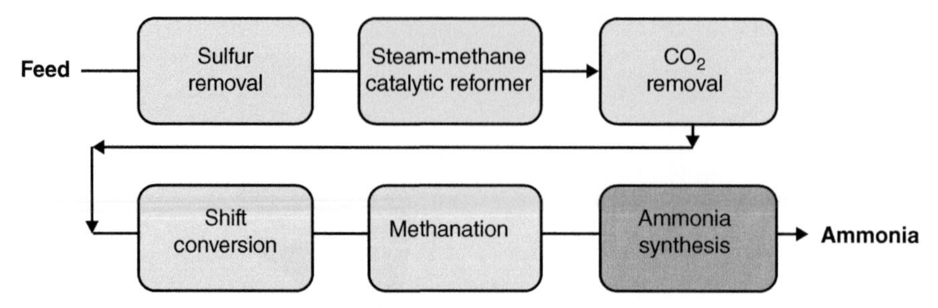

Figure 8.7 A summary of the processes needed to prepare H_2 gas for its conversion to ammonia in the Haber–Bosch process, based on conventional steam-reforming of natural gas into ammonia. *Source*: Adapted from Wikipedia (2020a).

The key to the conversion of H_2 to ammonia in the Haber–Bosch process is the use of high pressures of 60–180 bar, depending on the commercial process, and certain catalysts (based on magnetite). From the standpoint of process complexity, it seems far simpler to avoid all the steps needed for sulfur and carbon-based gas removals in the process using fossil fuels (Fig. 8.7) and to use H_2 directly for ammonia production. However, the high temperatures and pressures used in the Haber–Bosch process still make it a very energy-demanding process. At present, there are no commercially viable alternatives to the Haber–Bosch process for direct ammonia production. Although electrochemical methods are being examined to directly produce ammonia from N_2 gas the yields are barely in the detection range, and thus have a long way to go to reach the high production rates of the chemical processes.

There are several different ways to produce H_2 gas that would not release CO_2 as discussion that include: water splitting using renewable or nuclear power, fossil fuels combined with carbon capture (producing solid carbon or CO_2 injection into the ground), renewable methane (biogas), and biological-based processes (fermentation and microbial electrolysis cells). Among these many approaches green H_2 gas production from water splitting (water electrolysis) is already commercially available, although it is not as economical as H_2 production using natural gas in steam reforming without carbon capture (Chen et al., 2018). Water electrolyzers can achieve energy efficiencies of around 70–80% depending on the specific technology, with a theoretical minimum (100% efficient) electrolyzer using 39.4 kWh/kg H_2 (142 MJ/kg). If the H_2 gas is further compressed for storage and transport, that could consume an additional 50 kWh (54 MJ). This energy would only be for the H_2 production, and further energy would be needed for ammonia production via the Haber–Bosch process. The production of renewable H_2 is addressed in more detail in Chapter 6 (from biological-based processes) and Chapter 7 (using renewable electricity for water splitting).

The low cost of solar and wind electricity is stimulating commercial investment in ammonia production from hydrogen gas produced using water electrolyzers. The world's largest producer of ammonia, Yara, began development in 2018 of a 2.5 MW solar array to produce H_2 gas that could be used for ammonia production, which was claimed to reduce CO_2 emissions by half (Service, 2018). Additional efforts are being made in the Netherlands, Spain, and Germany for water electrolyzer-based hydrogen or ammonia production as well. Recently it was announced that Saudi Arabia would invest $5 billion in a plant to produce renewable H_2 gas (Scott, 2020). This H_2 would then be converted into NH_3 for transport to a destination, with the intended use being its conversion back into H_2. However, given the large amount of global ammonia use it is not clear that it would be more profitable to convert back to H_2 rather than just using the ammonia.

Recovery of Ammonia from Wastewater Streams

Instead of producing ammonia from hydrogen gas, another approach is to recover ammonia from domestic and animal wastewaters. However, ammonia in wastewaters is rarely recovered. In the United States, about 80% of all ammonia emissions into the atmosphere are associated with livestock production (Wikipedia, 2020b). Much of the wastewater from pig farms is sent to lagoons where 80% of that ammonia can be volatilized rather than treated to avoid its release or recovered for re-use for example as fertilizer.

Much of the nitrogen in our food ends up in domestic used water, but most treatment plants do not recover ammonia as a potential resource. At most used water treatment plants ammonia is first converted to nitrate, and then the nitrate is removed through its conversion to N_2 gas. This conversion of ammonia to nitrate takes energy, although it is well integrated into many treatment systems and thus energy consumption for this is relatively low considering energy consumed in other processes at these plants. However, even a small concentration of ammonia has a high energy value compared to the energy used for wastewater treatment. A typical used treatment plant consumes ~0.6 kWh/m^3 of wastewater treated (McCarty et al., 2011). An ammonia concentration in this wastewater of 30 mg/L would have an energy value of 0.43 kWh/m^3 based on the amount of energy it would take to make that ammonia using a Haber–Bosch process.

There are currently no well-accepted methods for recovery of ammonia from used waters. One effective method for high concentrations of ammonia is struvite production, which accomplishes both nitrogen and phosphorus removal (Perera et al., 2019). Struvite can be used directly as a fertilizer, although the economical production of struvite depends on farmers willing to pay for fertilizer in this form rather than industrially produced fertilizers. Typically, Mg^{2+} salts are added into the water at concentrations needed to produce the chemical struvite, but this addition of the salt increases the cost of the product. Methods are also being examined to concentrate ammonia in dilute wastewaters, using desalination or adsorption technologies so that its recovery is more economical than destruction of the ammonia through conversion to N_2 gas (Perera et al., 2019, Kim et al., 2018). Other alternatives are using ammonia in water to produce proteins which can be fed to animals, thus recycling this ammonia within the food chain (De Vrieze et al., 2020). However, such approaches to date have resulted in very little nitrogen recycling from human or animal wastewaters.

8.6 USING THE ENERGY UNIT D FOR OUR DIET

If food were rated in units of D rather than Calories, we would never be confused about what percentage of our needed daily energy were in a food item. For example, if a side dish is 375 Calories, and a sandwich is 720 Calories, what percentage of 2000 Calories is that? You can do the math, likely with a calculator, but wouldn't it just be easier to see that the side dish is 19% D, and the sandwich 36% D? Combined, these would meet 55% of your daily energy needs (likely without the need for a calculator).

Items prepared by restaurants often contain far more energy than we really need for a single meal. Here are some examples of processed foods at restaurants for a day of eating out for three meals:

- **Breakfast**: Denny's Ham and cheese omelet = 39% D
- **Lunch**: McDonald's Big Mac, fries, medium drink = 55% D

- **Dinner**: Outback Charbroiled Ribeye (12 oz), baked potato and house salad = 76% D; and with two drinks (medium sodas or beers) = 15% D; the dinner total = 91% D.

Based on these three meals, the total for the day is ***equal to 185% of your daily required calories or 185% D (1.85 D)***. Also, substituting a 720 Calorie sandwich for the burger lunch probably would not have helped all that much in the total, although avoiding the fries and medium drink could help a lot! The dinner alone was almost all the energy you needed for the day at 91% of the daily energy needed (based on 2000 Cal/d). Of course, your specific energy needs can vary quite a lot, as explained in Chapter 2, on factors such as height, weight, and activity levels. You might set your own personal D at 0.9 if you are shorter and have a lower weight than average, or 1.4 D if you are very active.

Our bodies have developed methods to store that excess energy. While some can go to protein (muscles) much of it is likely to end up as fat. A pound of body fat contains 1.75 D (3500 Calories). Think of it this way: to lose one pound in a week, you would need to reduce your food intake by 25% (0.25 D or 500 Calories a day). Clearly reducing our food by that amount per day would be a challenge if you were eating out. Also, it is not that simple to just estimate that you would lose 500 Calories a day over many days of dieting. Your body will reduce its baseline energy consumption as you diet, so it can take longer than a week to lose that pound of fat unless you exercise to maintain a good metabolism level (for example by walking or running a few miles, or other exercise). Nutrition experts caution us to be careful to maintain a healthy and balanced diet when we want reduce calories. The bottom line is that we should eat less and waste less food, and make smarter and more healthy choices in what we eat every day. Thinking about your food in D units can help!

8.7 FOOD WASTE AND OTHER FOOD-RELATED CO_2 EMISSIONS

Energy use and CO_2 emissions vary widely across the United States, and as expected the most populous cities have the highest CO_2 emissions. A more interesting analysis is CO_2 emissions is based on per capita use for food systems, where we can see some relatively intensive emissions localized in the Midwest and northern parts of Texas in the United States (Fig. 8.8). Of the 10 counties with the highest per-person CO_2 emissions 5 counties are in Kansas, 3 are in Texas, and the remaining 6 counties have populations totals in the bottom 10% in terms of population. All 10 are in areas in the bottom 20% based on population. These counties are high in CO_2 emissions related to food processing due to being fossil fuel intensive food-processing or farming areas. The CO_2 emissions for these counties based on the daily unit C are 0.62 C (5000 lb/cap-d) up to 99 C or more (72,000 lb/cap-d of CO_2 emissions). From the analysis above, the average is 8.1 C (5900 lb/y) based on total energy consumption related to food systems.

Project Drawdown

Solutions to reducing CO_2 emissions related to food, agriculture, and land use were summarized into 14 areas by Project Drawdown in their 2020 report (Wilkinson, 2020). These areas focused on aspects different from those of Canning (2017) as some of these include agricultural activities that are not part of the food system but these processes do impact CO_2 emissions through either carbon capture or release. For example, wetland protection does not have a direct link to a food system but it is listed in the land use activities by Project Drawdown that can impact CO_2 emissions. These releases are in gigatons (Gt) of CO_2 equivalents, and so impacts are translated into tons of CO_2

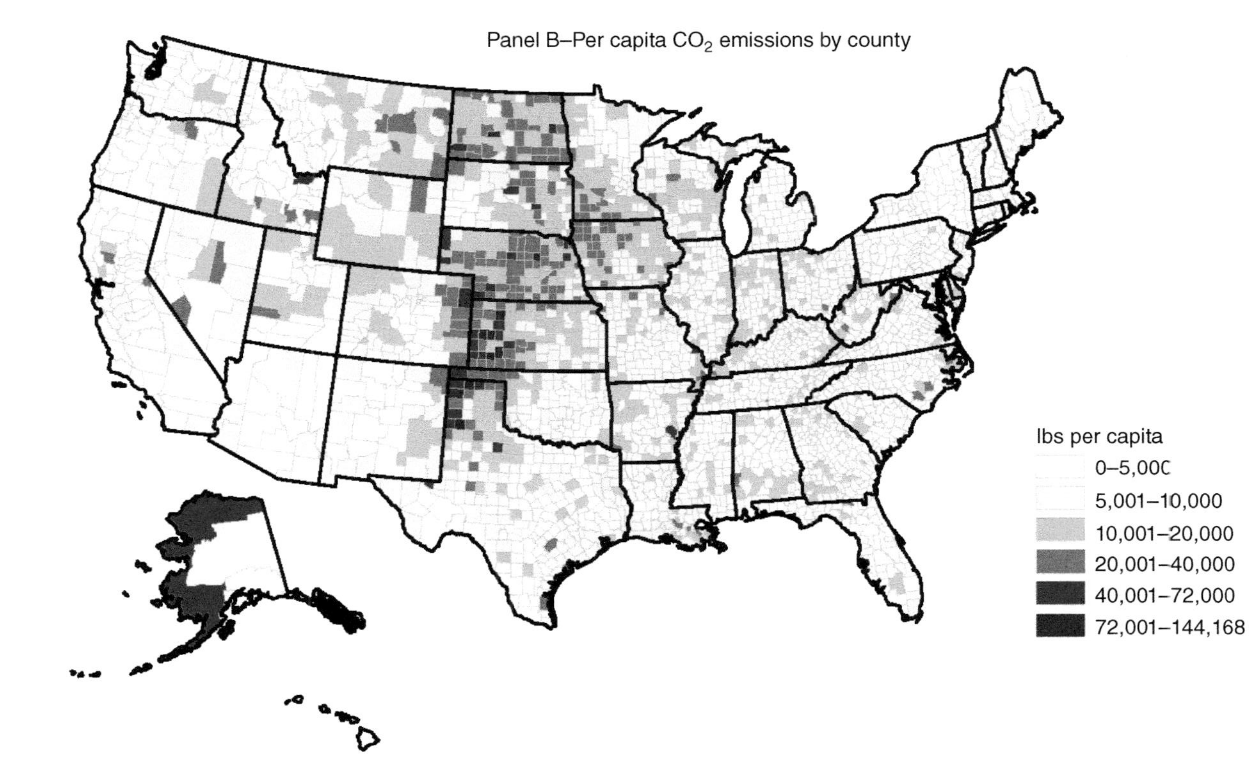

Figure 8.8 Annual CO_2 emissions due to the food supply energy consumption in lb per person based on counties in the United States, *Source*: Canning et al. (2017).

Table 8.4 Possible reduction in CO_2 emissions over a 30-year period (2020–2050) (Wilkinson, 2020) normalized to 9.4 billion people on a daily basis in units of carbon emissions C, where C is shown with a negative sign as these are carbon reductions.

No.	Solution	CO_2 (10^9 tons)	−C
1	Reduced food waste	87.4	0.93
3	Plant-Rich Diets	65.0	0.69
11	Silvopasture	26.6	0.28
17	Managed grazing	16.4	0.18
19	Perennial staple crops	15.5	0.17
20	Tree intercropping	15.0	0.16
21	Regenerative annual cropping	14.5	0.15
22	Conservation agriculture	13.4	0.14
23	Abandoned farmland restoration	12.5	0.13
25	Multi-strata agroforestry	11.3	0.12
29	Improved rice production	9.4	0.10
50	System of rice intensification	2.8	0.03
54	Nutrient management	2.3	0.02
65	Farm irrigation efficiency	1.1	0.01
	Total	**293.2**	**3.13**

Source: Data from Wilkinson (2020)

releases although they can occur due to emissions of other gases such as methane. The list of items by Project Drawdown most relevant to food systems and land use are summarized in Table 8.4, with a conversion to C values based on a global population estimate of 9.4 billion people in 2050 (Lewis and Nocera, 2006), and the total amount of reduced CO_2 emissions over a thirty-year period of 2020 to 2050 based on 2020 estimates by Project Drawdown (Wilkinson, 2020).

The results in Table 8.4 show the top way to reduce CO_2 emissions is to reduce food waste, which is also the #1 most impactful method of those assessed, followed by a plant-rich diet, and peatland protection. The first two of these items translate to a savings of 1.62 C in terms of impact, which is 51.8% of the total of 3.13 C for all these changes. What these solutions do not include are operations outside the food-agriculture system, for example the impact of all harvesting machines operating on solar- or wind-produced electricity, and the food systems outside the systems, for example in retail and processing aspects of the food chain.

Another notable aspect of items listed in Table 8.4 is that the top two items are things that we can help contribute to in our daily lives. Reducing food waste and a plant-rich diet are easy to imagine being incorporated into our eating habits at home or at restaurants. The other notable aspect of this table is that there are things we can do in our homes, and our daily activities, that can impact CO_2 emissions more profoundly than some of these large-scale activities. For example, farm irrigation efficiency is listed as 1.1 Gt of CO_2 over the 30-year period, equivalent to 0.012 C. The electricity use by an average home is 3.5 C (Fig. 4.2), so 0.12 C is just 3.43% of our daily energy use for our home. This demonstrates the importance of reducing our personal carbon footprint relative to our lives.

Example 8.1

In Table 8.4, it was estimated that farm irrigation efficiency could reduce CO_2 emissions by 1.2 gigatons (Gt) in 30 years. (a) Show the conversion for that number into units of C using an estimated 9.4 billion people in 2050. Compare this amount of C to (b) a single lightbulb turned on for 8 h a day, and (c) watching TV for 6 h a day. Assume the lightbulb uses 9 W (equivalent to a 60 W incandescent bulb) and that the TV is an LCD model that uses 0.39 D, and electricity is supplied by a coal fired power plant with 35% delivered efficiency.

(a) To convert this amount over a 30-year period, to a daily C unit, we have

$$\frac{1.2\,\text{Gt}}{9.4\times 10^{9}\,\text{cap}}\;\frac{1}{30\,\text{y}}\;\frac{10^{9}\,\text{ton}}{\text{Gt}}\;\frac{10^{3}\,\text{kg}}{\text{ton}}\;\frac{2.2\,\text{lb}}{\text{kg}}\;\frac{1\text{C}}{2\,\frac{\text{lb}}{\text{d cap}}}\;\frac{1\,\text{y}}{365\,\text{d}} = 0.012\,\text{C}$$

(b) For the lightbulb, we first calculate kWh, convert that to D, and use a D:C ratio of 1.2 for subbituminous coal (Table 4.2).

$$9\,\frac{\text{W}}{\text{cap}}\;\frac{8\,\text{h}}{\text{d}}\;\frac{1\,\text{kWh}}{1000\,\text{Wh}}\;\frac{1}{0.35}\;\frac{1\,\text{D}}{2.32\,\frac{\text{kWh}}{\text{d cap}}}\;\frac{1\,\text{C}}{1.2\,\text{D}} = 0.074\,\text{C}$$

(c) For the TV, we make a similar calculation, but we start with 0.39 D, as

$$0.39\,\text{D}\;\frac{1}{0.35}\;\frac{1\,\text{C}}{1.2\,\text{D}} = 0.93\,\text{C}$$

From these results we can see that there a greater impact on C emissions that could be achieved by just turning off the lights or the TV than those possible from changes to farm irrigation (normalized to one person). A single lightbulb operated for just 1.3 h would result in the same release of CO_2 as the proposed farm irrigation efficiency. The real "win" would come from saving energy in all these cases and using electricity that is renewable.

With a total impact of 2.18 C estimated from these changes, a lot more would need to be done to reduce the total CO_2 emissions, but many of these other actions that are required are listed in different sections of the Project Drawdown analysis (and are mentioned in other chapters in this book). One additional set of recommendations that is not included in other chapters relates to changes in land use that do not directly relate to food systems (Table 8.5). While these are not dedicated to food production, they seem to best fit in here as our land use is sometimes a trade-off for growing crops and using land for other purposes. The greatest benefits for reducing CO_2 emissions based on items listed in Table 8.5 are tropical forest restoration (0.58 C) and peatland protection (0.28 C) as these provide excellent carbon sinks due to CO_2 uptake into plant and carbon materials. Peat is made up of decomposing matter and it is made slowly over hundreds to thousands of years. Peatlands hold an estimated 500–600 Gt of carbon, so they hold about twice the carbon of the world's forests (Wilkinson, 2020). The topic of carbon storage in soils is further discussed in Chapter 14.

Table 8.5 Possible reduction in CO_2 emissions due to improvements in land use, calculated over a 30-year period (2020–2050) (Wilkinson, 2020) normalized to 9.4 billion people on a daily basis in units of carbon reductions C (negative sign).

No.	Solution	CO_2 (10^9 tons)	–C
5	Tropical forest restoration	54.5	0.58
12	Peatland protection and rewetting	26.0	0.28
13	Tree planation (on degraded land)	22.2	0.24
14	Temperate forest restoration	19.4	0.21
31	Bamboo production	8.3	0.09
41	Forest protection	5.5	0.06
49	Grassland protection	3.3	0.04
55	Biochar production	2.2	0.02
57	Composting	2.1	0.02
62	Sustainable intensification for smallholders	1.4	0.01
66	Recycled paper	1.1	0.012
68	Coastal wetland protection	1.0	0.011
71	Coastal wetland restoration	0.8	0.009
	Total	**147.8**	**1.58**

Source: Data from Wilkinson (2020)

Global Food System Emissions

Most of the above discussion has focused on CO_2 emissions for food systems, and except for the Project Drawdown calculations, has been centered on the United States. A recent study of global food systems by Clark et al. (2020) present a more comprehensive view of the food system relative to climate change. Their analysis suggests that the global food system accounts for 30% of all GHG emissions. They calculate that the major sources of GHG emissions are due to: land clearing and deforestation, which release nitrous oxide (N_2O) as well as CO_2; use of agrichemicals in addition to ammonia which release CO_2, N_2O and CH_4, fermentation-based processes in animals (cows, sheep and goats) which can produce CH_4; and manure from these animals which emit net N_2O and CH_4. The main conclusion is that even if all other GHG emissions were immediately stopped that the food system alone would result in emissions that would exceed the <1.5°C target of the Paris Climate Change agreement by 2050. Clark et al. (2020) suggest changes that could reduce GHG emissions by an estimated 14–48% by 2100, most of which are included in the different activities of Project Drawdown. The continued large impact of GHG emissions to climate change highlight the importance of including methods for carbon sequestration in plans for minimizing climate change impacts. The subject of carbon sequestration is further addressed in Chapter 14.

References

AGPRO. 2011. *Corn and soybeans water use* [Online]. Available: https://www.agprofessional.com/article/corn-and-soybeans-water-use#:~:text=There%20are%20roughly%2027%2C000%20gallons%20of%20water%20in,be%20stressed%20somewhat%20more%20and%20require%20less%20water [Accessed October 4 2020].

ANDREWS, C. 2019. *July fourth food! How many hot dogs and hamburgers are consumed in your state?* [Online]. Available: https://www.usatoday.com/story/money/2019/06/17/july-4th-hot-dog-and-hamburger-consumption-by-state/39580323/ [Accessed May 31 2020].

BIROL, F. 2019. The future of hydrogen: Seizing today's opportunities. https://www.iea.org/reports/the-future-of-hydrogen.

CANNING, P., CHARLES, A., HUANG, S., POLENSKE, K. R. & WATERS, A. 2010. Energy use in the U.S. food system. Economic Research Report Number 94. https://www.ers.usda.gov/webdocs/publications/46375/8144_err94_1_.pdf?v=0.

CANNING, P., REHKAMP, S., WATERS, A. & ETEMADNIA, H. 2017. The role of fossil fuels in the U.S. food system and the American diet. USDA, US Department of Agriculture. Economic Research Report Number 224. www.ers.usda.gov/publications.

CHEN, J. G., CROOKS, R. M., SEEFELDT, L. C., BREN, K. L., BULLOCK, R. M., DARENSBOURG, M. Y., HOLLAND, P. L., HOFFMAN, B., JANIK, M. J., JONES, A. K., KANATZIDIS, M. G., KING, P., LANCASTER, K. M., LYMAR, S. V., PFROMM, P., SCHNEIDER, W. F. & SCHROCK, R. R. 2018. Beyond fossil fuel–driven nitrogen transformations. *Science*, 360, eaar6611.

CLARK, M. A., DOMINGO, N. G. G., COLGAN, K., THAKRAR, S. K., TILMAN, D., LYNCH, J., AZEVEDO, I. L. & HILL, J. D. 2020. Global food system emissions could preclude achieving the 1.5° and 2°C climate change targets. *Science,* 370, 705–708.

DE VRIEZE, J., VERBEECK, K., PIKAAR, I., BOERE, J., VAN WIJK, A., RABAEY, K. & VERSTRAETE, W. 2020. The hydrogen gas bio-based economy and the production of renewable building block chemicals, food and energy. *New Biotechnology,* 55, 12–18.

HELLER, M. C. & KEOLEIAN, G. A. 2000. Life cycle-based sustainability indicators for assessment of the U.S. food system. University of Michigan.

INDUSTRIAL EFFICIENCY TECHNOLOGY DATABASE. 2020. *Ammonia: Benchmarks* [Online]. Available: http://www.iipinetwork.org/wp-content/Ietd/content/ammonia.html#benchmarks [Accessed May 28 2020].

KIM, T., GORSKI, C. A. & LOGAN, B. E. 2018. Ammonium removal from domestic wastewater using selective battery electrodes. *Environmental Science & Technology Letters,* 5, 578–583.

LEWIS, N. S. & NOCERA, D. G. 2006. Powering the planet: Chemical challenges in solar energy utilization. *Proceedings of the National Academy of Science*, 103, 15729–15735.

MCCARTY, P. L., BAE, J. & KIM, J. 2011. Domestic wastewater treatment as a net energy producer – Can this be achieved? *Environmental Science & Technology,* 45, 7100–7106.

MONFORTI-FERRARIO, F., DALLLEMAND, J.-F., PINEDO PASCUA, I., MOTOLA, V., BANJA, M., SCARLAT, N., MEDARAC, H., CASTELLAZZI, L., LABANCA, N., BERTOLDI, P., PENNINGTON, D., GORALCZYK, M., SCHAU, E., SAOUTER, E., SALA, S., NOTARINCOLA, B., TASSIELLI, G. & RENZULLI, P. A. 2015. Energy use in the EU food sector: State of play and opportunities for improvement.

PEACHMAN, R. R. 2019. Meat gets a makeover. *Consumer Reports*, 42–47.

PERERA, M. K., ENGLEHARDT, J. D. & DVORAK, A. C. 2019. Technologies for recovering nutrients from wastewater: A critical review. *Environmental Engineering Science,* 36, 511–529.

PIMENTEL, D., WILLIAMSON, S., ALEXANDER, C., GONZALEZ-PAGAN, O., KONTAK, C. & MULKEY, S. 2008. Reducing energy inputs in the US food system. *Human Ecology,* 36, 459–471.

SCOTT, A. 2020. *Tensions arise as clean hydrogen projects spread, C&EN*, American Chemical Society.

SERVICE, R. F. 2018. Liquid sunshine. *Science,* 361, 120–123.

STATISTICA. 2020. *Ammonia production in the United States from 2014 to 2019* [Online]. Available: https://www.statista.com/statistics/982841/us-ammonia-production/#:~:text=U.S.%20ammonia%20production%202014%2D2019™text=The%20United%20States%20is%20one,facilities%20in%20the%20United%20States [Accessed October 2 2020].

WIKIPEDIA. 2020a. *Ammonia production* [Online]. Available: https://en.wikipedia.org/wiki/Ammonia_production [Accessed May 30 2020].

WIKIPEDIA. 2020b. *Anaerobic lagoon* [Online]. Available: https://en.wikipedia.org/wiki/Anaerobic_lagoon [Accessed May 31 2020].

WILKINSON, K. 2020. *The drawdown review: Climate solutions for a new decade*, Project Drawdown.

WOOLSTON, C. 2020. Healthy people, healthy planet. *Nature,* 588, 854–856.

CHAPTER 9

HEATING AND BUILDINGS

9.1 HEATING AND INSULATION

The energy used to heat or cool a home depends in on its size, the amount, and effectiveness of insulation, exposure to sunlight, the number of windows, and the specific type and methods for operating the heating system (for example programmable thermostats). Even under nearly identical infrastructure conditions, the overall energy use also depends on how many people live there, temperature preferences, number of hours people work away from home, and sleep habits, as these can impact the length and duration of heating or cooling cycles. In Chapter 2, we discussed energy use for appliances and the overall energy needs for a home. In this chapter, we focus on heating a home that is a house that you own.

To minimize the cost of building a new home, less expensive and inefficient heating systems are often installed to maximize profits for the builder. For many people, the house you live in was likely already built and so you are using a heating system that was originally installed when it was built or modified by a previous owner. Heating systems will eventually need to be replaced, so at some point, if you stay in that home you will need to decide on a replacement for parts or all of the heating system for your home. Even before you need to replace heating system components, you may want to save on your heating bills and lower your CO_2 emissions, so there are some changes that you can easily make in the existing infrastructure without overhauling your existing one.

An energy audit can reveal important shortcomings that can directly impact your bill and your carbon emissions, and it is worth getting one if you will be in that home for more than a few years. The energy auditor will suggest several immediate changes that could reduce energy use for heating. For example, sealing cracks around windows or doors, getting a new (smart) thermostat, or adding additional insulation can really save you money and benefit the environment by reducing energy consumption and carbon emissions. Preventing air leaks and providing additional insulation will also improve the overall atmosphere of your home by minimizing cold spots or chilly areas. Due to the low cost of heating oil or electricity for many years, homes were often built with little insulation as it was cheaper to heat using inexpensive fuels rather than make a large payment for insulation. However, with climate change, we now know it is imperative to reduce our use of fossil fuels, and thus, good insulation is essential to reduce your carbon emissions. Energy audits can help you decide if your house can be improved through investment in additional insulation or if sealing cracks and fixing doors or windows will make a large difference with little cost. Some states will perform energy audits for free, while others can give rebates or tax breaks on charges made by hiring companies for energy audits.

Daily Energy Use and Carbon Emissions: Fundamentals and Applications for Students and Professionals,
First Edition. Bruce E. Logan.

The newest generation of "smart" thermostats are wise investments to reduce energy consumption and heating bills. Thermostats can be programmed to lower the temperature at night and raise it in the morning before you get up so that it is warm while you get ready for work and lowered if the house is unoccupied during the day. Many of these smart thermostats can be set differently for workdays and weekends. Thus, you can delay turning on the heat on weekends if you get up later in the morning on those days and if you are home on the weekends but not weekdays. Some thermostats can "sense" when you are home, so if you are in a small apartment or the thermostat is located in an area of the house that you will frequently go by, you can have the thermostat heat when it knows you are around and go to a default lower temperature when you are not at home. Many smart thermostats can be connected to the Internet using your home Wi-Fi, and therefore, you can turn down the heat if you forgot to do that even after going on vacation, or warm up the house before you get home after a long absence. These modern smart thermostats linked to the Internet can also switch between heating or cooling options depending on the outside temperatures. If you have dual heating options, the smart thermostat can be programmed to switch from using a heat pump on cooler temperatures outside to heating using a gas furnace for the much colder days, which can overall save you money. For many reasons, it is worth investing in one of these new thermometers. If you want to discard an older thermometer, make sure that you safely dispose it if it has mercury in the sensor (as seen by the silvery liquid in a tube).

The focus of the remainder of this chapter is on aspects of the home that directly impact heating energy use and costs and the carbon emissions from different types of heating systems. While improving insulation or timing your heating (or cooling) schedule to reduce energy consumption, you will have an infrastructure in your home that will be based on one or more heating options and therefore you need to understand which of these options can help you save money and reduce carbon emissions. First, consider whether electricity or gas is a good option, and second, whether geothermal energy might be helpful in your area. Third, consider the type of hot water heater that you use as water heaters are usually the third largest energy use in your home (following building heating and cooling).

9.2 COMPARING HEATING SYSTEMS BASED ON CARBON EMISSIONS

The heating system that you use in your home can have a large impact on your overall CO_2 emissions, probably second only to your choices about your car and other methods of transportation. Many older homes used oil for heating, while later on homes transitioned to electrical heating for many reasons including ease of adding heating into different rooms or locations in the house, the rapid response time of electrical heating, and not having to have a big, bulky oil tank in or near your home. Today, the choices for homes are mostly whether you will use gas or electric for heating, and to what extent you will pay for more energy efficient units compared to (generally) cheaper but less energy efficient units.

What Is the Best Way to Heat My Home?

To answer this question, we will consider heating an average-sized US house using the following technologies: natural gas, oil, electricity by direct (baseboard or floor) heating, and electricity using a heat pump. However, to make this comparison fair based on the fuels used, we must also consider the amount of energy, and more importantly fossil fuel energy, that goes into the energy used for heating. Does "best" to you mean the least expensive, most reliable, or lowest carbon emissions?

Here we focus on carbon emissions, but we will initially consider the costs of the energy needed to heat a house using these different technologies. An average house size in the United States is about 1700 ft^2 (158 m^2), but the sizes of homes and heating energy vary quite a lot and they are larger than average sizes in many other countries and regions. Here, we assume as a baseline electric heating (using baseboard heating) of 1200 kWh/mo or about 39 kWh/d. This is larger than the average US electric use per home over a month (~30 kWh) but this will help to focus on colder climates where heating is needed for a large part of the year. This base condition will then be compared to heating with oil, natural gas, and electricity using a heat pump. Next, we will evaluate carbon emissions for these different heating technologies to identify the heating methods that can minimize CO_2 emissions. The reliability or a life cycle analysis of heating systems will not be considered here. The calculation method used to evaluate energy demand and carbon emissions is shown in the example below.

Example 9.1

Calculate (a) the amount of primary fuel used, in units of total energy D and the fossil fuel energy used in D_{ff} for electricity use of 1200 kWh/mo for a home with one person. (b) What would be the carbon emissions in units of C based on the average for the US grid?

(a) Table 2.5 showed that the energy in the primary fuels for the US energy grid was 35 EJ to produce a total of 14.2 EJ of electricity, or a ratio of D:D_e = 2.46 based on all fuels (same energy units in number and denominator; this ratio is equal to an energy efficiency of 40.6%) or D_{ff}:D_e = 1.64 based only on the fossil fuels used as a part of these electricity production. The amount of energy used in units of D (using an average of 30.4 d/mo) in terms of the electricity energy (D_e), fossil fuel energy (D_{ff}), and total primary energy (D) is:

$$\text{Electricity: } \frac{1200\,\text{kWh}}{\text{mo cap}}\ \frac{1\,\text{mo}}{30.4\,\text{d}}\ \frac{1\,\text{D}}{\frac{2.32\,\text{kWh}}{\text{d cap}}} = 17.0\,\text{D}_\text{e}$$

$$\text{Fossil fuel: } 17.0\,\text{D}_\text{e}\ \frac{1.64\,\text{kWh}}{\text{kWh}} = 27.9\,\text{D}_\text{ff}$$

$$\text{Primary: } 17.0\,\text{D}_\text{e}\ \frac{2.46\,\text{kWh}}{\text{kWh}} = 41.8\,\text{D}$$

(b) In Example 4.4, the ratio of primary energy to carbon emissions was calculated as D:C = 2.40, and D_{ff}:C = 1.59 based only fossil fuels. Using the ratio based on total primary energy, the carbon emissions would be:

$$41.8\,\text{D}\ \frac{1\,\text{C}}{2.40\,\text{D}} = 17.4\,\text{C}$$

From this calculation, we see that when comparing energy use within the context of total carbon emissions, it is important to calculate the emissions from the fuels that were used to produce that electrical energy.

Table 9.1 The energy used and costs for heating a typical house and carbon emissions based on primary energy (D) used to produce the heat or electricity.

Type	Efficiency	kWh/mo	$/y	D	D_{ff}	C
Natural gas (older)	80	1500	564	21.3	21.3	9.2
Natural gas (newer)	95	1263	475	17.9	17.9	7.7
Oil	70	1714	1633	24.3	24.3	14.9
Electric	100	1200	1728	17.0	41.9	17.5
Heat pump (7.5 HSPF)	218	552	794	7.8	19.3	8.0
Heat pump (8.8 HSPF)	255	470	677	6.7	16.4	6.8
Heat pump (12.5 HSPF)	363	331	477	4.7	11.6	4.8

See the example calculation in Example 9.1 above.

Prices for the different energy sources used to heat homes vary widely, but for this comparison, we assume US national averages of $0.12/kWh for electricity, $0.95/CCF for natural gas, and $2.90/gal of oil. For the different scenarios shown in Table 9.1, we can calculate the kWh used for each month based on the method of heating and then convert these data to D assuming the average days a month (30.4 = 365/12). For older natural gas furnaces, the efficiencies were low (70–80%) but modern furnaces can reach heating efficiencies of 97%. Electric baseboard heating efficiency is assumed to be 100% as all energy in the electricity is eventually dissipated as heat. Older heat pumps can have heating seasonal performance factor (HSPF) ratings that can be low relative to those made today, with minima of 6.8 in 1992 and 7.7 in 2006. New heat pumps must have at least an HSPF of 8.2 in the United States (since 2015), and this will increase to 8.8 in 2023, but you can already buy a heat pump with a rating of 12.5. HSPF ratings are further discussed in the sections below. Based on the results shown in Table 9.1, the lowest costs are obtained the same using the very highest efficiency natural gas system of $475/y or the very best heat pumps available today ($477/y). Thus, for the given system efficiencies, *the costs for natural gas and heat pumps appear to be about the same.* To address climate change, however, we need to further examine these systems in terms of fossil fuel energy use and carbon emissions.

When we compare these systems based on primary energy and carbon emissions, we can see much clearer advantages for choosing a heat pump over natural gas. The high efficiency heat pump (12.5 HSPF) has the lowest fossil fuel energy requirement with 11.6 D_{ff}, compared to the natural gas requirement of 17.9 D_{ff}. This lower energy requirement is due to the much greater efficiency of a heat pump than other systems. While fuels can never exceed an efficiency of 100% based on heat generated compared to the heat value of the fuel, heat pumps are capable of transferring more heat to inside the home than the energy it took to do that. The efficiency of a heat pump is further examined below.

The carbon emissions from heat pumps with the minimum allowable HSPF rating of 8.8 that can be sold today, and the most efficient heat pumps (HSPF = 12.5), both produce much less net carbon emissions compared to the other options. Natural gas does have the best energy to fuel ratio, with a D:C ratio of 2.23, while for oil it is 1.63. The electrical grid energy to carbon ratio is even better due to the use of renewables and carbon-neutral energy sources (D:C of 2.40), and it will continue to improve if the amount of electricity produced using renewables is increased. Electric baseboard heating, although rated with 100% energy efficiency, produces 17.5 C due to its reliance on electricity from the grid compared to natural gas, which has a lower D:C value. Heat pumps, although they use electricity from the grid (which includes coal and natural gas), have much lower C values because of their greater energy efficiencies. Thus, when the amount of fuel used and the carbon emissions are considered, heat pumps can produce heat for this scenario with as little as 4.8 C. Therefore, based on energy costs, total fossil fuel use, and CO_2 emissions, *a high efficiency heat pump clearly is the best option for heating your home.*

9.3 ENERGY RATINGS

Furnaces

The efficiency of a furnace is rated in terms of efficiency, which in the United States is on an Energy Star rating expressed as annual fuel utilization efficiency (AFUE), or annual fuel utilization efficiency which indicates it efficiency at converting the fuel energy into heat energy (Fig. 9.1a). The higher the rating the better, but the AFUE can never be >100%. The minimum AFUE for commercial sale in the United States is 78%, with average values of 80–85%. The very best high efficiency systems average above 90% and some units can achieve 95–97%. States in northern (cold) climates must now have AFUE ratings of 90% or more. Some older furnaces (20–25 y or more) are likely only 60% efficient, meaning that 40% of the energy in the fuel is wasted in terms of heating your home.

Heat Pumps

Why do heat pumps work better than electric baseboard or floor heating? Compared to electric heating and furnaces, which are limited to 100% efficiency based on a ratio of the heat transferred

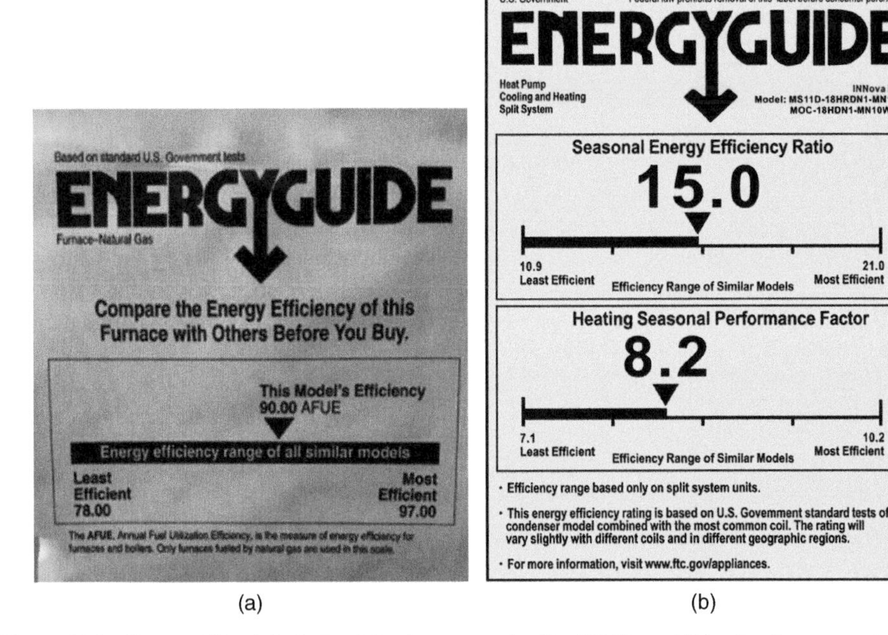

(a) (b)

Figure 9.1 Energy Star labels for (a) a furnace rated in AFUE and (b) a heat pump rated in HSPF for heating and seasonal energy efficiency ratio (SEER) for cooling.

to the energy used, heat pumps can be more than 100% efficient in terms of heat delivered relative the energy used!

To explain how heat pumps can be so efficient, we first need to discuss heat pump ratings and throw in a little bit of thermodynamics. When you purchase a heat pump, it is rated for heating in terms of HSPF, or heating season performance factor (Fig. 9.1b). The HSPF is a ratio of the heat the pump can produce in units of Btu, to the electricity it uses in units of Wh. Therefore, the HSPF is calculated as

$$\mathrm{HSPF} = \frac{\text{Heat produced (Btu)}}{\text{Electrical energy used (Wh)}} \times 100\% \tag{9.1}$$

However, the rating is better viewed as the *rate* heat is produced, such as Btu/h, divided by the electrical *power* used being fed into the heater in units of kW, which are both power. The number comes out the same whether the calculation is done as a ratio of energy or power since the unit of time cancels out.

The HSPF would be better presented as a "dimensionless" or unitless number as the numerator and denominator are both in energy units, but different units of energy are used. A dimensionless HSPF will indicate energy efficiency rather than a scaled efficiency because the numerator and denominator in a scaled efficiency comparison are in different energy units. By using 1 Btu = 0.29 Wh (or 1 Wh = 3.412 Btu), we can redefine a HSPF as a dimensionless ratio of heat energy to electrical energy where both the numerator and denominator have the same units, $HSPF_d$, as:

$$\mathrm{HSPF}_d = \mathrm{HSPF}\frac{(\mathrm{Btu})}{(\mathrm{Wh})}\,\frac{1\,\mathrm{Wh}}{3.412\,\mathrm{Btu}} = 0.29\,\mathrm{HSPF} \tag{9.2}$$

By converting the units from Btu to Wh, the HSPF_d number is now a ratio of Wh/Wh, and therefore, a true measure of heat energy to electrical energy as both the numerator and denominator are now in the same units. In Table 9.1, one of the heat pumps was indicated to have a rating of 7.5 HSPF, so the dimensionless ratio for this heat pump, or the heating efficiency, is

$$\mathrm{HSPF}_d = \mathrm{COP}_h = 0.29\,\mathrm{HSPF} = 0.29\,(7.5) \times 100\% = 218\% \tag{9.3}$$

where the term HSPF_d is more widely known by experts as the coefficient of performance, $\mathrm{COP_h}$, and the subscript h is used here to indicate that it is calculated for heating (as opposed to cooling). The above result indicates that the unit is 2.2 times more efficient at heat transfer into the home than the electricity used by the unit to accomplish this transfer of heat. On this basis, the unit has an efficiency of 218%! The upper limit in performance of a heat pump is set by the temperature difference between the heat source and destination and the laws of thermodynamics, so no true thermodynamic efficiency can be larger than 100%. The reason a heat pump appears to be more than 100% efficient is that it is moving heat, not generating heat from a fuel.

The maximum performance that any heating or cooling unit can accomplish, known as the maximum coefficient of performance, $\mathrm{COP}_{h,\max}$, is limited to a maximum by thermodynamics. For heating, it is:

$$\mathrm{COP}_{h,\max} = \frac{T_H}{T_H - T_C} \tag{9.4}$$

The efficiency of a heat pump relative to stated HSPF and the $\mathrm{COP}_{h,\max}$ can therefore be calculated as:

$$\eta = \frac{0.29\,\mathrm{HSPF}}{\mathrm{COP}_{h,\max}} \times 100 \tag{9.5}$$

Using Eq. (9.5), it is therefore possible to calculate the upper limit to heat pump efficiencies based on heat moved to electricity used, as shown in the example below.

Example 9.2

(a) Calculate a heat pump efficiency with an HSPF rating of 7.5 assuming an indoor temperature of 22°C and an outdoor temperature of 10°C. (b) Estimate the improvement in electricity use if a 7.5 HSPF unit is replaced with a 12.5 HSPF.

(a) First, we calculate the maximum possible efficiency for these two temperatures based using an absolute temperature in Kelvin (adding on 273 K to each temperature in degrees Celsius) as

$$COP_{h,\max} = \frac{T_H}{T_H - T_C} = \frac{(295\,K)}{(295\,K) - (283\,K)} = 24.6$$

The thermodynamic efficiency, which is based on this upper limit of the COP, is therefore

$$\eta = \frac{0.29\,HSPF}{COP_{h,\max}} \times 100 = \frac{0.29\,(7.5) = 2.2}{(24.6)} \times 100 = 8.8\%$$

(b) For the 12.5 HSPF rating, we have

$$\eta = \frac{0.29\,HSPF}{COP_{h,\max}} \times 100 = \frac{0.29\,(12.5) = 3.63}{(24.6)} \times 100 = 14.7\%$$

This increase in efficiency translates to a 67% improvement in performance.

The above example shows that while the heat pump can be quite efficient, moving heat at 2.2–3.6 times the energy used, it has an overall thermodynamic efficiency (theoretical maximum) of ~15%. Based on progress in the past years in improving heat pump efficiencies, it seems likely that HSPF ratings will continue to improve further over time. One limitation of the heat pump is that performance decreases with colder outdoor temperatures, and so a HSPF rating requires tests over a range of temperatures. Thus, the actual composite efficiency would require comparisons at each temperature difference, but for the example here a single temperature is used as an illustrative example. The performance of most heat pumps substantially decreases as the temperatures get colder as it is increasingly more difficult to extract heat from the colder temperatures. Therefore, most heat pumps will switch to an alternative heat source (resistance electric heating or natural gas) as temperatures approach or go below 0°C.

The reduced carbon emissions by using a heat pump compared to other heating options should provide the best incentive for switching from an electric or natural gas system. As shown in Table 9.1, the C values for the heat pump with the HSPF rating of 12.5 were 4.8 C, compared to the 9.2 C for the older natural gas furnace, or 17.5 C for baseboard electric heating. Using a heat pump with electricity generated by photovoltaics on your roof will not only eliminate CO_2 emissions from your home, but a 4.7 D rating for the heat pump will also greatly reduce the number of solar panels needed (11 panels) compared to a 17 D needed for electric baseboard heating (42 panels) for this home.

9.4 GEOTHERMAL HEATING

With heat pumps providing a clear advantage over other heating methods, it seems very straight forward that by installing photovoltaic solar panels on your house that you can reduce your home CO_2 emissions by using heat pumps. The one challenge for using a heat pump is that the greatest need for heating energy is in the winter months when both heat pump performance drops and solar output by photovoltaic panels is at the lowest electricity generation level for the year. Since heat pumps become less efficient at temperatures below freezing temperatures, a secondary system such as a gas furnace might be necessary if temperatures routinely go below 10–25°F (−12 to −4°C). However, you could avoid the use of a gas furnace by installing a geothermal or ground-based heat pump system.

Home geothermal energy systems are based on a simple concept: the upper 10 ft of the earth remains a nice constant 50–60°F (10–16°C) throughout the year (The Climate Reality Project, 2018). Home geothermal systems tap into this temperature based on heat pump technologies: the heat is either moved from the soil and into your house in the winter, or heat from your house is transferred into the soil during the summer to cool your house. Therefore, you do not need to live in an area that has hot underground heat sources, you only need to have sufficient land and soils suitable for creating a heat exchanger system in the ground.

The performance of a geothermal heat pump, or any heat pump, is limited by the temperature difference and the smaller the difference, the better the performance as shown above for the heat pumps. For example, using an indoor desired temperature of 72°F (T_H = 22°C = 295 K) with soil temperature of 55°F (T_C = 13°C = 286 K), the $COP_{h,max}$ would be 33. The higher the $COP_{h,max}$ the better your system will work. If you were using the heat pump in the winter with an outside temperature of 32°F (T_C = 273 K), then the $COP_{h,max}$ would only be 13, showing the decreased performance in colder temperatures. Therefore, by using the ground as your source of heat you can gain access to a temperature close to that of your home and thus achieve good heat pump efficiencies. Note that for cooling there are similar efficiencies that can be calculated as done above. For more on this approach using heat pumps for cooling, see the discussion in Chapter 10.

Geothermal systems are not cheap, but they will save money and reduce your carbon footprint! Home systems can cost between $10,000 and $25,000 by some estimates (The Climate Reality Project, 2018) to $20,000 to $25,000 by others (Consumer Reports, 2019), so it is very important to get an accurate estimate for your home before making a decision to install one of these systems. A key aspect is how much digging and drilling would be needed for your property. However, there can be federal, state, or local tax benefits that may help reduce your net costs. Once installed, typical savings are 30–70% on heating and 20–50% on cooling, so you need to assess savings relative to capital investment. A goal would be a payback in about 10 y, although these systems have very long lifetimes and so payback would occur long after your initial payment was covered.

9.5 WATER HEATERS

Gas and Electric Water Heaters

Heating water is the third most energy-consuming process in your home (following space heating and cooling, Table 3.4). Estimates for the energy used on average varies, with water heating estimated to be 12% (U.S. Energy Information Administration, 2020) to 14% (Direct Energy, 2020) of home energy use, and 17% for Pennsylvania. If we assume a total daily home energy use of 13 D and 12%, then D = 1.56 (3.6 kWh/d) for the water heater. However, if we use estimates that a water heater runs 3 h/d and uses 4.5 kW, then that translates to 5.8 D (13.5 kWh/d).

The actual need will depend on the number of people in the home, length of showers, insulation of piping and the heater, type of fuel, and age of the heater. No matter which number comes closest to the actual value for your home, the water heater clearly consumes a substantial percentage of home energy.

What type of water heater should you use? Most water heaters use direct electric heating and they are the least expensive to purchase. We expect that carbon emissions from water heaters would follow the same trends for home heating that was shown in Table 9.1, and that is the situation. Electric heaters release the most CO_2, and natural gas heaters cost less to operate since they use less primary energy and have a lower carbon footprint due to the use of natural gas compared to electricity for electric baseboard heaters. The most advantageous home heating was to use a heat pump, and heat pumps are also an option for water heaters as well (depending on the location in your home of the water heater).

Conventional natural gas or electric water heaters were evaluated for many years in terms of their energy factor (EF) rating, which was calculated as the amount of water heated per Btu. The EF is a ratio of the energy in the heated water to that actually used, or

$$\mathrm{EF} = \frac{\text{Energy in the hot water}\left(\dfrac{\mathrm{Btu}}{\mathrm{gal}}\right)}{\text{Energy used}\left(\dfrac{\mathrm{Btu}}{\mathrm{gal}}\right)} = \frac{E_w}{\mathrm{E}_{\mathrm{used}}} \tag{9.6}$$

where the energy in the water is based on the definition of a Btu, which is the energy needed to raise 1 lb of water 1°F. If we assume the water is being heated from 58 to 135°F, and we use the density of water at 58°F which is 8.34 lb/gal, then we can calculate the energy needed to heat the water as:

$$E_w = \frac{1\ \mathrm{Btu}}{1\ \mathrm{lb}\,^\circ\mathrm{F}}\frac{8.34\ \mathrm{lb}}{\mathrm{gal}}(135-58)\,^\circ\mathrm{F} = 642\,\frac{\mathrm{Btu}}{\mathrm{gal}} \tag{9.7}$$

The Energy Star ratings highlight cost of operation of the unit and the EF rating used to be included in the text below the annual cost. The EF may be in the manufacturer's specifications for the water heater you currently have, so it is worth considering how that number is calculated for comparisons to newer units. The annual energy use is based on the performance of the system and hot water use of 64.3 gal/d, which was assumed to be the use in a household of three people.

Example 9.3

Calculate the annual water use for a water tank for a gas heater with an EF = 0.57, assuming a natural gas fuel price of P = $6.19 /mmBtu, and a water daily use of 64.3 gal/d.

The actual energy used is $E_{used} = E_w/EF$, and so the annual cost is based on a daily use of $Q = 64.3$ gal/d the price of the fuel is calculated as:

$$\mathrm{Cost} = \frac{E_w}{EF}PQ$$

$$\mathrm{Cost} = \frac{642\,\dfrac{\mathrm{Btu}}{\mathrm{gal}}}{(0.57)}\left(\frac{\$6.19}{\mathrm{mm\,Btu}}\right)\left(\frac{64.3\ \mathrm{gal}}{\mathrm{d}}\right)\frac{1\ \mathrm{mm\,Btu}}{10^6\mathrm{Btu}}\frac{365\ \mathrm{d}}{\mathrm{y}} = \$163/\mathrm{y}$$

The EF ratings are now being phased out in favor of a system based on using multiple criteria that more comprehensively describes energy performance. These evaluations include the "first-hour rating" or the energy needed to heat water up in the first hour, and the uniform energy factor (UEF) which is based on a more complex method of analysis of heat energy relative to energy use. The UEF method is not further described here as none of the UEF method information is posted on the Energy Star label on water heaters. Energy guide labels highlight the estimated cost for using the water heater for 1 y in an average home. Each of these labels gives the cost and the range for similar water heaters (same fuels and similar first-hour water heating values) (Fig. 9.2). The cost of the fuels can account for large differences in costs. For example, the natural gas unit uses 202 therms (THM) a year while the propane unit uses less energy over a year (188 THM), but costs more than twice to operate for the same amount of water heated. The labels do state the average cost for the energy source, but they do not have any listed energy efficiencies.

With the implementation of this new method of evaluating water heaters, the overall efficiency can no longer be directly calculated. However, we can get a sense of the efficiency of these different systems based on the energy used for 1 y, although these energy units are different for the gas-powered systems (reported in therms) and the electrical power systems (reported in kWh). Another way to view the energy use is in terms of the energy in the water that would be

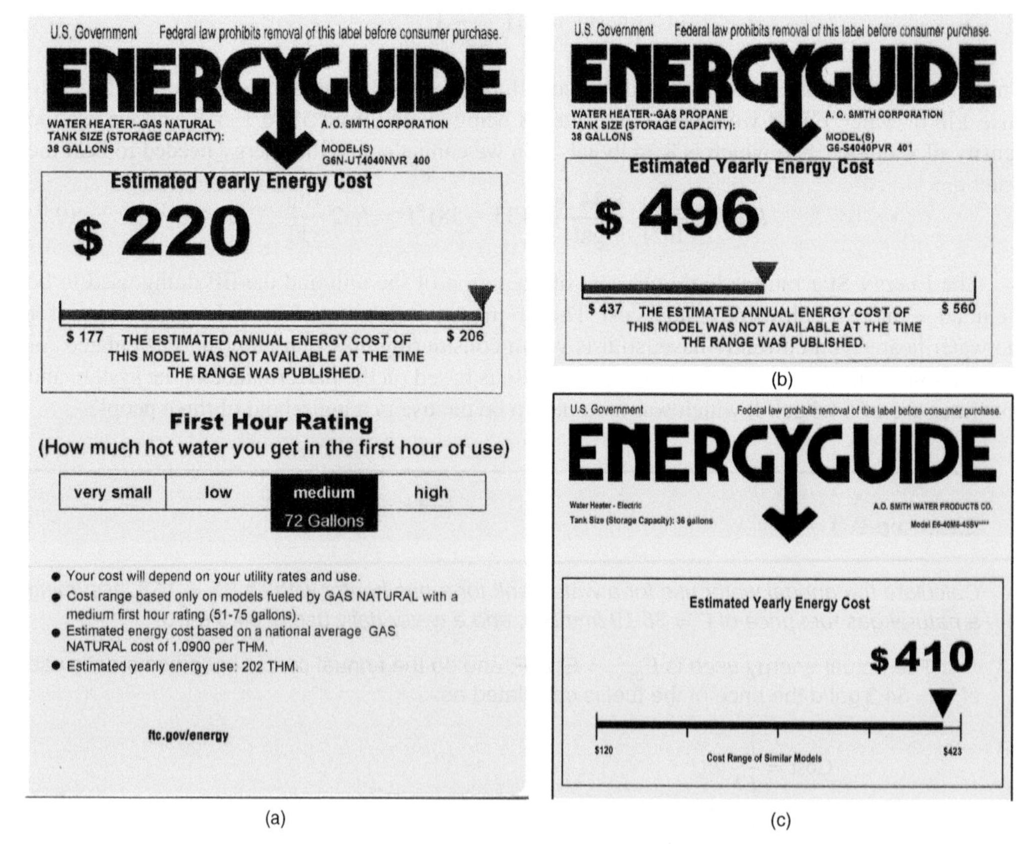

Figure 9.2 Examples of Energy Star labels for water heaters fueled by: (a) natural gas (38 gal capacity, 202 THM), and partial labels for heaters fueled by (b) propane (38 gal, 188 THM) and (c) electricity (36 gal, 3419 kWh).

heated up using that energy compared. The new method does not have strict requirements of using 64.3 gal/d (64.3 w) as in the previous method, but the comparison of the water heated by the energy consumed in a day to this 64.3 gal will give us a sense of overall energy efficiency. This approach is shown in the example below.

Example 9.4

The current methods of estimating energy use are more complex than previous methods but one way to compare relative performance is to calculate the amount of water that could be heated for the given annual energy consumption. For the three water heaters in Figure 9.2 compare the volume of water per day based on the stated annual energy consumption. Assume that one gallon of water requires 642 Btu (to go from 58 to 135°F)

For the natural gas water heater, we see that 202 THM are used per year. Therefore, we can calculate the actual energy use and efficiency of the water heater for these test conditions, η_{wh}, as:

$$\frac{202\ \text{THM}}{\text{y}}\ \frac{10^5\ \text{Btu}}{\text{THM}}\ \frac{1\ \text{y}}{365\ \text{d}}\ \frac{\text{gal}}{642\ \text{Btu}} = 86\frac{\text{gal}}{\text{d}} = 86\ \text{w}$$

For the propane water heater, we can calculate this in one step as:

$$\frac{188\ \text{THM}}{\text{y}}\ \frac{10^5\ \text{Btu}}{\text{THM}}\ \frac{1\ \text{y}}{365\ \text{d}}\ \frac{\text{gal}}{642\ \text{Btu}} = 80\frac{\text{gal}}{\text{d}} = 80\ \text{w}$$

For the electrical water heater, we need to convert the energy used in kWh to Btu, and then, we make the same calculation as:

$$\frac{3419\ \text{kWh}}{\text{y}}\ \frac{3412\ \text{Btu}}{\text{kWh}}\ \frac{1\ \text{y}}{365\ \text{d}}\ \frac{\text{gal}}{642\ \text{Btu}} = 50\frac{\text{gal}}{\text{d}} = 50\ \text{w}$$

From this comparison, we see that the electric water heater is meeting the water needs for a year with the energy equivalent to heating 50 w, while the propane heater would use more energy as it is equivalent to 80 w, with the natural gas heater consuming the most energy of 86 w. These results are different from the EF calculation based on efficiencies relative to the use of 64.3 w. With the new method, it is not possible to calculate an efficiency just based on a starting and ending temperature as the efficiency of the electric heater would then be estimated to be larger than 100%, which is not possible using this technology.

Heat Pump Water Heaters

One final alternative to the above water heaters is a heat pump water heater, which also works well in combination with electricity produced by a solar photovoltaic system. Heat pump water heaters are just like heat pumps used to heat your home: they pull heat from the surrounding air and transfer it into the water in the tank. Since they do not directly use electrical energy to heat the water, they can be as much as two to three times more efficient in terms of heating than direct electric heating (US Energy Information Administration, 2020a). The cost of a heat pump water heater is $2500–$4000 for all new components with installation, or roughly twice what a conventional and new electric water heater system would cost ($1100–$2200). However, a heat pump water heater can save up to

61% of your electricity cost, with a typical savings of $3400 over the lifetime of a typical system (Trout, 2020). They also last longer (13–15 y) compared to electric heaters (8–12 y) which means fewer units over the long term.

The main concern with a heat pump water heater is where it will be located, as that can impact performance and temperature comfort in your home. Since the heat pump moves heat from the room into the tank, and it moves more heat than it generates, it will effectively be cooling the room that it is placed in. That cooling can be beneficial if it is located next to your dryer. However, if you have a cool basement, it will make the basement colder, requiring additional heating if you want to make that room more comfortable for use for example in the winter months. The performance the water heater is similar to that of any heat pump or geothermal system, with the temperature of the surroundings improving efficiency if it is closer to that of the heater.

Comparison of Water Heaters

Energy Star labels should provide estimates of carbon emissions based on using the water heater so that consumers can make decisions that include how much CO_2 will be emitted each year. The energy used can be converted from the different annual energy units into daily energy units, and then using the D:C ratio for the fuel, the daily carbon emissions in terms of C can be obtained for each device. In the case of electricity, the factor 2.46 can be used to calculate the D in the fossil fuels based on fuels used to make the electricity, and a D:C ratio of 1.59 for CO_2 emissions resulting from the mix of fuels used to generate electricity.

In addition to the three systems examined above, we also consider an "on-demand" or tank-less flow water system and a water heater using a heat pump. The amount of energy used as electricity of 4.04 De for the electric water heater is lower than that of the gas systems which are all above 6 D (Table 9.2). However, when we account for the primary energy used to make that electricity the electrical water heater has a similar consumption of fossil fuel use (D_{ff}). The heat pump system uses the least energy of all these systems with only 1.79 D_{ff}. When we consider CO_2 emissions, we see that the heat pump water heater has a C value of only 1.1, with the other systems around 3–4 C.

While these D and C values for the heat pump water heater are low, the placement of the water heater will be essential in determining its overall impact on home energy use and climate emissions. If the water heater is placed in a location that is heated, then it will just draw heat from that is produced by another method of heating. If it is placed in a cold basement, its efficiency will also not be similar to conditions estimated for determining the energy use. However, the placement of all these different water heaters will be important relative to the use of heating and cooling for specific houses and tank locations. Tank location should therefore be carefully considered in

Table 9.2 The energy used for different water heaters based on primary energy (D) and carbon released based on the daily CO_2 emissions (C).

Fuel type	Energy/y	$/y	D	D_{ff}	D_{ff}:C	C
Natural gas-38 gal	200 THM	220	6.92	6.92	2.23	3.1
Natural gas-4.7 gpm	180 THM	199	6.23	6.23	2.23	2.8
Propane-38 gal	188 THM	496	6.51	6.51	1.82	3.6
Electric-36 gal	3419 kWh	410	4.04	6.56	1.59	4.1
Heat pump-46 gal	950 kWh	114	1.12	1.79	1.59	1.1

See related calculations in Example 9.1 above.

house construction (or during a renovation) to minimize the impact of the water heater on overall energy demands.

Solar Thermal Water Heaters

If you live in Australia, chances are that your water heater accounts for about 28% of energy use, compared to about 17% (1.6 D) in Pennsylvania and many other states in the United States (Wikipedia, 2020). That relatively high percentage of energy use is why the Australian government provides incentives to install solar water heaters in houses, although only ~3–4% of houses have them (mostly new houses). Solar water heaters are not limited to warm climates although they may be more economical, easier to integrate into houses, and have a faster payback compared to other climates.

The water heater is an opportunity for capturing solar power directly as heated water. There are many types of solar water heating systems which are classified as active or passive (US Energy Information Administration, 2020b). Active water heating systems use pumps to either directly circulate water in climates where freezing temperatures are rare, and a heat transfer fluid in climates more prone to very cold temperatures that will experience freezes. Passive systems are less expensive, do not have pumps, and are therefore cheaper but they are typically not as efficient at capturing solar heat as the active systems. One consideration for solar heating is how much roof space you have for panels, where your water heating system is located, and many other structural factors.

A solar water heater typically consists of a flat panel on your roof where the sun heats water or it heats another liquid which is usually a mixture of glycol (typically ethylene glycol) and water. Some systems use hydrocarbon oils, while other systems can use refrigerants such as chlorofluorocarbons (CFCs). From an environmental perspective, these other fluids come with risks, especially the CFCs which are potent greenhouse gases (GHGs) if they leak into the environment. Solutions with glycols may need to be replaced every 3–5 y, and their proper disposal or recycling is essential to the environment.

A solar water heating system costs on average ~$3500 (Angie's List, 2020). The components are the tank, collector, switches and plumbing system ($1000–$4000), and labor for installation. Unless you live in a sunny, warm climate all year you likely would want an active water heating system that has backup power. However, you may get tax credits (26%, the same as photovoltaics) which could reduce this cost to ~$2500.

Although the cost of a solar water heating system will be more than a conventional one, once it is installed your water heating bill will decrease by half or more (50–80%). If you are buying a new home, that is a great time to install one as the cost can be put into your mortgage and thus you can derive additional tax benefits over the lifetime of the mortgage. If you spend more than $15 per month, you may find the system to immediately be profitable due to tax benefits. If you have an electric water heater your savings will likely be greater than if you have a gas heating system.

Water Heaters Powered by Photovoltaics

An alternative to a solar thermal water heater is to just power your electric water heater using electricity from your solar panels. For a home use of 1.6 D of electricity for your water heater and an average of 0.58 D per solar panel (Example 6.1), you would need about three panels. At ~$700 for an installed panel after tax discounts of 26%, that is about $2100, and those panels will last for 25+ y. The added advantages are no pipes running between the tank and collectors, and no fluids needed as you just use electricity. Therefore, if you have a limited amount of space on your roof for optimal location of the solar water heater, and you want to have photovoltaics, you will have to do some location-specific calculations to see what is more economical.

9.6 HOME AND BUILDING ENERGY ANALYSIS FROM DRAWDOWN

Project Drawdown has identified several activities relevant to buildings and heating systems that could make significant contributions to reducing GHG emissions. There are 10 activities that could reduce GHGs by 71.6 gigatons (Gt) of CO_2-equivalent emissions by 2050 (Table 9.3). When these are normalized to the expected global population of 9.4 billion, this translates to an overall change of –0.77°C over the 30-y period. Insulation and widespread adoption of light-emitting diode (LED) lighting provide the two largest reductions in emissions, followed by high-performance glass and smart thermostats. In terms of capital expenses, LED lightbulbs and smart thermostats are the easiest and most expedient ways to reduce consumption of fossil fuels and therefore to reduce CO_2 emissions.

It is surprising, given the very high efficiencies of heat pumps and the large amount of energy used for home heating and water heaters, that heat pumps rank only 46th on this list with an impact similar to the use of solar hot water. However, a complete retrofit of heating systems to the use of heat pumps would be expensive and place a burden on CO_2 emissions through increased manufacturing. Certainly, it is worth considering a replacement of direct electric heating in the future for any home or business that would need a system replacement.

Table 9.3 Possible reductions in CO_2 emissions (GT) over a period of 30 y (from 2020 to 2050) due to activities most related to buildings (selected from 76 activities).

No.	Solution	CO_2 (10^9 tons)	–C
16	Insulation	17.0	0.18
18	LED lighting	16.1	0.17
27	High-performance glass	10.0	0.11
36	Smart thermostats	7.0	0.07
37	Building automation systems	6.5	0.07
38	District heating	6.3	0.07
46	High-efficiency heat pumps	4.2	0.04
48	Solar hot water	3.6	0.04
73	Green and cool roofs	0.6	0.006
74	Dynamic glass	0.3	0.003
	Total	71.6	0.77

The value of C is shown as a negative sign as these are carbon reductions normalized to a global population of 9.4 billion people in 2050.

Source: Data from Wilkinson (2020).

References

ANGIE'S LIST. 2020. *How much does a solar water heater cost?* [Online]. Available: https://www.angieslist.com/articles/how-much-does-solar-water-heater-cost.htm#:~:text=The%20average%20solar%20water%20heater,to%20%245%2C500%20for%20the%20installation [Accessed June 4 2020].

CONSUMER REPORTS. 2019. *Heat pumps* [Online]. Available: https://www.consumerreports.org/cro/heat-pumps/buying-guide/index.htm#:~:text=The%20minimum%20federal%20HSPF%20rating,15%20SEER%20and%208.5%20HSPF [Accessed June 5 2020].

DIRECT ENERGY. 2020. *What uses the most electricity in my home?* [Online]. Available: https://www.directenergy.com/learning-center/energy-efficiency/what-uses-most-electricity-in-my-home [Accessed May 19 2020].

U.S. ENERGY INFORMATION ADMINISTRATION. 2020a. *Heat pump water heaters* [Online]. Available: https://www.energy.gov/energysaver/water-heating/heat-pump-water-heaters [Accessed May 19 2020].

U.S. ENERGY INFORMATION ADMINISTRATION. 2020b. *Solar water heaters* [Online]. Available: https://www.energy.gov/energysaver/water-heating/solar-water-heaters [Accessed May 19 2020].

THE CLIMATE REALITY PROJECT. 2018. *How exactly does geothermal energy work?* [Online]. Available: https://climaterealityproject.org/blog/how-exactly-does-geothermal-energy-work#:~:text=Geothermal%20energy%20is%20largely%20used,through%20a%20looping%20pipe%20system [Accessed June 3 2020].

TROUT, J. 2020. *Are heat pump water heaters worth the cost?* [Online]. Available: https://www.consumeraffairs.com/homeowners/heat-pump-water-heater-value.html [Accessed June 4 2020].

U.S. ENERGY INFORMATION ADMINISTRATION. 2020. How is electricity used in U.S. homes?. https://www.eia.gov/tools/faqs/faq.php?id=96&t=3 [Accessed May 10 2020].

WIKIPEDIA. 2020. *Solar hot water in Australia* [Online]. Available: https://en.wikipedia.org/wiki/Solar_hot_water_in_Australia [Accessed June 4 2020].

WILKINSON, K. 2020. *The drawdown review: Climate solutions for a new decade*, Project Drawdown.

References

CHAPTER 10

COOLING AND REFRIGERATION

10.1 WHY ENERGY FOR COOLING IS INCREASINGLY IMPORTANT

People want to be more comfortable in the hot weather by staying cooler, but the world is getting hotter. In the United States, the energy used for cooling in the average home has already reached and exceeded (on average) the amount of energy used for heating (Table 3.4) (US Energy Information Administration, 2020). By 2100, worldwide power consumption for cooling is predicted to increase 33 times compared to that used today. In the United States, residential (4%) and commercial (5%) sectors used about 9% of the electricity produced for cooling (377 billion kWh in 2018). In only 15 y, the number of air-conditioned homes in the United States increased from 64 to 100 million. In 2010, 50 million air conditioner (AC) units were sold in China. In the Middle East, cooling makes up a large proportion of energy use. For example, 40% of electricity used in Mumbai, India was for cooling, and 50% of electricity use in the summer in Saudi Arabia is for cooling.

Cooling is not just be needed for comfortable living in some homes and workplaces, it could be essential to save lives. When the temperatures get too hot and humid, the body can no longer cool itself to maintain needed temperatures. We have already discussed that each person uses about 1 D, or ~100 W, and a large part of that is dissipated as heat that needs to be removed from the body. If it is too humid, water cannot evaporate fast enough on skin to cool the body, and a person will overheat. While some climate models suggest dangerously high ambient temperatures will need to be dealt with in the mid-21st century, there is already evidence of extreme heat and humidity in some coastal subtropical locations (Raymond et al., 2020). These extremes occur when skin temperatures exceed about 35°C (95°F), and thus, it is not possible to maintain a normal body temperature of 36.8 ± 0.5°C except by water evaporation. When the wet-bulb temperature (TW) reaches above 35°C, cooling by sweating fails to maintain proper body temperature. Warm temperatures in very humid regions therefore can already be deadly without sufficient cooling.

To understand energy use for cooling, two topics are addressed in this chapter: energy use for home refrigerators and energy use by different types of air conditioning units. The way performance of these appliances and units is reported is different, and thus, the amount of energy used can be a little difficult to relate to our other energy-consuming activities. Therefore, for each topic, we start out first with the units used to characterize each appliance or device and then convert these to kWh and D units. The discussion of energy use related to cooling proceeds to a more in-depth analysis of the energy efficiency of cooling systems. Finally, the importance of refrigerants is discussed within the context of climate change.

Daily Energy Use and Carbon Emissions: Fundamentals and Applications for Students and Professionals,
First Edition. Bruce E. Logan.

10.2 ENERGY USE FOR REFRIGERATORS

When you look at a new refrigerator in the U.S., you will usually see the Energy Star tag (Fig. 10.1) displayed (often inside the door of the unit) showing how much money it will cost on average to use the refrigerator each year. The tag will also indicate where that particular model ranks compared to other models of a similar size and configuration in terms of annual expenses per year, and how many kWh/y it would use for a typical household. The cost of the electricity can be based on national or local averages, so it is always important to consider the kWh used so that you can be sure to calculate the cost for operation based on the price of electricity in your area.

Example 10.1

A 27 ft³ refrigerator Energy Star label indicated the annual cost was $77 based on operation that consumed 637 kWh/y for one person. Calculate the price of electricity used for this estimate and the energy used per day in kWh and D. Compare your results to those given in Table 3.4.

Using the given information, we calculate the cost and energy in kWh/d or D as:

$$\text{Electricity price} = \frac{\$77}{\text{y}}\ \frac{\text{y}}{637\ \text{kWh}} = \frac{\$0.12}{\text{kWh}}$$

$$\text{Energy use} = \frac{637\ \text{kWh}}{\text{y cap}}\ \frac{1\ \text{y}}{365\ \text{d}} = 1.75\ \frac{\text{kWh}}{\text{d cap}}\ \frac{1\ \text{D}}{2.32\frac{\text{kWh}}{\text{d cap}}} = 0.75\ \text{D}$$

The electricity price matches the national average for 2020, which is also listed in the small print on the tag. Table 3.4 has 0.78 D for a modern refrigerator, which would be equivalent to 1.8 kWh/d, which are numbers in good agreement with those for this refrigerator.

The amount of energy used by refrigerators has changed a lot over the past decades with 2000 kWh/y (2.4 D for one person) in the early 1970s and 900 kWh/y (1.1 D) in the early 1990s for an average-sized refrigerator. The capacity of the refrigerator is an important factor in energy use, with smaller models (19 ft^3) rated at 420 kWh/y (0.50 D), and larger models using much more energy. For example, the label in Figure 10.1 shows a range in cost of $74–$104, which would translate to 620 kWh (0.73 D) to 870 kWh (1.0 D) based on the average price of electricity of $0.12/kWh. Therefore, while the *average* energy use may have decreased some of larger units consume quite a bit more electrical energy for their operation. In general, side-by-side models are less energy efficient than those with above and below refrigerator and freezer doors.

Changes in total energy use for home refrigerators over the past decades must also factor in the number of refrigerators in a home. In 1993, about 15% of homes had two refrigerators but by 2005 this had increased to 22% of homes. If we consider that cooling requirements for a single refrigerator were reduced by ~50%, but nearly a quarter of homes have two refrigerators, the gain in overall energy use is not as impressive. Also, many homeowners buy new refrigerators and then use the old one as a second unit, so that 15% of homes in the United States have fridges that are at least 20 y old, and thus much less energy inefficient than those made today (Mooney, 2014).

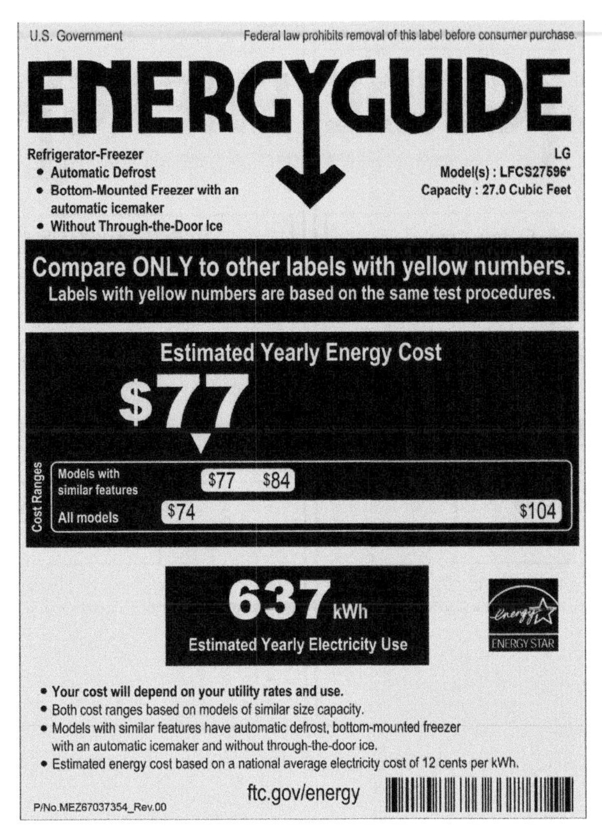

Figure 10.1 Example of an Energy Star tag for a 27 ft^3 refrigerator showing the average energy use of this model in terms of cost, where that cost ranks among similar models (the sliding scale), and the estimated kWh used per year.

10.3 ENERGY USE FOR AIR CONDITIONERS

For ACs, the cost for operation is also given on the Energy Guide tag as it is for refrigerators. However, we can see that information for kWh used per year is not listed on the AC tag (Fig. 10.2). Furthermore, the cost of the kWh may also be missing from the tag making it difficult to know how much energy is used when the unit is operated. The energy use and costs for a kWh should always be included for public information as this allows a simple calculation of energy use for operation of the AC unit. For example, if the cost is \$0.12 kWh and the energy cost is listed as \$143 (Fig. 10.2), then the unit uses 1190 kWh/y or 3.2 kWh/d = 1.4 D (on average). Also, if the unit is only used for 3 mo a year (during the summer), then it is more reasonable to calculate energy consumption for the period that it is used. For this case in Figure 10.2, the energy consumed would be 5.5 kWh/d or 2.4 D (over the 3-mo period). Adjusting the use to summer months helps us to better understand how daily use will impact our monthly electricity bill and overall energy use during the summer months, compared to energy costs for heating during the winter months.

A national average energy cost should be included on this AC Energy Guide tag, and it should be a national average not a local average, especially if electricity is inexpensive. The use of a very

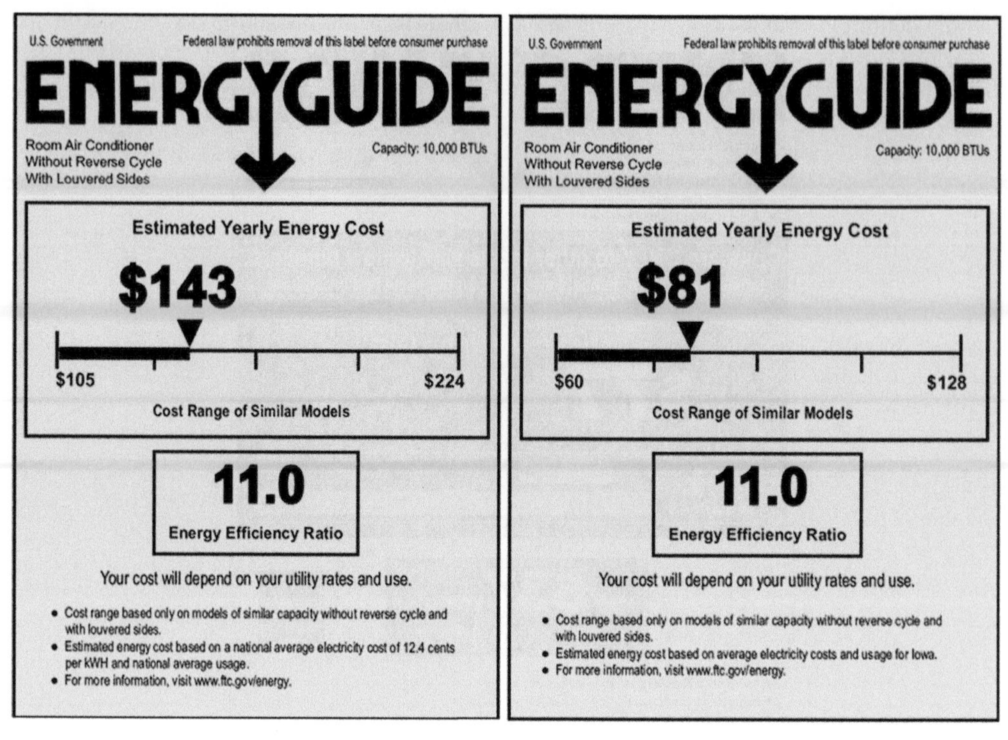

Figure 10.2 Example of an Energy Star tag for the same air conditioner, with the prices specified for the one on the left (12.4 cents/kWh), and the one on the right just saying average electricity costs for Iowa.

low estimated electricity cost will lead to consumers choosing less expensive models that could release more carbon into the air than more energy efficient models (Davis, 2014).

Power Used and Performance of AC Units

ACs have a completely different method of evaluation than other appliances as they have historically indicated the energy efficiency in terms of an energy efficiency ratio (EER) (Energy.gov, 2020), while more recently the units are rated in terms of seasonal energy efficiency ratio (SEER). The EER accounts for performance between two temperatures, typically 27°C indoors and 35°C outdoors, the SEER is a performance indicator over a range of temperatures. The EER number allows you to directly evaluate energy consumption, while the SEER cannot be used in the same way due to the range of operating temperatures. For both EER and SEER, the higher the number, the more efficient the unit.

We will examine the energy performance of air conditioning units here in terms of the EER to illustrate what the energy rating means for a single temperature difference. The EER rating for ACs, like the heating season performance factor (HSPF) for heating units, is a confusing mixture of English and metric units. The EER is defined in terms of performance in terms of the rate the unit can extract heat in Btu/h (English units) and power use in units of Watts (metric units) (Energy.gov, 2020) as:

$$\mathrm{EER} = \frac{\text{Cooling capacity } (CC) = 1\frac{\mathrm{Btu}}{\mathrm{h}}}{\text{Power consumed } (P) = 1\ \mathrm{W}} \tag{10.1}$$

What this definition means is that the unit for every Btu of heat removed in 1 h takes 1 W of power if it is rated at 1 EER. For the Energy Star label above, that means that the AC can remove 11 Btu/h per W of electricity consumed. Thus, the more heat that can be removed, the higher the EER. The second E in EER is "efficiency," but that is not an absolute efficiency. The EER is useful when comparing the performance of different units but it is not a thermodynamic efficiency.

The other number on the tag, in very small letters, is the unit rating in Btu, which for the one shown in Figure 10.2 is 10,000 Btu. For large AC systems, the units are sometimes rated in "tons," where 1 ton is equal to 12,000 Btu. What the label should say (but it does not!) is that this is the heat in Btu indicates *Btu that can be removed in 1 h*, so the rating for a 10,000 Btu system means that it can remove heat at a rate of 10,000 Btu/h. From that unit-specific information, and the EER, we can calculate the power and estimate the daily energy used for the AC unit as shown in the example below.

Example 10.2

An AC with an 11 EER rating is indicated to have a capacity of 10,000 Btu. Based on these data, (a) what is the power for the unit when operating? (b) If the unit runs 6 hours a day, how many kWh are consumed by the unit each day, and what is the energy consumed expressed as the daily energy unit D?

(a) Using Eq. (10.1) and using the given performance of 10,000 Btu/h, the power is

$$P = \frac{CC}{\text{EER}} = \frac{10\ \text{Btu/h}}{11\left(\frac{\text{Btu/h}}{\text{W}}\right)} = 910\ \text{W}$$

(b) If the unit runs at this power level for 6 h a day, then it will consume

$$910\ \frac{\text{W}}{\text{cap}}\ \frac{6\ \text{h}}{\text{d}}\ \frac{1\ \text{kW}}{1000\ \text{W}} = 5.45\frac{\text{kWh}}{\text{d cap}}\ \frac{1\ \text{D}}{2.32\frac{\text{kWh}}{\text{d cap}}} = 2.35\ \text{D}$$

This result of 2.35 D compares well to the estimate we have above based on cooling of 2.4 D if the AC unit energy demand for the year was spread out over a period of only 3 mo.

10.4 UNDERSTANDING ENERGY UNITS FOR COOLING

The Energy Star tags are useful for consumers to compare how much they might spend on a unit in a year, but they are not geared towards understanding energy use in terms of overall efficiency. The EER ratings also do not fit into thermodynamic efficiency ratings that are often used for converting fuels into work or power. For example, if a power plant has a 37% energy efficiency, then that means that 37% of the energy in the fuel ended up as electrical power. This energy efficiency is therefore always less than 100%. For refrigerators, ACs, and other cooling systems the energy is not used to do work, it is the energy that was needed to accomplish transfer of heat. Thus, in terms of heat removed for cooling, or heat added by a heat pump (heating), more energy can be moved than that used by the unit because moving heat and energy to do that are separate tasks. An air conditioning unit can therefore have a *rating that makes it appear that it has a thermodynamic efficiency of larger than 100% efficient* (which it does not!). However, the unit can remove more energy in the form of heat than it uses in energy in the electricity to move that heat.

Consider this example (which is not a perfect analogy): If you carry a bag of groceries from your car to your table, it may contain 500 Cal of food, but you certainly did not use that many Calories to bring in those groceries! If used 10 Cal to move 500 Cal of food into your house, your efficiency would appear to be $500/10 \times 100 = 5000\%$. Similarly, an AC can move more heat energy from inside your house to outside your home than the AC consumed while doing that! The AC (and heat pumps) work on moving heat around, but they do not directly consume that heat, which allows the energy moved to be less than the energy consumed – or the power moved less than the power consumed, where time is how long it took to do that. This concept was explored for heat pumps in Chapter 9 to move heat. Here we can apply it to cooling systems.

To see how energy movement for cooling works, we start with converting the EER rating, which is calculated in units of Btu/h and Watts, to have the same units of Watts in both the numerator and denominator. If we convert the heat or energy in Btu first to Wh, using 0.29 Wh per Btu, we can then get Watts for the numerator, as:

$$\mathrm{EER_d} = \frac{\mathrm{EER}\left(\frac{1\ \mathrm{Btu}}{\mathrm{h}}\right)}{P\,(\mathrm{W})}\frac{0.29\ \mathrm{W}}{\frac{\mathrm{Btu}}{h}} = 0.29\ \mathrm{EER}\,\frac{\mathrm{W}}{\mathrm{W}} = 0.29\ \mathrm{EER} \tag{10.2}$$

When the EER is converted to units of W/W, then we have the $\mathrm{EER_d} = 0.29\,\mathrm{EER}$. This conversion provides an $\mathrm{EER_d}$ in units known as a dimensionless (or unitless) number since the units in the numerator and denominator cancel each other out. To indicate this is a dimensionless number we have added a subscript onto the EER, which therefore indicates the units are the same in the numerator and denominator. The $\mathrm{EER_d}$ is also known as the coefficient of performance (COP) (Wikipedia, 2020).

If we use our example from above that a unit has an EER of 11, then the $\mathrm{EER_d}$ is as follows:

$$\mathrm{EER_d} = 0.29\,\mathrm{EER} = 0.29\,(11) = 3.19\frac{\mathrm{W}}{\mathrm{W}} = 0.319 \tag{10.3}$$

We have calculated that the AC unit is transferring heat out at a rate of 3.19 W per 1 W of power used by the AC unit, or 3.19 times faster than the power used by the AC unit. With an EER rating of 11, this AC unit is relatively efficient if you consider that back in the 1970s AC units typically had EER ratings of only 5, meaning that heat was removed at a rate of only 1.45 W for every W of electricity used.

Maximum Performance and Efficiency

The efficiency of a refrigeration process used to move heat is ultimately limited by the laws of thermodynamics. Thus, the maximum efficiency of the system cannot exceed that based on the Carnot cycle. Therefore, the maximum $\mathrm{EER_d}$ ($\mathrm{EER_{d,max}}$) (Wikipedia, 2020), is as follows:

$$\mathrm{EER_{d,max}} = \frac{T_C}{T_H - T_C} \tag{10.4}$$

where T_C is the cold temperature, and T_H is the hot temperatures relative to absolute zero, or in degrees Kelvin. For example, using the temperatures typically used to calculate the EER, the indoor desired temperature would be $T_C = 300$ K (27°C, or 80°F) with an outdoor temperature of $T_F = 308$ K (35°C, 95°F) (Wikipedia, 2020). The efficiency of an AC unit relative to the maximum based on the Carnot cycle, η, is therefore

$$\eta = \frac{\mathrm{EER_d}}{\mathrm{EER_{d,max}}} \times 100 \tag{10.5}$$

Using Eq. (10.5), we can therefore translate a given EER posted on an Energy Star label into an efficiency for a specific range of indoor and outdoor temperatures.

Example 10.3

Calculate the efficiency of an AC with an 11 EER rating, assuming an indoor temperature of 27°C and an outdoor temperature of 35°C.

First, we calculate the maximum possible efficiency for these two temperatures based using absolute temperatures in Kelvin as

$$\mathrm{EER_{d,max}} = \frac{T_C}{T_H - T_C} = \frac{(300\,K)}{(308\,K) - (300\,K)} = 37.5$$

$$\eta = \frac{\mathrm{EER_d}}{\mathrm{EER_{d,max}}} = \frac{0.29\ \mathrm{EER}}{\mathrm{EER_{d,max}}} = \frac{0.29\,(11)}{(37.5)} \times 100 = 8.5\%$$

While the EER rating of 11 may at first seem impressive, this calculation shows that the rate heat is moved is only 3.19 times as much as the power and that the AC unit overall is achieving an efficiency of only 8.5% of that possible based on the thermodynamic limit. Clearly, there is room for improvement in the efficiency of these units with only an EER rating of 11!

The performance of an individual AC unit can be calculated as shown using the EER for any specific indoor and outdoor temperatures. However, outdoor temperatures vary considerably during a single day, as well as over the summer months, and therefore, AC's are now rated by the SEER. While the EER is calculated for a fixed indoor temperature of 27°C and an outdoor temperature of 35°C, the SEER is based on performance over a range of outdoor temperatures (18 to 40°C). The SEER rating is not specific to a region, so the performance of a unit in any given location will likely be different than the SEER conditions due to variations in temperatures due to location and weather. One estimate aimed at relating the SEER to the EER (Wikipedia, 2020) is as follows:

$$\mathrm{EER} = -0.02\ \mathrm{SEER}^2 + 1.12\ \mathrm{SEER} \tag{10.6}$$

The actual relationship between EER and SEER, however, depends on the specific AC unit and actual climate conditions and thus cannot be accurately predicted. For example, using Eq. (10.6), a SEER of 13 is approximately equal to an EER of 11. Commercial unit specifications that list both show actual ratios of these two ratings that can be much different. For example, for one unit the EER was given as 11, but the SEER = 18. For another AC, the indicated metrics were SEER = 40 and EER = 15, compared to 12.8 calculated for the EER using Eq. (10.6) for the stated SEER. Thus, the Eq. (10.6) provides some estimate of the relationship between the EER and SEER, but specific data are needed to compare these two energy ratings. The US government has passed regulations on the minimum SEER, with values increasing from SEER = 10 in 1992, to 13 in 2005, and 14.5 for units that are Energy Star certified. These are just minima, so some units could have higher ratings. For example, some ductless AC systems are now available that have SEER ratings of up to 42.

10.5 COOLING OPTIONS

The three main refrigerant-based or chilled air conditioning options are central ACs, wall and window units, and ductless units. In the dry southwestern states, houses are frequently cooled using evaporative cooling systems, which do not use refrigerants. In an evaporative cooler water is trickled down over porous materials with air drawn through that wetted material, resulting in water evaporation and cooling of the air. This cooling method is much more efficient than chilled air systems using a refrigerant, but on hot days the amount of cooling possible is limited. When the whole home is to be cooled using a chilled air system, central AC units are usually used avoiding hot and cold room air exchange, but generally more money is spent as the whole house is cooled. For example, in the hot and humid states in the United States (Fig. 10.3) about $500 is spent annually to cool a home using central air compared to ~$400 spent in homes that use individual units. Wall and window units cool less efficiently and therefore the total cost can be more than twice that of the central air unit on a per square foot basis (Fig. 10.3).

Central AC units require ducts in the house to carry air from the chiller into the rooms and separate ducts to return cooled air. These systems use two separate units with the outdoor unit venting the heat and the indoor unit dispersing the cool air to different rooms. Window units combine both processes into a single unit, but they also restrict the use of windows for allowing the inflow of air on days when chilled air cooling is not needed. Ductless AC units, popular in Asia for many years, are now becoming more common in the United States. Ductless AC units convey refrigerant in tubes to units located inside a home that then release cold air. They have the advantage of not

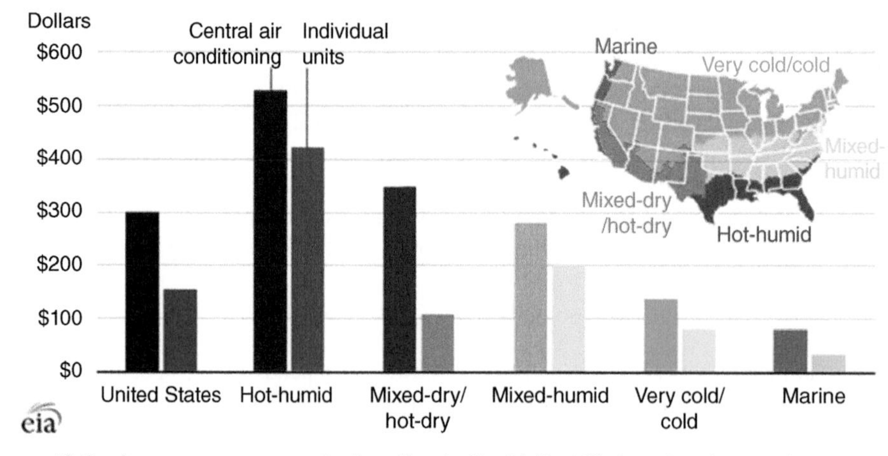

Figure 10.3 Average energy use by location in the United States showing costs per square foot. When using individual units, likely only parts of a home are cooled. *Source*: EIA (2018).

needing a complex open duct system inside the home. Although they can be expensive to purchase to retrofit a home that did not previously have a centralized AC system, such systems are usually less expensive to install into an existing home than a series of air ducts to convey these air streams with more conventional systems.

10.6 REFRIGERANTS AND GHGs

Liquid refrigerants used in ACs and other cooling devices are critical to efficiency and performance, but the use of these different refrigerants has important implications for human safety and global climate change. Early refrigerators used toxic or flammable gases such as ammonia, methyl chloride, and propane, which were considered to be dangerous if they leaked out of the system. These chemicals were therefore replaced by chlorofluorocarbons (CFCs) which were safer to handle and use. However, CFCs were banned in the 1990s due to their impacts on depleting the ozone layer. CFCs were replaced by hydrochlorofluorocarbons (HCFCs), but these are being phased out for the same reason.

CFCs and HCFCs were largely been replaced by hydrofluorocarbons (HFCs) as these chemicals do not adversely impact the ozone layer. However, HFCs are extremely potent greenhouse gases (GHGs), ranging from being 1000–9000 times as potent as CO_2 as a GHG. The most common HFC is 1430× more damaging to climate change than CO_2 (Climate & Clean Air Coalition, 2020). Thus, although they are only 1% of total GHG emissions, their impacts can be far greater than CO_2 on climate change.

The huge impact of HFCs and other refrigerants still in use has led to their classification as being extremely important for drawing down GHG emissions by Project Drawdown (Table 10.1). Refrigerant management is the #4 most impactful method for drawing down GHGs, with the use of alternative refrigerants that do not contribute to climate change listed as #7. Taken together, these two activities could lead to a CO_2-equivalent reduction in GHGs of 101.2 Gt by 2050. This is about 10% of the total of 997.2 Gt of CO_2-equivalent emissions listed by Project Drawdown that need to be removed could be reduced between 2020 and 2050. Even if HFCs and their predecessors are no longer being sold, it is critical that we avoid the release of these existing chemicals that are currently present in old ACs, refrigerators, freezers, and other cooling devices into the environment.

If we convert the total amount of CO_2 removed over 30 y into units of C, based on a world population in 2050 of 9.4 billion, then this translates to 1.08 C for these two refrigerant GHG sources. It will be difficult to achieve a reduction in use and releases of refrigerants given that the

Table 10.1 The number in the first column corresponds to the solution listed by Project Drawdown.

No.	Solution	CO_2 (10^9 tons)	–C
4	Refrigerant management	57.7	0.62
7	Alternative refrigerants	43.5	0.46
	Total	101.2	1.08
—	Building/Heating solutions	71.6	0.77

The value of C is shown as a negative sign as these are carbon reductions normalized to a global population of 9.4 billion people in 2050.
With the number indicated to be in the order for that solution from highest to lowest.
Source: Adapted from Wilkinson (2020).

production, consumption, and emissions of these gases are growing at a rate of 8% per year. There are other factors related to buildings that will impact energy use for cooling, such as insulation and the use of smart thermostats. These and other solutions were listed in Chapter 9 in the discussion on heating and buildings. If summed up, these represent a further potential reduction of 71.6 Gt of CO_2, or 0.77 C based on the estimate for the population in 2050 (Table 10.1).

What are possible alternatives to HFCs? These include two of the chemicals originally used as refrigerants, propane, and ammonia, which are safe to use as long as they do not leak out of the system. Another alternative class of chemicals that is being investigated are hydrofluoroolefins (HFOs) which are alkenes containing fluorine but not any chlorine molecules. While HFOs do not appear to contribute to ozone depletion and have little potential to be GHGs, they are based on the use of highly fluorinated compounds. There is now widespread recognition of the danger of per- and poly-fluorinated compounds (PFAS) to the environment (NIH, 2020). While HFOs fit within the definition of PFAS compounds, HFOs and a number of other highly fluorinated substances have not yet been considered to be PFAS compounds (Mass.gov, 2019). The production and use of these types of highly fluorinated compounds will be a very large concern in decisions in moving forward with these chemicals as refrigerants compared to non-fluorinated chemicals.

The refrigerants used in home refrigerators in the United States is now changing to isobutane (also known as methyl propane), a four-carbon hydrocarbon (C_4H_{10}) that has been extensively used in Europe, China, and much of the rest of the world. The chemical does not cause depletion of ozone in the atmosphere nor is it a GHG. Isobutane has been used in some small refrigerators in the United States. In 2019, the US Environmental Protection Agency approved the use of larger amounts of isobutane in a single unit, thus enabling it to be used in larger home refrigerators. It is now expected that within a few years isobutane will completely replace all other refrigerants used in home refrigerators. The main disadvantage of this chemical is that it has an explosion risk but there have been extremely few reports of such events. Certainly, no chemical suitable for use as a refrigerant will be perfect in all aspects. The best choice of refrigerants for the present and the future will therefore be a compromise in some aspect of health, safety, and the environment, but for home refrigerators, the future looks bright for avoiding the use of chemicals that contribute to global warming.

References

CLIMATE & CLEAN AIR COALITION. 2020. *Hydrofluorocarbons (HFCs)* [Online]. Available: https://www.ccacoalition.org/fr/slcps/hydrofluorocarbons-hfc [Accessed May 21 2020].

DAVIS, L. 2014. *Better yellow labels* [Online]. Available: https://energyathaas.wordpress.com/2014/10/27/better-yellow-labels/ [Accessed May 20 2020].

EIA. 2018. *Air conditioning accounts for about 12% of U.S. home energy expenditures* [Online]. Available: https://www.eia.gov/todayinenergy/detail.php?id=36692 [Accessed October 13 2020].

U.S. ENERGY INFORMATION ADMINISTRATION. 2020. *Room air conditioners* [Online]. Available: https://www.energy.gov/energysaver/room-air-conditioners [Accessed May 19 2020].

MASS.GOV. 2019. *Per- and poly-fluorinated alkyl substances (PFAS): policy analysis (draft)* [Online]. Available: https://www.mass.gov/doc/draft-pfas-policy-analysis-november-12-2019/download [Accessed May 21 2020].

MOONEY, C. 2014. *Why it's not okay to have a second refrigerator* [Online]. Washington Post. Available: https://www.washingtonpost.com/news/wonk/wp/2014/11/26/why-its-not-okay-to-have-a-second-refrigerator/ [Accessed May 26 2020].

NIH. 2020. *Perfluoroalkyl and polyfluoroalkyl substances (PFAS)* [Online]. Available: https://www.niehs.nih.gov/health/topics/agents/pfc/index.cfm [Accessed May 22 2020].

RAYMOND, C., MATTHEWS, T. & HORTON, R. M. 2020. The emergence of heat and humidity too severe for human tolerance. *Science Advances,* 6, eaaw1838.

US ENERGY INFORMATION ADMINISTRATION. 2020. *How much electricity is lost in electricity transmission and distribution in the United States?* [Online]. Available: https://www.eia.gov/tools/faqs/faq.php?id=105&t=3 [Accessed May 14 2020].

WIKIPEDIA. 2020. *Seasonal energy efficiency ratio* [Online]. Available: https://en.wikipedia.org/wiki/Seasonal_energy_efficiency_ratio [Accessed May 20 2020].

WILKINSON, K. 2020. *The drawdown review: Climate solutions for a new decade*, Project Drawdown.

[illegible] *[illegible] and polyfluoroalkyl substances* (PFAS) [illegible]
[illegible] Accessed May 22 [illegible]

[illegible] MATTHEW, [illegible] M. 2020. The [illegible]

[illegible] INFORMATION [illegible]

[illegible]

[illegible]

CHAPTER 11

CARS

11.1 WHY CARS MATTER FOR CLIMATE CHANGE

The energy use by our own cars and other vehicles is amazingly large based on all our other energy uses. Cars with internal combustion engines (ICEs) consume a lot of energy inefficiently, and they can consume it quickly as they have large engines. Both ICEs and electric vehicles (EVs) need a source of energy that is large relative to its weight (high energy density), and engines (ICEs) or motors (EVs) that can produce a lot of power. In Chapter 3, we discussed energy use by sector in terms of the unit D, and we compared the ratio of energy that went into our intended use (services) to the energy that was wasted (rejected). Among 5 categories, transportation had the lowest conversion efficiency with only 21% overall conversion of energy used into services. In the next chapter, transportation as a complete topic will be discussed in terms of how we can evaluate our transportation infrastructure for improving energy efficiency and reducing carbon emissions.

In this chapter, the focus is on cars because just about everyone in the United States will choose to buy one, or more likely several, over their lifetime. For some people, a car is just a method of transportation so that you can go to work, run errands, or see your family or friends, and the specific vehicle is not that important. For many other people, however, choosing a car is more than just finding a vehicle for transportation: It might be about making a personality statement, or presenting yourself to your friends or others by the image conveyed when you are driving that vehicle; or it could be that you just like driving powerful or fast vehicles. These varied interests can lead to cars that are excessive in size, weight, or other properties that are far in excess of basic transportation needs.

The cost of buying and operating a vehicle is a concern for just about everyone. The total cost of owning a car is only partly due to its fuel consumption efficiency and how far you drive every year. The greater proportion of the cost of owning a car is due to a combination of factors that include insurance, depreciation, repairs, taxes, and fees. So why should a person care about a car in terms of fuel efficiency? Because there are a lot of cars out there and they are all very inefficient (but to different extents) in terms of fuel use for miles driven. Therefore, cars are a key factor in reducing global CO_2 emissions, and thus addressing climate change, as well as improving local air quality. In the sections below, fuel efficiency is examined in terms of the carbon emissions to see where the best path is to lowering CO_2 emissions and getting to a goal for zero use of fossil fuels. Spoiler alert on the bottom line: We have a long way to go!

Daily Energy Use and Carbon Emissions: Fundamentals and Applications for Students and Professionals, First Edition. Bruce E. Logan.

11.2 INTERNAL COMBUSTION ENGINES AND CARBON EMISSIONS

One way to understand the large amount of energy in the gasoline that we use to power most of our cars is to compare the energy in gasoline that we use in the US every day, to the daily electrical energy used for our home. The total amount of gasoline used in the United States in 2019 was 142 billion gallons, with 92% of that used for light duty vehicles defined as cars, sport utility vehicles, motorcycles, and light trucks (US Energy Information Administration, 2020). Assuming 328.2 million people, that is equivalent to 1.19 gal/d per person, or 18 D (assuming 15.2 D for a gallon of gasoline). For comparison, the average electricity use for a home is 13 D_e of electricity, or with 2.52 people on average in a home, 5.2 D_e per person. Thus, our energy use for gasoline is over three times the energy in the electricity used for a home (on a per person basis). The primary energy for our home would be 2.46 times that of the electricity, or 12.8 D_p, which is still smaller than the gasoline D value.

There is little that we can do to impact gasoline use overall in the United States, but we can consider our own gasoline use based on miles driven to go to work, or our how many miles we drive each year for our own car. On average, a person in the US commutes 16 mi/d (Harris, 2020). In 2018, the average miles per gallon for a new car was 25.1, and it was projected to increase to 25.5 in 2019 (U.S. Environmental Protection Agency, 2020). Using these averages, that would mean energy use of 9.7 D just for your daily commute to work. If we use the US annual average of 13,500 mi for our car, that is 22 D. These numbers for average gasoline use of 18 and 22 D are a bit different, but the estimate for of 25.5 mpg is for new cars not all light duty vehicles. Thus, it is best to use the average miles that you accumulate on your car every year, and you can calculate your energy use and carbon emissions for your own specific lifestyle.

Gasoline and Petrol, mpg and L/100 km

Cars in the US tend to be larger than those in many countries in the European Union (EU) and have much higher estimated gas consumption. To get a wider view on energy use for vehicles around the world, we also need to consider that most cars in the US currently use gasoline, while those in the EU use petrol (gasoline) or diesel. The amount of diesel used in some EU countries can be a substantial portion of fuel use for automobiles, as shown in Figure 11.1. Most countries in the EU, as well as others, remain highly dependent on either gasoline or diesel fuels. The use of alternative fuels, mostly biofuels, is apparent only in Poland, Italy, and Turkey for the listed countries. The number of cars using electric vehicles, which are included in alternative fuels, will need to drastically change across Europe in order to reduce carbon emissions.

Another consideration in evaluating fuel use for cars is understanding the way that fuel use is reported, as this varies for different countries. For example, in the United States, the fuel consumption is reported based on the distance you can go per volume of fuel used or in miles per gallon (mpg) of gasoline (English units). However, in the EU, the inverse is reported as a comparison of volume of fuel used per distance traveled, and metric units are used, so the car is rated in terms of liters of fuel used per 100 km (L/100 km). To convert between these mileage units, you can use a conversion factor of 235, as shown in Example 11.1 below.

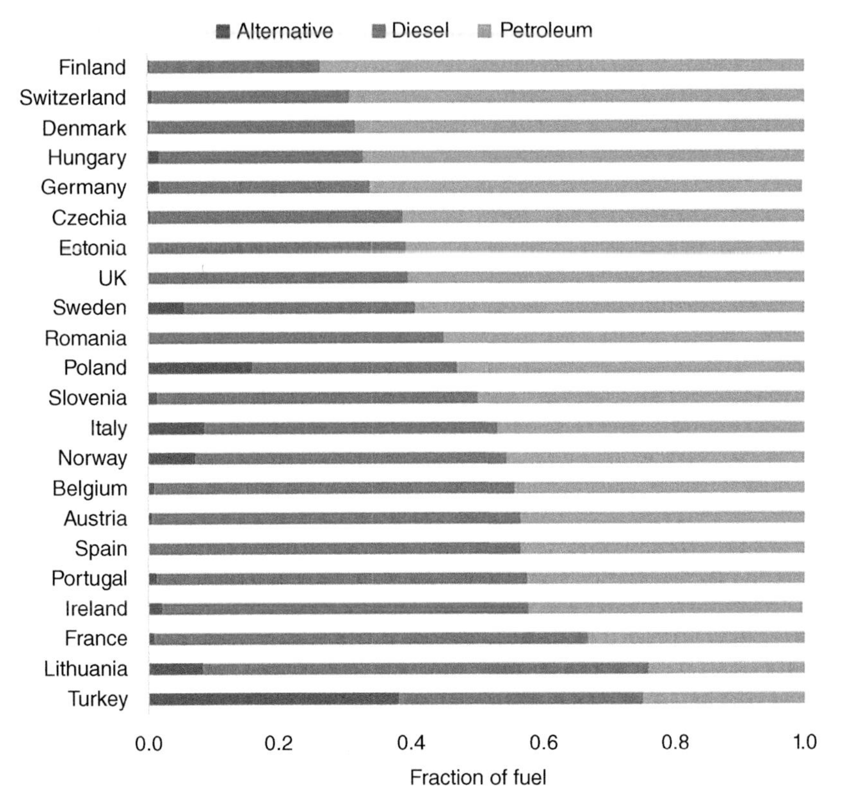

Figure 11.1 Comparison of energy sources for cars in the US. Alternative fuels include electricity, natural gas, biofuels, and others. *Source*: Adapted from Eurostat (2020).

Example 11.1

Fuel efficiency of cars is calculated in terms of mpg in the United States, but the EU and many other countries use L/100 km. Derive an equation to convert the units from mpg to L/100 km and then use that to convert 25.5 mpg to L/km.

We will set up the equation to use V_L, defined as the volume of fuel in liters to go 100 km, based on M, the mpg rating for the car.

$$V_L(L) = \frac{1}{M\frac{\text{mi}}{\text{gal}}} \frac{3.78\ \text{L}}{\text{gal}} \frac{1\ \text{mi}}{1.61\ \text{km}} 100\ \text{km} = \frac{235}{M\ (\text{mpg})}$$

Based on this equation, a car that gets 25.5 mpg would use 9.2 L to go 100 km. One quick way to remember this conversion factor of 235 is to approximate it as 234 so the numbers are in sequence. The difference between the two numbers will not impact the significance of the calculation.

In the EU, CO_2 emissions from cars are reported in units of g/km, and therefore, it is not the mileage that is reported on the sticker when the car is for purchase from a dealer but instead how much CO_2 is emitted. This reflects an attitude in the EU that the fuel is not as important as the overall CO_2 emissions. However, it follows that the lower the amount of CO_2 emitted the better the mileage rating. There are strict methods to measure CO_2 emissions in the United States and Europe, and some automobile manufacturers have tried to manipulate conditions to make overall emissions appear lower. For example, there was a large scandal in 2015 when it was discovered that company Volkswagen was using programming and methods to effectively manipulate how the vehicle responded in those tests so that the car was rated lower than the actual emissions. In 2015, the goal for CO_2 emissions was to be <130 g/km of CO_2, and recently, this has been updated to set a target of 95 g/km of CO_2. In Example 11.2 below, we explore how these numbers relate to fuel efficiency based on average properties of gasoline and diesel fuel.

Example 11.2

The European Union had a goal for CO_2 emissions in 2015 of 130 g/km, and for 2021 the target is 95 g/km. What is the fuel consumption to meet these goals for gasoline and diesel fuel in L/km and mpg?

From Table 4.2, we see that gasoline produces 19.6 lb/gal and diesel 22.4 lb/gal of CO_2. We can use these data and some unit conversion to get the fuel consumption. For gasoline, we have:

$$130\frac{\text{g}}{\text{km}}\,\frac{\text{gal}}{19.6\,\text{lb}}\,\frac{2.2\,\text{lb}}{1000\,\text{g}}\,\frac{3.78\,\text{L}}{\text{gal}}\,100\,\text{km} = 5.5\frac{\text{L}}{100\,\text{km}},\ \text{or 43 mpg}$$

$$95\,\text{g/km}\,\frac{\text{gal}}{19.6\,\text{lb}}\,\frac{2.2\,\text{lb}}{1000\,\text{g}}\,\frac{3.78\,\text{L}}{\text{gal}}\,100\,\text{km} = 4.0\frac{\text{L}}{100\,\text{km}},\ \text{or 58 mpg}$$

The mpg values are calculated from the L/100 km results using the equation in Example 4.1. Similarly, for diesel, we have 4.8 L/100 km = 49 mpg for the goal of 130 g/km, and 3.5 L/100 km and 67 mpg for the goal of 95 g/km.

The fuel consumption values calculated based on the EU standards seem extraordinarily high compared to those for the United States, where an average car gets 25.5 mpg. Why are the goals for the United States and the EU therefore so different? First, the EU has a set of methods and conditions for measuring CO_2 emissions that are different from those used by the US Environmental Protection Agency (EPA) to evaluate fuel use. The averages reported by the EU and the United States are based on tests that include a variety of operational conditions including speed, starts, stops, and so forth, all of which will impact emissions and fuel use. Second, cars in the EU tend to be smaller, and consequently, they have smaller engines than those in the United States. Another factor to consider when looking at mpg ratings in EU publications is that the mpg is often based on using an imperial gallon. While a gallon in the United States is equal to 3.78 L, the imperial gallon is 4.55 L. Based on these definitions, a conversion equation based on imperial liters is $V_L = 282/M$. For example, a Prius rated at 52 mpg in the United States would be equivalent to 4.5 L/100 km, or 62 mpg in imperial gallons. For these reasons, EU estimates of mileage are often about 20% larger than those in the United States, with the US estimates considered to be more reflective of actual mileage (Ingram, 2014).

11.3 UNDERSTANDING ENERGY USE BY ELECTRIC CARS

The *performance* of electric vehicles (EVs) is evaluated in several different ways, but the preferred units are kW for power and kWh for energy storage. Cars with ICEs are rated in hp for power. These values for ICEs can be compared to EVs by using the conversion factor that 1 hp = 0.746 kW (1.23 hp = 1 kW). The typical US car has a 120 hp engine, equivalent to 90 kW. A hybrid Prius that gets 52 mpg (2019) has an engine 71 hp or 53 kW, with a combined (engine and electric motor) equal to 121 hp, or about the average power of a US car sold today. An electric vehicle like the Nissan leaf rated at 80 kW would have a corresponding 107 hp motor, while a Tesla with a performance package of 615 kW could reach an amazingly high power output of 750 hp! For comparison, a Chevrolet Corvette Stingray has around 495 hp (and the Tesla is faster in a 0–60 mph acceleration test).

The *range* of an ICE vehicle is a combination of the fuel tank capacity and the fuel use. The Prius, which has a 9-gal tank and averages 52 mpg, therefore has an estimated range of 468 mi. The actual distance before needing to refuel a car depends on driving speed, terrain, starts/stops and so forth. For EVs and hybrid vehicles, the range is more difficult to evaluate as energy can be recaptured during driving. Thus, a car such as the 2016 Nissan Leaf which has a 30 kWh battery and rated at 30 kWh/100 mi, has an advertised range of 107 mi rather than 100 mi. A 2018 Hyundai Ionic with a 28 kWh pack was given a 25 kWh/100 mil rating but a range of 124 mi (rather than 112 mi based on the average mileage rating and battery energy capacity).

Apparent Efficiency Based on mpg-e

The fuel efficiency for an EV is also given in terms of mpg-e, or the miles that the vehicle can travel in EPA tests using the same energy contained in one gallon of gasoline. This rating of a vehicle in terms of mpg-e depends on the energy defined for a gallon of gasoline, and this energy content varies. The US EPA defines 1 gal of gas to have 115,000 Btu, or 33.7 kWh. However, the value given by the US Energy Information Agency (EIA) in 2019 for 1 gal of gasoline was 120,333 Btu or 35.3 kWh. This latter definition of 35.3 kWh is what we have been using so far in this book for a gallon of gasoline. While the EIA value can change from year to year, the EPA has a fixed number for all years to maintain consistency in the ratings from year to year. Thus, the mpg-e values will vary a little depending on the Btu estimate for the fuel. For the calculations below using EPA ratings, 33.7 kWh is used for a gallon of gasoline compared to 35.3 kWh used in other calculations in this book.

Some examples of fuel efficiencies based on mpg-e ratings and kWh/100 mi using the EPA ratings are shown in Table 11.1. One of the first and most popular EVs, a Nissan Leaf, has a rating of 30 kWh/100 mi and an equivalent of 112 mpg-e based on energy in 1 gal of gasoline (2016 vehicle). A new Tesla Model 3 (2020) has an impressive 141 mpg-e, while the Tesla Model S is rated at only 89 mpg-e. New automobiles with combustion engines in the US average 25.5 mpg, with the Toyota Prius hybrid having one of the best fuel ratings by EPA of 52 mpg. If we convert these into kWh/100 mi equivalents, then the average car would achieve 132 kWh/100 mi and the Prius would be 64 kWh/100 mi.

Actual Fuel Efficiency Based on mpg-eE

While these mpg-e values are impressive relative to ICE vehicles, the mpg-e rating lacks consideration of the *primary energy* needed to fuel these vehicles, as well as energy losses when charging

Table 11.1 Fuel equivalent ratings for electric vehicles in kWh and mpg equivalents (mpg-e) based on a gallon of gas having 33.7 kWh (the value assumed by the US EPA).

Automobile	kWh/100 mi	mpg-e	mpg-eE
Tesla Model 3-standard range (2020)	24	141	45
Hyundai Ionic electric (2017)	25	136	44
BMW i3 (60 Ah, 2016)	27	124	40
Tesla Model 3-Long range (2020)	28	121	39
Chevrolet Bolt (2017)	28	119	38
BMW i3 (94 Ah, 2017)	29	118	38
Volkswagen Golf (2016)	29	116	37
Nissan Leaf (2016)	30	112	36
Tesla Model S (85 kWh, 2015)	38	89	29
Toyota RAV4 (2014)	44	76	24

The mpg-eE is the back-calculated primary energy used based on efficiency of electricity production from power plants (40.6%) and transmission losses (7.1%) and a charging loss of 15% for an overall efficiency of 32.2%.
Source: Data from Wikipedia (2020).

the batteries. As was discussed in Chapter 2, the average efficiency of the electrical grid based on primary fuels is 40.6%, with 32.2% for coal plants and 43.1% for natural gas. The electricity once produced needs to be transmitted to the user, and these distribution and transmission losses are 0.94 EJ (not including onsite use), so that 13.28 EJ of electricity is delivered for 14.22 EJ produced by the electrical power industry (Table 2.6). Therefore, for every 1 kWh consumed to charge an electric vehicle 1.071 kWh needs to be produced. Using the average efficiency of a power plant of 40.6%, that means 2.64 kWh (=1.071 kWh/0.406) was consumed in primary energy (for fossil fuels). Therefore, we can consider the overall efficiency of fuel use to point of consumption as 37.9%. The other issue to consider with electric vehicles is charging efficiency. Batteries can provide direct electrical power storage although losses result in charging/discharging (round trip) efficiencies of 85% for lithium batteries (Mongird et al., 2019). If we include this addition 15% loss, then the efficiency is further reduced to 32.2%

The mpg-eE is defined as the energy efficiency of the car based on using the primary energy sources from the US electrical grid and accounting for energy losses including battery charging. With an overall energy efficiency of 32.2%, we can recalculate the mileage efficiencies for electric cars corrected for these losses, by multiplying the mpg-e values by 0.322, as shown in Table 11.1. With this revised efficiency, the best electric vehicle at only 45 mpg-eE equivalent, which is the same as a Toyota Prius hybrid vehicle. All the other vehicles fall below this 52 mpg-eE value. However, all these mpg-eE values are better than the average new combustion engine vehicle in the United States which has an EPA rating of 25.5 mpg. If charging losses are ignored then the equivalent mpg-eE for the Tesla Model 3 would be 53, or about the same as a Prius. Improved efficiencies in battery charging (which is possible by a slower charge cycle) still only brings the best electric vehicle to be equivalent to the Prius.

Another factor for ICE vehicles is the energy used to produce the gasoline or diesel. Refineries are ~90% efficient, and energy is needed to transport the fuels from refineries to fuel stations. The energy use at a refinery is a mixture of fuel use and some electricity. If we assume the 90% refinery efficiency with 15% of the energy loss due to electricity used (Levi, 2011), then the 33.7 kWh assumed by the US EPA for a gallon of gas requires about 38.0 kWh, or an overall 13% loss of

4.6 kWh/gal of gasoline. However, this calculation is further complicated by adding 10% ethanol into the gasoline, which has a much different energy demand. Given uncertainties in these numbers, and an unknown about of energy for transportation of the fuels, it seems reasonable to conclude approximately an overall loss of 15% of primary energy for producing a gallon of gasoline, or an overall energy efficiency for gasoline to point of use of about 85%. Thus, the Prius number of 52 mpg-eE would be reduced to 44 mpg-eE. This percentage loss is therefore not negligible although it is quite a bit smaller than that of the 32.2% energy losses for electricity to an EV.

We conclude from this analysis that on average EVs are more energy efficient than average cars sold in the United States that use combustion engines, but that the most efficient combustion engine car (Prius, rated at 52 mpg-eE but 44 mpg-eE with refinery losses) has about the same efficiency as the best EV (Tesla Model S, long range, 45 mpg-eE) running on electricity produced by power plants in the United States. These efficiencies would be different in other countries depending on the makeup of the fuels and energy sources for the grid. For example, if only coal is used then the efficiency based on delivery to the battery decreases to 26.8% ($0.322 \times 0.969 \times 0.86$), or for the Tesla, this translates to 38 mpg-eE. However, ICEs are currently tied to fossil fuels while EVs have the potential to have their energy source be completely provided by renewable energy sources that produce electricity.

11.4 CARBON EMISSIONS FROM CARS WITH DIFFERENT FUELS

The analysis of fuel efficiency for the different ICE cars or EVs addresses relative energy use, but it does not fully address carbon emissions from a perspective of fuel to final CO_2 emissions. As we saw in Chapter 4, one gallon of gas is equivalent to 9.8 C in terms of carbon emissions, while subbituminous coal had 12.6 C for the same energy, and natural gas was only 6.82 C. For electricity from the US grid, the D:C ratio is 2.40 based on the fossil fuels used to produce that electricity. Electricity production for the electric vehicle will vary considerably depending on the energy source used to produce that electricity. Another factor is the efficiency of electricity production, with fossil fuel plants averaging 32.2% compared to 40.8% for the average based on the energy sources for the electrical grid. As noted above, transmission and charging efficiencies will lower these overall efficiencies. In the example below, we explore a range of possibilities for carbon emissions based on different vehicles and energy efficiencies.

Example 11.3

(a) Calculate CO_2 emissions in terms of the unit C for an average new car in the United States using gasoline (25.5 mpg), and a Tesla 3 (24 kWh/100 mi) electric vehicle, assuming a person drives 37 mi/d (13,500 mi annually) and electricity produced by the US grid with an overall delivered efficiency of 32.2%. (b) What would be the CO_2 emissions for the Tesla 3 from a high-efficiency natural gas plant, assuming a 63.5% fuel efficiency and transmission (1.071%) and charging (14%) losses.

(a) For the gasoline-fueled car, assuming the EPA defined energy content of gasoline, and 9.8 C for a gallon of gas (Table 4.5), we have

$$E = 37 \frac{\text{mi}}{\text{d cap}} \frac{\text{gal}}{25.5\ \text{mi}} \frac{33.7\ \text{kWh}}{\text{gal}} = 48.9 \frac{\text{kWh}}{\text{d cap}}$$

$$\text{Daily energy unit} = 48.9 \frac{\text{kWh}}{\text{d cap}} \frac{1\ \text{D}}{2.32\ \frac{\text{kWh}}{\text{d cap}}} = 21.1\ \text{D}$$

$$\text{Daily } CO_2 \text{ emissions: } 37 \frac{\text{mi}}{\text{d cap}} \frac{\text{gal}}{25.5\ \text{mi}} \frac{9.8\ \text{C}}{\frac{\text{gal}}{\text{d cap}}} = 14.2\ \text{C}$$

(b) For the Tesla 3, with D_e defined as the energy used in a day based on electricity used, and D_p the primary energy used to produce that electricity included, we have

$$\text{E} = 37 \frac{\text{mi}}{\text{d cap}} \frac{24\ \text{kWh}}{100\ \text{mi}} = 8.9 \frac{\text{kWh}}{\text{d cap}}$$

$$\text{D}_\text{e} = 8.9 \frac{\text{kWh}}{\text{d cap}} \frac{1\ \text{D}}{2.32\ \frac{\text{kWh}}{\text{d cap}}} = 3.9\ \text{D}_\text{e}$$

$$\text{D}_\text{p} = \text{D}_\text{e} \frac{1}{\eta} = (3.9) \frac{1}{0.322} = 11.9\ \text{D}$$

$$\text{Daily } CO_2 \text{ emissions} = 11.9\ \text{D}_\text{p} \frac{\text{C}}{2.4\ \text{D}_\text{p}} = 5.0\ \text{C}$$

(c) If the Tesla 3 is charged using electricity from a high-efficiency natural gas plant, we first need to calculate the delivered efficiency with the two losses. Then, we can calculate the carbon emissions using a D:C = 2.23 ratio for natural gas as:

$$\eta = (0.635)\,(1 - 0.071)\,(1 - 0.14) = 0.507$$

$$\text{Daily } CO_2 \text{ emissions} = (3.9\ \text{D}_\text{e}) \frac{1}{0.507} \frac{\text{C}}{2.23\ \text{D}_\text{p}} = 3.4\ \text{C}$$

The calculations presented in Example 11.3 show that there can be a wide range of carbon emissions based on these two scenarios of an average vehicle producing 14.2 C from daily use of the gasoline vehicle, to as little as 5 C using US grid electricity and 3.4 C if that electricity was supplied by a high-efficiency natural gas turbine plant. However, it should not be concluded that all electric vehicles are superior to gasoline engines in terms of carbon emissions. The results of additional calculation are provided in Table 11.2 for some other conditions to provide greater context on what other vehicles could produce. For example, a Toyota Prius hybrid with a 52 mpg average fuel consumption is calculated to have 13% less carbon emissions than a high-performance Model S (7.9 C) if the electricity is derived from US grid. Even the standard Tesla 3, which has the same equivalent fuel rating as the Prius (52 mpg), would have greater CO_2 emissions (11.9 C) than both of these cars if electricity was derived only from a coal fuel power plant. However, if the Tesla 3 is charged using electricity from the US grid, then it would have a 5.0 C, or about 29% less

Table 11.2 Emissions in units of C based on 37 mi/d (13,500 mi/y).

Automobile	Energy source	kWh (gal)	D_e	D_p	C
Average car	Gasoline	48.9 (1.45)	—	21.1	14.2
Prius	Gasoline	23.9 (0.71)	—	10.3	7.0
Tesla S (85 kWh)	Electricity/US grid	14.1	6.1	18.9	7.9
Tesla 3-standard	Electricity/Coal PP	8.9	3.8	14.3	11.9
	Electricity/US grid	8.9	3.8	11.9	5.0
	Electricity/NG-HEPP	8.9	3.8	7.6	3.4

For the electric vehicles, D is calculated in terms of electricity or primary fuel indicated, with % production for a coal plant to charged battery, 32.2% using US grid energy, and 50.7% efficiency for a high-efficiency combined cycle natural gas electricity plant. For the gasoline energy sources, D and C values would be ~15% higher if gasoline production and delivery energy losses were included.

carbon emissions. If electricity that powered the Tesla S was produced by a high-performance gas turbine that had an efficiency of 63.5%, for a delivered overall efficiency of 50.7%, then the carbon emissions would be the lowest of all cases with 3.4 C.

Solar Power for Your Car

If you can obtain solar power for your home, either by contracting with a supplier of renewable energy to your electric grid or through your own home solar panels, then the decision should be to drive an electric vehicle for all D values of the electric vehicles. However, the car with the lowest D_e would need the fewest photovoltaic panels on your home. The daily energy output for a solar panel varies depending on your location in the country (as well as how well the panel is oriented toward the sun), but for a 320 W panel in the northeastern United States, you can expect around 0.58 D per panel (4.2 h of equivalent maximum sunlight, Fig. 6.1) (US Energy Information Administration, 2020). For the Tesla 3, you would therefore need ~7 panels to supply electricity every day for your car, compared to ~11 panels for the Tesla S. These would be photovoltaic panels in addition to those needed to supply your home with electricity (~30 panels). With a current installed cost of ~ $1000 per panel, that is a large investment, but the lifetime of the panels (25 y) can help to justify this independently of the environmental benefits!

Battery Storage for Electric Vehicles and the Grid

One big challenge for powering a car directly from electricity produced by our home solar panels is that your car is likely to be where you work during the day and not charging at home when the sun is out. However, if you are putting electricity into the grid while you are at home and charging your car at work or from the grid when you get home, overall, this can average out (but it might not be as economical as charging it at home during the day). One way to avoid the need to use grid energy is to have a home battery system that can be charged by your solar panels, although that does add significantly to the cost of your system. Alternatively, you might be able to fully charge your car on weekends, depending on the amount of use of the car during the week versus weekends, and therefore, you might not need additional electrical power storage.

The battery in your car, and batteries in your home should you decide to use them, provide another opportunity to help balance energy use by the grid. In the future, you might be able to use your car battery as a storage battery to power your home. While not possible today, the possibility of a network of car batteries could help to provide a more stable electrical grid. Balancing the electrical load of users and energy supply is a big challenge even today, and it will become an enormous challenge in the future as more intermittent energy production sources are added into the grid.

11.5 HYDROGEN FUEL CELL VEHICLES (HFCVs)

In 2003, President George W Bush said that with a national commitment "our scientists and engineers will overcome obstacles to taking these (hydrogen fuel cell) cars from laboratory to showroom, so that the first car driven by a child born today could be powered by hydrogen, and pollution-free." However, the program established for this goal quickly collapsed in the United States when a new Secretary of Energy, Steven Chu, declared his lack of support for hydrogen-powered vehicles (Wald, 2009) by cutting funding for research and infrastructure development. Ironically, Chu admitted in 2012 that the program had merit and fuel cell vehicles had potential for impact. But by then the damage had been done to research and advancement of a U.S. hydrogen infrastructure for transportation. The rest of the world, however, did not stand still and continued to develop and promote hydrogen fuel cell vehicles (HFCVs).

Today, hydrogen fuel cells are used in forklifts, buses, and several automobile manufacturers sell, lease, or have developed prototype hydrogen fuel cell cars. The three currently commercially available HFCVs are the Toyota Mirai, which is mass produced only in the HFCV version, as well as the Hyundai Nexo and the Honda Clarity. Because a hydrogen fuel cell produces electricity, these are electric vehicles powered by an onboard electrochemical engine rather than stored energy in batteries. These cars have a notable advantage of being able to be refueled in minutes, as opposed to hours for battery electric vehicles. By the end of 2019, Toyota had sold over 10,000 Mirai vehicles. This car can be refueled with hydrogen gas in as little as 5 min, and it has a range of over 300 mi, which is better than most battery electric vehicles but comparable to some of the newer Tesla models.

The main issue with HFCVs is the same as that for the battery models: What is the primary source of energy? Most H_2 gas is produced from natural gas by an endothermic (energy required) reaction, and thus, HFCVs would remain just as dependent on fossil fuels as electric cars if the source of the hydrogen was not changed. Photovoltaic production of hydrogen gas via water electrolysis would remove that dependence on of fuel cells fossil fuels, but the same can be said for electricity production to charge battery electric vehicles. HFCVs can be very efficient with about a 60% energy efficiency in conversion of hydrogen to electricity, relative to gasoline automobiles. A water electrolyzer is about 80% efficient in converting electricity to H_2, so the overall efficiency from solar production to H_2 (neglecting gas compression) would be about 42%. However, this would not be as good as an electric vehicle as only ~7% electrical power is lost in transmission to a site, and typically ~15% of the electrical power is lost in charging the batteries. Solid oxide hydrogen fuel cells have efficiencies that can reach 90% (Fuel Cells Bulletin, 2019) but they are so far not practical for vehicles due to the high operating temperatures.

Will HFCVs overtake electric vehicles in the United States? As things look now for light duty vehicles (cars and small trucks), most carbon-neutral plans are showing a very small growth rate for HFCVs compared to electric vehicles. For example, in the state of California, only a small percentage of the ~35 million vehicles are expected to be HFCV by 2045, while about half of those vehicles will be electric (Mahone et al., 2020) (Fig. 11.2). For heavy duty vehicles, such as trucks, however, slightly more HFCVs are expected to be on the road compared to battery electric vehicles.

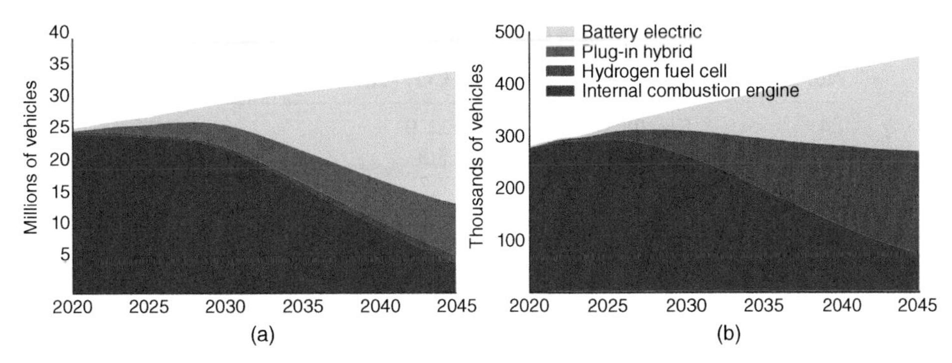

Figure 11.2 Planning document for the state of California on the fuels used for (a) light duty vehicles and (b) for heavy duty vehicles from 2020 to 2050. *Source*: Adapted from Mahone et al. (2020).

One clear disadvantage of the HFCVs is the limited number of fueling stations, with only 45 in the United States in 2018, compared to more than 20,000 electric charging stations at that time. Overall, HFCVs remain a technology that will continue to have important advantages in energy storage capacity and refueling rates that make them worth continuing to examine and develop as an alternative to combustion engine vehicles.

11.6 AUTOMOBILES OF THE FUTURE

As the world population grows and more people live in cities, the need for personal automobiles should diminish on a per-person basis. There are also increasing trends in new car purchases for electric and hybrid vehicles, but without strong curbs on emissions, carbon taxes, and efforts to curtail the sale of combustion-based vehicle emissions will continue to be a problem for both air quality (in terms of combustion by products) and the CO_2 concentrations in the air. The UK, France, and India announced in 2017 plans to eventually ban sales of new gasoline and diesel cars. Since then, many more countries especially in Europe have also announced plans to eliminate sales of new gasoline and diesel and other methods to limit the use of combustion vehicles in favor of electric or other zero emission vehicles. However, as discussed above, these efforts will not be sufficient to curtail CO_2 emissions unless the electricity grids in these countries also become carbon neutral.

Project Drawdown has listed many energy-related topics that can reduce CO_2 emissions through changes made that are related to transportation. Surprisingly, electric cars are listed in the category of energy and not transportation. Electric cars are listed here in Table 11.3 as cars are the focus this chapter. Also included are hybrid cars and car polling, which are also listed in Table 12.2 but discussed in the next chapter. For electric and hybrid car advances and further use, including carpooling, it was estimated by Project Drawdown that 27.5 Gt of CO_2 could be avoided over the period of 2020–2050. Normalizing that by the world population produces a value of −0.19 C. Hopefully, this is an underestimate of the true impact that could be made by banning sales of new automobiles that burn fossil fuels as many countries continue to advance plans to avoid their sales. Automobile manufacturers will also reduce options for cars that use ICEs as the markets shrink. The use of biofuels and hydrogen for vehicles could also have impacts that do not directly appear in the list of changes on the list of topics by Project Drawdown. Changes in the preferences of buyers toward zero emission vehicles, tax incentives, governmental regulations, and technology advances will all need to play a role in reducing the emissions from automobiles going into the future.

Table 11.3 Three Project Drawdown solutions related to cars.

No.	Solution	CO_2 (10^9 tons)	−C
24	Electric cars	11.9	0.13
33	Hybrid cars	7.9	0.08
34	Carpooling	7.7	0.08
	Total	**27.5**	**0.19**

The value of C is shown as a negative sign as these are carbon reductions normalized to a global population of 9.4 billion people in 2050. Note that these three items are also listed in Chapter 12 under the general topic of Transportation.
The number in the first column corresponds to the solution listed by Project Drawdown.
Source: Adapted from Wilkinson (2020).

References

EUROSTAT. 2020. *Most EU cars run on petrol* [Online]. Available: https://ec.europa.eu/eurostat/web/products-eurostat-news/-/DDN-20191024-1 [Accessed May 13 2020].

FUEL CELLS BULLETIN 2019. VTT's reversible fuel cell for highly efficient hydrogen production. *Fuel Cells Bulletin,* 2019, 12, 14-15.

HARRIS, D. 2020. *How far do Americans drive to work on average?* [Online]. Available: https://itstillruns.com/program-vue-keyless-entry-remote-7296360.html [Accessed May 13 2020].

INGRAM, A. 2014. *Why European gas-mileage ratings are so high -- and often wrong* [Online]. Available: https://www.greencarreports.com/news/1091877_why-european-gas-mileage-ratings-are-so-high--and-often-wrong [Accessed May 14 2020].

LEVI, M. 2011. *Do gasoline based cars really use more electricity than electric vehicles do?* [Online]. Council on Foreign Relations. Available: https://www.cfr.org/blog/do-gasoline-based-cars-really-use-more-electricity-electric-vehicles-do [Accessed February 15 2021].

MAHONE, A., SUBIN, Z., MANTEGNA, G., LOKEN, R., KOLSTER, C. & LINTMEIJER, N. 2020. Achieving carbon neutrality in California: PATHWAYS scenarios developed for the California Air Resources Board (Draft, August 2020).

MONGIRD, K., FOTEDAR, V., VISWANATHAN, V., KORITAROV, V., BALDUCCI, P., HADJERIOUA, B. & ALAM, J. 2019. Energy storage technology and cost characterization report.

U.S. ENVIRONMENTAL PROTECTION AGENCY. 2020. *Automotive trends report: Highlights of the automotive trends report* [Online]. Available: https://www.epa.gov/automotive-trends/highlights-automotive-trends-report [Accessed May 13 2020].

US ENERGY INFORMATION ADMINISTRATION. 2020. *Gasoline explained: Use of gasoline* [Online]. Available: https://www.eia.gov/energyexplained/gasoline/use-of-gasoline.php [Accessed May 13 2020].

WALD, M. L. 2009. *U.S. drops research Into fuel cells for cars* [Online]. Available: https://www.nytimes.com/2009/05/08/science/earth/08energy.html?auth=link-dismiss-google1tap [Accessed May 15 2020].

WIKIPEDIA. 2020. *Electric car EPA fuel economy* [Online]. Available: https://en.wikipedia.org/wiki/Electric_car_EPA_fuel_economy [Accessed May 14 2020].

Wilkinson, K. 2020. *The drawdown review: Climate solutions for a new decade*, Project Drawdown.

CHAPTER 12

TRANSPORTATION

12.1 MY ENERGY USE FOR TRANSPORTATION

Before we consider how much energy different modes of travel consume, we first start with the average energy used for a person to walk or run a mile. Walking is good exercise at any age both for body and mind, especially if you walk in a nice place. Running is even better as you are more active, and it helps your heart and mind as well, but you will not burn as many Calories as you might expect running versus walking. While you are running, you are burning calories approaching twice what you do while walking, so if you cover the same distance you use more power to get there by running compared to walking. An average person with a weight of 160–180 lb will use about 100 Cal to walk 1 mi, with less Calories used by a smaller person (65 Cal/mi for a 120 lb person), over an average period of time of about 15 min. In the same period, you can use 15–30 Cal (60–130 Cal/h) sitting to 25–50 Cal (100–200 Cal/h) standing. Using these numbers, we can calculate our mpg equivalent based on primary energy, or mpg-eE, for the same energy as used in a gallon of gasoline, as shown in the example below.

Example 12.1

Assume a person expends 100 Cal to walk a mile. (a) Calculate the mpg-e for a person to use the same energy as a gallon of gasoline based on units of D. (b) For a 140 lb person, it is estimated that you use 7.6 Cal/min walking compared to 13.2 Cal/min running. Compare the energy used in 15 min to run a mile in 9 min to that needed to walk a mile in 15 min.

(a) Since 100 Cal is a fraction of the 1 D of energy that we eat to power the human body, and there is 15.2 D in a gallon of gas, then by the ratio of these numbers, we have

$$\frac{1\ \text{mi}}{100\ \text{Cal}}\ \frac{2000\ \text{Cal}}{1\ \text{D}}\ \frac{15.2\ \text{D}}{\text{gal}} = 300\ \text{mpg-e}$$

From this result, walking 1 mi consumes 100/2000 = 0.05 D, so therefore a human could (theoretically) walk 300 mi using energy equivalent to 1 gal of gasoline. This result is slightly lower than another estimate based on a 150 lb person of 360 mi (Wikipedia, 2020a).

Daily Energy Use and Carbon Emissions: Fundamentals and Applications for Students and Professionals, First Edition. Bruce E. Logan.

(b) The energy needed to walk or run a mile is calculated as:

$$\text{Walking: } \frac{7.6\,\text{Cal}}{\text{min}}\,15\,\text{min} = 114\,\text{Cal}$$

$$\text{Running: } \frac{13.2\,\text{Cal}}{\text{min}}\,9\,\text{min} = 118\,\text{Cal}$$

The Calories for running are a little higher than the ~100 assumed for part a due to different assumptions. However, we also note that the time is not equal for the runner, as only 9 min are used leaving 6 min for another activity. Assuming that when the runner is done that person sits for the next 6 min using ~100 Cal/h, we then need to add:

$$\text{Sitting after running: } \frac{100\,\text{Cal}}{60\,\text{min}}\,6\,\text{min} = 10\,\text{Cal}$$

Therefore, the totals for the same amount of time are as follows:

$$\text{Walking: } 114\,\text{Cal} = 0.057\,\text{D}$$

$$\text{Running, sitting: } 128\,\text{Cal} = 0.064\,\text{D}$$

The result is that the runner does consume slightly more Calories or D, which in this case is calculated to be 12%. Compared to just sitting (25 Cal = 0.0125 D), walking is 4.6x larger and running is 5.1x larger.

12.2 ENERGY USE FOR TRANSPORTATION OPTIONS

When you are in a city, you can spend a lot more time walking around or standing while you run errands, shop, and visit different locations in the city. As we saw in the previous example, walking consumes about 4.6× more D than sitting, but even standing consumes ~50% more energy than sitting. To get to the city, or to move around in the city over longer distances than you might want to walk, you will need to consider other transportation options. The best option for getting around cities from both an energy and safety perspective is a bicycle (assuming there is sufficient and safe room from cars). Assuming 550 Cal for 10 mi (1.8 h for an average person) (Ketchum, 2019), this can be calculated to be 550 mpg-e.

The best-motorized option after walking or biking in a city is intercity rail, based on the mpg-e equivalents for many different modes of powered transportation, assuming a certain number of people for each method (Table 12.1). The D value indicates how much energy is consumed for a person to travel on the equivalent of a gallon of gasoline, for just one mile. The motorized or electric methods of transportation span 0.27–0.59 D, meaning that 1 D of energy will get you about 1.7–3.7 mi in a mechanical device, compared to 20 mi walking or 36 mi riding a bicycle. Kick scooters (see below) can have the highest energy efficiencies, but they are not considered to be a safe mode of travel.

The number of people per vehicle greatly impacts the energy efficiency of these different modes of traffic. For example, the top hybrid gasoline-powered vehicle that gets 52 mpg is surpassed only by intercity rail, with other rail forms (including transit and commuter rail) having higher energy demands (Table 12.1). However, a car with an average fuel rating of about 26 mpg

Table 12.1 Calculated D values for different modes of transportation based on miles per equivalent energy in a gallon of gasoline (mpg-e) and the assumed number of people during transit.

Vehicle type or travel mode	mpg-e	Persons	kWh (mi^{-1})	D (mi^{-1})
Bicycling[a]	550	1	0.064	0.028
Walking[a]	300	1	0.12	0.051
Intercity rail	2.76	20.5	0.62	0.27
Car-best hybrid/elec.[a]	52	1	0.68	0.29
Airlines (domestic)	0.46	113.2	0.68	0.29
Transit rail	1.99	25.8	0.69	0.30
Motorcycles	42.20	1.2	0.72	0.31
Commuter rail	1.42	31.6	0.79	0.34
Cars-avg (1.6 people)	25.62	1.6	0.89	0.38
Light trucks	19.57	1.8	0.98	0.42
Transit buses	3.29	9.1	1.17	0.51
Cars-avg (1 person)[a]	25.62	1	1.38	0.59
Demand response	6.01	1.4	4.13	1.78

[a] Calculated here for the stated assumptions of mpg-e.
Source: Data from US Department of Energy (2020).

with only 1 person is at the bottom, with 0.59 D/mi. Another surprising result is that airplanes flown on domestic routes can be as energy efficient as other modes of travel. The high D rating requires very full planes, but this has been true in the US except for times after the 911 terrorist attacks and certain times during the COVID 19 pandemic. The relationship between driving and flying is further explored in the next section.

The last category, referred to as demand response by the US Department of Energy (DOE), is travel to destinations defined by the travelers when they ask (or demand) travel assistance and not travel to only specific destinations at defined times. While on demand travel can include taxis, it also includes certain shuttle buses and other vehicles. One important form of demand response travel is for areas where normal bus travel is not considered to be economical or for passengers with disabilities who require special assistance or more time to get into or out of the vehicle. These special transport conditions may require a larger bus with special lifts. This on demand category can also include special airport shuttles with variable numbers of passengers from different locations going to the same destination. The nature of this travel means that the vehicle will likely travel empty or not go by the most direct route for any single passenger. Therefore, demand response travel is the most energy intensive of all methods, with a rating of 1.78 D.

Ride Hailing

Demand response travel includes many categories of travel, but one form that is used by many travelers these days is ride hailing companies, such as Uber and Lyft. Does transport using these companies help transportation congestion and energy use? The simple answer is that they do not: This form of travel increases overall energy use and leads to greater urban traffic congestion. Data in Table 12.1 suggest that Demand Response uses 3.7 times more energy than a car with 1.6 people, and 2 times more energy than a car with one person. A study by the Union of Concerned Scientists (UCS) on ride hailing by Uber and Lyft showed that ride hailing trips produced 47% more carbon emissions on average than those same trips in a private car. Furthermore, when considering the range of options for these trips, such as walking, biking, buses, and others, they found that ride-hailing trips were 69% more polluting than the other options they could choose. The main reason for the added pollution was the time the cars spent driving around looking for a fare or going from a current location to a fare location.

Ride hailing can also contribute to traffic congestion, leading to more energy consumption and pollution in getting from place to place (Liang, 2020). An analysis of ride hailing companies, also known as transportation network companies, showed that weekday vehicle hours increased traffic delays by 62% compared to a prediction of 22% delays in the absence of such transport options in San Francisco between 2010 and 2016 (Erhardt et al., 2019). Clearly, ride hailing is a convenience and a benefit to many people, but it comes at an increased energy and pollution cost compared to other travel options. Electrification of on-demand provider vehicles can help (Bauer et al., 2020), but not if the electric vehicles have mpg-eE ratings equivalent to mpg ratings for internal combustion engine cars. In addition, charging stations would need to be conveniently located to avoid long distances to recharge the vehicles.

Electric Kick Scooters

One new and quite popular mode of transport in cities in the United States is electric kick scooters. These are estimated to have an energy efficiency of several hundred to a thousand mpg-e. By one estimate, 1 kWh was needed to go 82.8 mi (Tillemann and Feasley, 2018) or 2900 mi on the energy in 1 gal of gas (35.3 kWh). Converting from electrical to primary for the US grid, that would be 1200 mpg-eE = 0.013 D_p/mi. This energy requirement places kick scooters much higher in terms of energy efficiency than either walking or riding a bicycle. However, riding the scooter requires greenhouse gas (GHG) emissions based on making additional electricity and the exercise of walking and riding a bicycle bring additional health benefits. These energy efficiencies for kick scooters also do not include pickup and recharging of the scooters, so the actual energy benefits will be lower than those just based on a fully charged scooter.

While the energy advantages of an electric scooter may be clear relative to other gas or electric vehicles, safety is a very important issue for all these devices. Scooters are dangerous to ride, with 2.4 – 18× more people injured using scooter sharing than bike sharing. In a study in Portland, OR, the main cause of non-fatal injury was falling (83%) compared to collisions with a car (14%) or a pedestrian or another scooter (3.4%) (Electric scooter guide, 2020). Most people riding an electric kick scooter do not wear helmets, which greatly increases the possibility of serious injuries.

Smart Transportation

Reducing traffic congestion can improve mileage efficiency in a car and extend vehicle range and reduce air pollution. Timing light changes, restricting the rate cars flow onto crowded highways or enter cities, and using google maps to avoid congestion areas and accidents can all reduce energy consumption and improve the quality of the trip. What else can be done? Simply a more relaxed attitude while driving can help. A study by Oak Ridge National Laboratory showed that aggressive driving, with excessive speeding and breaking, can reduce gas mileage by 10–40% in stop-and-go traffic, and 15–30% in highway traffic (Thomas et al., 2017). Intelligent traffic signals that respond to traffic flow and greater vehicle safety devices through onboard cameras that avoid accidents, could continue to reduce congestion. The use of smart systems in cities will not only reduce CO_2 emissions, their use will also ease the minds of commuters and help with overall safety hopefully by decreasing aggressive driving.

12.3 AIR TRAVEL AND HIGH-SPEED RAIL

Long-distance domestic travel can include several options, with cost and convenience often the major considerations. However, due to a need to reduce carbon emissions additional consideration should be given to energy use and carbon emissions from these different methods of travel. In Table 12.1, different travel options were compared based on mpg-e, and it was calculated that D values for air travel could be similar to those based on travel in a car. In this section, we can delve further into that comparison based on looking at how a specific trip would impact D use and C emissions for a car versus a plane, and then examine longer travel options by airplanes relative to emissions and options for rail travel.

Travel by Car or Plane for Short Distances?

If you are traveling across the country, you will likely fly to your destination. However, for shorter trips that could take 3–12 h by car, you might consider carbon emissions for air travel compared to your car. Carbon emissions are often reported for an air ticket. If these emissions are not listed on your ticket or reservation, you can use a carbon emissions calculator to estimate CO_2 emissions for your trip, such as the International Civil Aviation Organization (ICAO) (2020) calculator. The emissions from that website are based on averages for routes, plane sizes, and typical numbers of seats occupied, with data given for travel distance, and estimated CO_2 emissions based on total fuel use. CO_2 emissions for your automobile can be estimated by miles traveled (for example from google maps), the mpg for your car, and assuming CO_2 emissions for your vehicle based on 9.8 C for a gallon of gasoline. As shown in the example below, the number of people in the vehicle and the proximity of the airports to your destinations can have a big influence on the CO_2 emissions. It may also be useful to make a few additional calculations to check on whether the results seem reasonable, as shown in Example 12.2.

Example 12.2

You are planning a trip between University Park, PA, and Washington DC. Estimate the CO_2 emissions for your travel in terms of C units for travel by (a) plane, and (b) car using a typical 25.5 mpg car or a 52 mpg-eE vehicle.

(a) Using a CO_2 emissions calculator (International Civil Aviation Organization [ICAO], 2020), travel between the University Park (SCE [State College airport code]) airport and Washington DC airport (IAD [Washington Dulles airport code]) which has a direct flight, indicated: 133 mi, 2248 lb of fuel, and 83 lb of CO_2 (converted from km and kg to mi and lb). Using these data, the total C for a one-way trip is as follows:

$$83\frac{\text{lb CO}_2}{\text{d cap}}\ \frac{1\ \text{C}}{2\frac{\text{lb CO}_2}{\text{d cap}}} = 42\ \text{C}$$

We can provide a rough check on this number using the given fuel rating of 2248 lb and the density of jet fuel (6.71 lb/gal), by first calculating the total C for the flight of *n* people:

$$2248\ \frac{\text{lb}}{n}\ \frac{\text{gal}}{6.71\ \text{lb}}\ \frac{21.1\ \text{lb CO}_2}{\text{gal}}\ \frac{1\ \text{C}}{2\frac{\text{lb CO}_2}{\text{d cap}}} = 3534\ \frac{\text{C}}{n}$$

The number of passengers on this flight would therefore be

$$n = \frac{3534\ \text{C}}{42\ \text{C}} = 84$$

The capacity of the planes between these two designations varies from about 50 to 68 depending on the configuration, so it is not possible to fit 84 people on the plane. If we assume that the flight had 50 people, then based on the total C that would result would be 71 C for the flight.

(b) Travel estimated by car between the same two airports is 203 mi, so the total C values for these cars with two different mpg ratings are as follows:

$$\text{Car (25.5 mpg)}: 203\ \text{mi}\ \frac{\text{gal}}{25.5\ \text{mi}}\ \frac{9.8\ \text{C}}{\text{gal}} = 78\ \text{C}$$

$$\text{Car (52 mpg)}: 203\ \text{mi}\ \frac{\text{gal}}{52\ \text{mi}}\ \frac{9.8\ \text{C}}{\text{gal}} = 38\ \text{C}$$

This calculation suggests that the average car emissions (78 C) would only be slightly higher than that of the plane assuming the value of 71 C (not 42 C). Therefore, the car and airplane have about the same CO_2 emissions, which is different than the conclusion based on data in Table 12.1 that the airplane would use less fuel. In addition, if a 52 mpg car is used, the overall carbon emissions (38 C) are about half that of the plane. If there are two people in the high mpg car that would reduce carbon emissions per person to 19 C.

The results of this analysis suggest it is worthwhile to consider the specific trip and the performance of your car when planning a trip, and that the estimates provided by the carbon emissions calculator need to be carefully evaluated as well.

High Speed Rail Versus Planes

Trains that can travel at speeds of up to 250 kph or 160 mph are considered to be high speed trains, with the first train that reached these speeds operated in Japan in 1964. Today, there are high-speed trains running between many major European cities and in other countries, with China leading the way in the size of the network for these trains with over 50% of the installed miles (more than 18,500 mi in total). It is difficult to evaluate carbon emissions for these trains due to the different fuel sources that supply the electrical power. When evaluated based on energy use per person, trains are expected to use less energy than cars. However, it is not clear how to calculate the carbon emissions except in a few cases where it is known what the power source is for that train system. For example, Eurostar trains are run primarily using electricity produced in France (which has a large percentage of electricity provided by carbon-neutral nuclear power), and so a London-Paris trip is estimated to have 90% less carbon emissions than flying between these two cities (Wikipedia, 2020b). Germany as a country has 38% of electricity from renewable sources, but many German trains run off grid energy supplied by dedicated plants which may use fossil fuels.

High-speed trains run on electricity so as the electric grid shifts more to renewable energy, it can be expected that the carbon emissions will be lowered. However, airplanes are likely to remain tied to jet fuel for many decades to come. Therefore, trains when available can be the better option compared to planes when their routes are reasonably consistent with travel plans and destinations.

12.4 ENERGY FOR PAVEMENT MATERIALS

Another factor to consider in energy use is the energy needed for road construction. The Federal Highway Administration estimated that 1% of total GHG emissions in the United States was due to pavement construction, maintenance, and rehabilitation (Mcelvery, 2021). It is estimated that 2–4 TJ are needed to construct 1 km of highway (single lane), although that can vary from as little as 1 TJ to up to 17 TJ depending on the complexity of the site and construction process (Pavement Interactive, 2020). About 75% of the energy use used in producing the pavement, with the remaining 20% for energy for transportation costs and 5% for construction. If we assume 3 TJ/km, that is equivalent to 1.34×10^6 kWh/mi of a single lane road. On average, it was estimated that constructing a road would consume about the same amount of energy as used by cars traveling on the road for the first 2 y, so that the energy D to make the road would become a small fraction of energy use over much longer periods of time. We show how this energy use relates to D units and the number of cars that would travel that road in the example below.

Example 12.3

Assuming 3 TJ/km to construct a single km of road, (a) relate this amount to D assuming the energy was spread out over 1 or 20 y. (b) For this energy, how many cars would travel over that 1 mi of pavement to equal that amount of energy in 1 y?

(a) We convert this to D units assuming 1 y and 2.32 kWh/d cap = 1 D, as

$$3\,\frac{\text{TJ}}{\text{km y cap}}\;\frac{10^6\,\text{MJ}}{\text{TJ}}\;\frac{1\,\text{kWh}}{3.6\,\text{MJ}}\;\frac{1\,\text{km}}{0.62\,\text{mi}}\;\frac{\text{y}}{365\,\text{d}}\;\frac{1\,\text{D}}{\frac{2.32\,\text{kWh}}{\text{d cap}}}=\frac{1590\,\text{D}}{n}\;\frac{\text{cap}}{\text{mi}}$$

If we average this 1 mi of road over 20 y, that is about 79 D.

(b) For the number of cars, using 25.62 mpg from Table 12.1, or 0.59 D/mi per car, we have

$$n_c=\frac{1590\,\text{D}}{(1\,\text{cap})}\;\frac{\text{cap}}{\text{mi}}\;\frac{1}{\frac{0.59\,\text{D}}{\text{mi car}}}=2690\text{ cars}$$

Based on this calculation, that means about 2690 people (one person per car) would have used 0.59 D on their way to work for gasoline-based vehicles considering the energy needed for building that road.

Asphalt can be partly made using recycled tires in the form of crumb rubber. There is no federal law that requires this approach, although some states such as California require the use of tire rubber in 35% of the roads in the state, with a content of 18–20% by weight of the binder (Mcelvery, 2021), with the binder consisting of about 5–8% of the final weight. Tires must be shredded, with the steel structure stripped out, and then the remaining material ground into small particles (0.2–2 mm in size). These small crumb rubber particles can then be mixed with the binder used for asphalt, typically a black and gummy mixture of hydrocarbons, and the rocky aggregate to form the asphalt. The rubberized binder may increase the total amount of binder used but overall, it can lower the amount of binder needed over the life cycle of the pavement due to the improved durability and therefore the extended lifetime of the material.

As road pavement ages, it needs to be repaired and repainted. Eventually, the pavement materials will either be removed and replaced with fresh asphalt, or receive an overlay of new asphalt which has an energy cost of 100–300 MJ/m^2. This new asphalt overlay roughly works out to be about 1/4th of the energy needed for construction of the original road, or about 400 D averaged over a year. While less energy is needed for this process of road repair by an asphalt overlay, the energy demands are quite substantial considering the number of roads that must be maintained for our modern transportation systems.

One advantage of using asphalt for roads rather than cement-based materials is that it is possible to nearly completely recycle asphalt pavements. In the United States, about 80–95% of pavements are reused. With materials costs accounting for 75% of building a road in the production of the original asphalt, recycling represents substantial savings in new road construction. The thickness of the layers and soil supports can impact the overall energy consumption, but a comparison of methods found that recycling asphalt consumed ~110 MJ/m^2 of surface, compared to 140 MJ/m^2 for overlay and 300 MJ/m^2 or reconstruction (Thenoux et al., 2007). Thus, we estimate that recycling can reduce overall costs by about one-third compared to reconstruction and thus also save money and reduce the environmental impact of road materials.

12.5 WHAT FUELS WILL BE USED IN THE FUTURE FOR TRUCKS, SHIPS, AND PLANES?

Unless plans are made to reduce the use of fossil fuels, CO_2 emissions will not decrease to the low levels needed over the next few decades. In the United States, California is leading in plans to reduce CO_2 and other GHG emissions. The most recent report on plans for achieving carbon neutrality in the state of California include three scenarios:

- High carbon dioxide reduction (High CDR)
- Balanced
- Zero-carbon energy

The High CDR plan would reduce GHGs by 80% by 2045, and it relies heavily on CO_2 removal technologies, allowing for higher carbon emissions than the other plants. The High CDR plan would require 80 million tonnes (80 billion kg or 0.080 Gt) of CO_2 removal from the atmosphere using economical technologies that do not yet exist. The zero-carbon plan is the most aggressive one for avoiding CO_2 emissions, relying less on CO_2 capture technologies, while the balanced plan represents a compromise plan between the other two scenarios.

For the transportation sector in CA, all scenarios are indicated in Table 12.2 to require 100% of sales of light duty vehicles (cars and small trucks) to be battery electric vehicles (BEVs). However, a more detailed analysis of these plans, presented in Figure 11.2, showed that a mix of vehicles would be in use by 2035 due to continued use of vehicles sold before that time, as well as zero-emission hydrogen fuel cell vehicles (HFCVs). As the plans become more aggressive in reducing CO_2 emission, or going from the High CDR to the zero-carbon plan, there is a greater increase in the use of electricity or other carbon-neutral fuels for medium duty vehicle (MDV) and high duty vehicle (HDV). In addition, there are plans for increasing electrification of rail and aviation travel. Currently, there are no large commercial electric aircraft so this plan would require

Table 12.2 Mitigation plans for reducing CO_2 emissions for transportation options in California.

Scenario	Assumptions for each plan to meet the different goals
High CDR	100% BEV sales for LDV by 2035 100% BEV sales for MDV by 2040 45%/48% BEV/CNG sales for HDV by 2040, 7% diesel sales remaining for long-haul 50% rail electrification, no aviation electrification
Balanced	100% BEV sales for LDV by 2035 100% BEV sales for MDV by 2035 45%/48% BEV/HFCV sales for HDV by 2035, 7% diesel sales remaining for long-haul 75% rail electrification, no aviation electrification
Zero-carbon energy	100% BEV sales for LDV by 2030 100% BEV sales for MDV by 2030 50%/50% BEV/HFCV sales for HDV by 2030 75%/25% rail electrification/hydrogen, 50% of in-state aviation electrified

Abbreviations: BEV = battery electric vehicle, LDV = light duty vehicle, MDV = medium duty vehicle, HDV = heavy duty vehicle, CNG = compressed natural gas.
Source: Mahone et al. (2020).

rapid advances in technology development and implementation. Biofuels are being used for a limited number of commercial aircraft flights, but the production of those fuels is not expected to be sufficient now, or in the future, to support their widespread use in aircraft travel.

The annual CO_2 emissions for aviation, international aviation, and international shipping vary, but most estimates show that these emissions are very large and comparable by themselves in magnitude to the emissions of highly industrialized countries. For example, all air travel was estimated to account for 2.5% of global CO_2 emissions (in 2019), prior to the COVID 19 pandemic (Tabuchi, 2019). Based on the International Energy Agency (IEA) estimate for 2019 of 33.3 Gt (IEA, 2020), that would equal 0.88 Gt of CO_2. For 2017, there are more data available on global emissions than in 2019. In 2017, the global CO_2 emissions were estimated as 37 Gt of CO_2 by one source (Wikipedia, 2020c), and 32.7 Gt by the IEA. Using 37 Gt and the 2.5% that would suggest 0.93 Gt/y for all air travel (Wikipedia, 2020c). However, CO_2 emissions from international air travel were listed in 2017 as 0.54 Gt/y, or about 1.43%, which is substantially less than the 2.5% estimate. Assuming 0.93 Gt/y, emissions from aviation would be comparable to those of 1.32 Gt for Japan, 0.80 Gt by Germany, and 0.67 Gt for South Korea. However, if the estimate for international air travel (as opposed to all air travel) of 0.54 Gt/y (1.43%) was accurate, then that would place air travel between emissions of Canada (0.617 Gt/y) and Indonesia (0.51 Gt). The estimated carbon emissions from world shipping were 0.68 Gt/y. (Wikipedia, 2020c), placing the levels between those of Germany and South Korea.

It is too soon to see how the COVID-19 pandemic will impact air travel in the future, but typically, air travel has rebounded after disasters, such as the 911 bombing of the two trade towers in New York City which temporarily completely halted global air travel. Therefore, if the time for countries around the world to be fully vaccinated can be used to estimate the time for air travel to resume to pre-COVID 19 levels, it may take until 2024 to see extensive international aviation resume.

Fuels for Future Aircraft

The aviation industry is strongly committed to reducing CO_2 emissions. One way to reduce these emissions is to replace fossil fuel-based jet fuels with biofuels. Currently, only around 0.1% of fuels used for commercial aircraft are biofuels. Air carries advertise their use of biofuels, and are working to increase this percentage, but it does not appear that sufficient biofuels could be produced to meet the fuel demands for global aviation. Therefore, other fuels and plane modifications are being examined to sustain the aviation industry.

Batteries alone do not appear to be able to power aircraft as they weigh too much for most planes to use as an energy source. Jet fuel is typically about 26–45% of the weight of the plane at takeoff (Wikipedia, 2021). For example, a Boeing 737–800 at takeoff weighs about 79 000 kg, with 27% of this weight used for the fuel and 27% for cargo. The best batteries produced for cars are about 50–100 times heavier than the same energy contained in gasoline or jet fuel (Chapter 6). Therefore, replacing the fuel with batteries would increase the weight of the plane more than ten times, and so the plane could never even leave the ground.

Greater attention has recently been given to fueling aircraft and large ships with hydrogen or ammonia fuels. Airbus announced in 2020 plans to develop three different types of aircraft all powered by liquid H_2, which they indicated could be produced for commercial use by 2035 (Frangoul, 2020). These aircraft include one that looks like a conventional jet in operation today, and a second turboprop plane that is longer than typical planes today and that has more flexible wings (Fig. 12.1). The third prototype, called a "blended-wing body," has a wider body which appears to incorporate the wings more into the body of the plane, with a series of smaller turbines running along the tail end of the plane. The transfer and storage of liquid H_2 on planes also present significant technical challenges.

Figure 12.1 Three prototype H-fueled airplanes announced by Airbus. *Source*: Reprinted with permission from Airbus.

Any new aircraft designs will take years to develop and implement, but it is a positive sign that commercial aviation is already considering changes to aircraft for future use. Planes typically will be in operation for 30 y. Therefore, if the H_2 or any new generation of planes is not available until 2035, it can be expected that planes will continue to be built the same way for years to come. Therefore, the continued use of existing planes running primarily on jet fuel would remain in operation until as late as 2065, would therefore continue to greatly add to global CO_2 emissions.

12.6 DRAWDOWN TRANSPORTATION RELATED SOLUTIONS

Transportation is one of the largest categories of energy use, so it is not surprising that many Project Drawdown solutions focus on aspects related to transportation ranging from cars and hybrid cars to the use of bicycles and more efficient trucks, planes, and methods of shipping cargo (Table 12.3). Overall, these transport-related topics sum to a possible reduction of 58 Gt of CO_2 over the period of 2020–2050. Some of solutions require nearly no investment, such as carpooling. Others, such as walkable cities can only work for a certain portion of the population.

The impact of "telepresence," or meetings over the internet, may be underestimated in terms of total carbon emissions going into the future. The COVID-19 pandemic for the first half of 2020 was indicated to reduce global CO_2 emissions by 17% due to several factors, but one of the main contributions to the reduced carbon emissions was certainly a lack of travel for meetings and work. As a result of the lockdown, and for travel restrictions that followed, nearly all meetings were restricted to be via the internet rather than in person. The reduction in travel for meetings, conferences, and commuting to work could likely have much greater long-term impacts than currently envisioned by the Project Drawdown team.

Table 12.3 Possible reduction in CO_2 emissions due to improvements in transportation-related travel, calculated over a 30-y period (2020–2050) and normalized to 9.4 billion people on a daily basis in units of carbon reductions C.

No.	Solution	CO_2 (10^9 tons)	−C
24	Electric cars	11.9	0.13
33	Hybrid cars	7.9	0.08
34	Carpooling	7.7	0.08
35	Public transit	7.5	0.08
39	Efficient aviation	6.3	0.07
44	Efficient trucks	4.6	0.05
45	Efficient ocean shipping	4.4	0.05
52	Bicycle infrastructure	2.6	0.03
60	Walkable cities	1.4	0.015
63	Electric bicycles	1.3	0.014
64	High-speed rail	1.3	0.014
67	Telepresence	1.0	0.011
75	Electric trains	0.1	0.001
	Total	**58.0**	**0.62**

Source: Data from Wilkinson (2020).

References

BAUER, G., ZHENG, C., GREENBLATT, J. B., SHAHEEN, S. & KAMMEN, D. M. 2020. On-demand automotive fleet electrification can catalyze global transportation decarbonization and smart urban mobility. *Environmental Science & Technology,* 54, 7027–7033.

ELECTRIC SCOOTER GUIDE. 2020. *Electric scooter safety report: An analysis from data just released from Austin & Portland* [Online]. Available: https://electric-scooter.guide/safety/electric-scooter-safety-report-austin-portland [Accessed June 15 2020].

ERHARDT, G. D., ROY, S., COOPER, D., SANA, B., CHEN, M. & CASTIGLIONE, J. 2019. Do transportation network companies decrease or increase congestion? *Science Advances,* 5, eaau2670.

FRANGOUL, A. 2020. *Airbus announces concept designs for zero-emission, hydrogen-powered airplanes* [Online]. Available: https://www.cnbc.com/2020/09/21/airbus-announces-concept-designs-for-hydrogen-powered-airplanes-.html#:~:text=European%20aerospace%20giant%20Airbus%20released,their%20primary%20source%20of%20power [Accessed October 18 2020].

IEA. 2020. *Global CO_2 emissions in 2019* [Online]. Available: https://www.iea.org/articles/global-co2-emissions-in-2019 [Accessed October 18 2020].

INTERNATIONAL CIVIL AVIATION ORGANIZATION (ICAO). 2020. *ICAO Carbon Emissions Calculator* [Online]. Available: https://www.icao.int/environmental-protection/Carbonoffset/Pages/default.aspx [Accessed June 16 2020].

KETCHUM, D. 2019. *How many Calories are burned after biking 10 miles?* [Online]. Livstrong.com. Available: https://www.livestrong.com/article/315685-how-many-calories-are-burned-after-biking-10-miles [Accessed June 14 2020].

LIANG, J. 2020. Ride-hailing: Convenience at what cost? *Union of Concerned Scientists.*

MAHONE, A., SUBIN, Z., MANTEGNA, G., LOKEN, R., KOLSTER, C. & LINTMEIJER, N. 2020. Achieving carbon neutrality in California: PATHWAYS scenarios developed for the California Air Resources Board (Draft, August 2020).

MCELVERY, R. 2021. Is the road to sustainable asphalt paved with tires? *Chemical & Engineering News.* American Chemical Society.

PAVEMENT INTERACTIVE. 2020. *Energy and road construction - what is the mileage of roadway?* [Online]. Available: https://pavementinteractive.org/energy-and-road-construction-whats-the-mileage-of-roadway [Accessed June 15 2020].

Tabuchi, H. 2019. *Worse than anyone expected': Air travel emissions vastly outpace predictions* [Online]. New York Times. Available: https://www.nytimes.com/2019/09/19/climate/air-travel-emissions.html#:~:text=Over%20all%2C%20air%20travel%20accounts,passenger%20cars%20or%20power%20plants [Accessed December 15 2020].

THENOUX, G., GONZÁLEZ, Á. & DOWLING, R. 2007. Energy consumption comparison for different asphalt pavements rehabilitation techniques used in Chile. *Resources, Conservation and Recycling,* 49, 325–339.

THOMAS, J., HUFF, S., WEST, B. & CHAMBON, P. 2017. Fuel consumption sensitivity of conventional and hybrid electric light-duty gasoline vehicles to driving style. *SAE International Journal of Fuels and Lubricants,* 10, 672–680.

TILLEMANN, L. & FEASLEY, L. 2018. *Let's count the ways e-scooters could save the city* [Online]. Available: https://www.wired.com/story/e-scooter-micromobility-infographics-cost-emissions [Accessed February 15 2021].

US DEPARTMENT OF ENERGY. 2020. *Alternative fuels data center, average per-passenger fuel economy by travel mode* [Online]. https://afdc.energy.gov/data/10311?page=2 [Accessed June 14 2020].

WIKIPEDIA. 2020a. *Energy efficiency in transport* [Online]. Available: https://en.wikipedia.org/wiki/Energy_efficiency_in_transport#Walking [Accessed June 15 2020].

WIKIPEDIA. 2020b. *High-speed rail* [Online]. Available: https://en.wikipedia.org/wiki/High-speed_rail [Accessed June 16 2020].

WIKIPEDIA. 2020c. *List of coutnries by carbon dioxide emissions* [Online]. Available: https://en.wikipedia.org/wiki/List_of_countries_by_carbon_dioxide_emissions [Accessed June 12 2020].

WIKIPEDIA. 2021. *Fuel fraction* [Online]. Available: https://en.wikipedia.org/wiki/Fuel_fraction [Accessed April 30 2021].

WILKINSON, K. 2020. *The drawdown review: Climate solutions for a new decade*, Project Drawdown.

CHAPTER 13

CONCRETE AND STEEL

13.1 ENERGY USE FOR BUILDING MATERIALS

The energy used for construction of large structures is a significant proportion of global energy use, and more of the world's population in the future will live in an urban environment. The carbon emissions from construction of large buildings that primarily use concrete and steel are particularly a concern relative to climate change as these materials can contribute enormous amounts of CO_2 into the atmosphere. While a comparison of energy for all building materials is beyond the scope of our analysis here, as it is not directly relevant to our daily energy use, the amounts of energy and CO_2 emitted from concrete and steel are important in our consideration of factors relevant to climate change. There continues to be an increase in cement production, particularly due to the urbanization of China. Global anthropogenic CO_2 emissions in 2018 were 33.1 Gt with cement and steel production accounting for 4.4 Gt, or about 13% of total emissions.

As the world population is estimated to increase from 7.5 to 9.4 billion people by 2050, the energy and carbon emissions related to buildings used for these additional people will have to increase if we continue to use the same construction and production methods for concrete and steel structures. Therefore, we should carefully consider alternative materials and methods that can be used to reduce global CO_2 emissions from these sources and ways to maximize reuse of these important contributors to overall CO_2 emissions.

13.2 CONCRETE AND CEMENT

Most concrete is made using Portland cement, with the main components of concrete consisting of cement (~15% of the final weight), sand, gravel, and water. To make Portland cement, limestone is heated with other materials like clays to a temperature of ~1450°C, producing what is known as "clinker," which must be ground up (typically with gypsum) to make the cement powder. Carbon emissions from cement production are primarily due to two sources: CO_2 released from the calcination of the limestone used to make the cement (~0.45 kg CO_2/kg of cement) and the thermal energy for the process from cement production to that of the final concrete (0.40 kg/kg). The remaining CO_2 emissions are associated with processing and transport of the ingredients, grinding of clinker, and packaging and shipping of the cement to its point of use. These other processes contribute to about 10% of the CO_2 emissions of concrete (~0.1 kg CO_2/kg of concrete) compared to ~0.13 kg CO_2 released per kg of concrete for only cement production (assuming 15% of the concrete is

Daily Energy Use and Carbon Emissions: Fundamentals and Applications for Students and Professionals, First Edition. Bruce E. Logan.

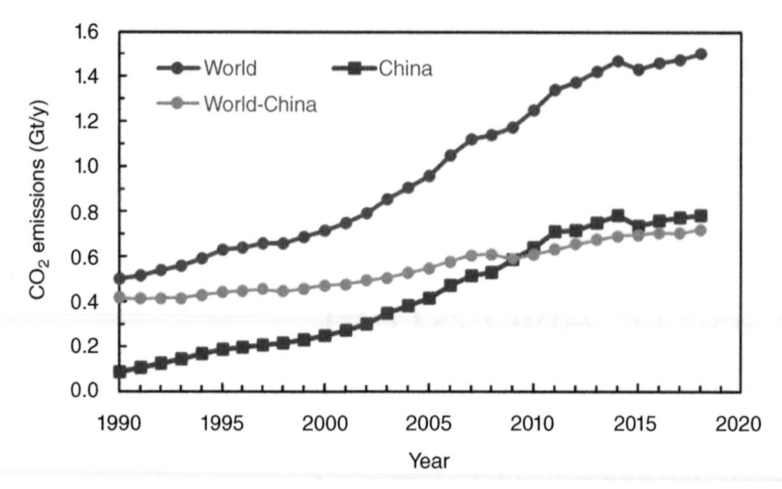

Figure 13.1 Growth in CO_2 emissions from cement production for the world, China, and the world without China. China's emissions grew by 850% over this period compared, contributing to a global increase of 200%. *Source*: Data from Zenodo (2020).

cement). Therefore, about half of the CO_2 released in concrete production cannot be avoided when using limestone ($CaCO_3$) as the CO_2 is released from the limestone when it is heated (calcined) to produce equal amounts of lime (CaO), the desired product, and CO_2, or

$$CaCO_3\ (+\ \text{heat}) \rightarrow CaO + CO_2 \tag{13.1}$$

Nearly as much CO_2 emissions (40%) arise from the use of fossil fuels to heat the kilns to temperatures needed to produce the lime as the CO_2 liberated with the heating of the limestone.

Over 30 billion metric tons of concrete were estimated to be produced in 2018 (Carbonbrief.org, 2020). Estimates for total annual CO_2 emissions only from the CO_2 released from the limestone (not including fuels) vary. For example, emissions reported were: 1.51 Gt (Andrew, 2018) to 2.08 Gt by CDIAC in 2014, and 1.44 Gt (Olivier et al., 2016) and 1.47 Gt (Andrew, 2018) in 2015. In 2018, emissions were estimated to be 1.50 Gt (Zenodo, 2020). When combined with CO_2 released from the fossil fuels used, these numbers would approximately double, resulting in about 3 billion tons of CO_2 in 2018 or about 8% of total global CO_2 emissions (Carbonbrief.org, 2020).

A comparison of CO_2 emissions from cement production to that of individual countries would suggest that cement production is the third largest "producer" on the planet compared to countries, second only to China and the United States (Carbonbrief.org, 2020). The enormous growth in carbon emissions is due primarily to China (Fig. 13.1). Without China, cement production emissions grew 72% between 1990 and 2018, but due to the 825% growth for China, global CO_2 emissions increased overall by 200% (Zenodo, 2020, Andrew, 2018). Over half of the cement production in 2018 was in China, which produced more cement between 2015 and 2018 than the United States did in the entire 20th century. The use of concrete is predicted to increase 25% by 2030 compared to its level of use today, and therefore, its production would have increased by four times compared to the amount used in 1990 (Carbonbrief.org, 2020).

Assuming a global population of 7.631 billion people in 2018 and ~1.5 Gt of CO_2, these carbon emissions released from the cement would translate to 0.6 C for CO_2 released from limestone, and a total of ***1.2 C for cement production*** including fuel consumption to produce that cement. The energy needed to produce cement can be estimated as shown in the example below. This estimated 1.2 C means that more CO_2 is released into the atmosphere each year just to produce cement than the amount of CO_2 exhaled each day by over 7.6 billion people.

Example 13.1

(a) Estimate the energy used for the world's supply of cement in 2015 in kWh assuming 1.435 Gt of CO_2 was released from the fuel used to make the cement (Olivier et al., 2016) using coal as the main fuel. (b) What would be the reduction in CO_2 emissions if only natural gas was used to produce that cement? (c) What is the amount of kWh/kg of cement produced and concrete made assuming a global production of 4.07 Gt of cement and a ratio of cement to concrete of 1:6.5 (by mass)?

(a) We can convert the CO_2 produced from any fuel into energy using the D:C ratios for that fuel. For coal, which has a D:C ratio of 1.2, we calculate the annual energy used as:

$$E = 1.435\frac{\text{Gt CO}_2}{\text{y}}\frac{10^9\text{tonne}}{\text{Gt}}\frac{10^3\text{kg}}{\text{tonne}}\frac{2.2\text{ lb}}{\text{kg}}\frac{1\text{ C}}{2\dfrac{\text{lb}}{\text{d cap}}}\frac{1.2\text{ D}}{\text{C}}\frac{2.32\dfrac{\text{kWh}}{\text{d cap}}}{1\text{ D}} = 4.4\times10^{12}\frac{\text{kWh}}{\text{y}}$$

Assuming a global population of 7.61 billion people, this is 0.67 D.

(b) Using this amount of energy and replacing the fuel with natural gas which has a D:C ratio of 2.23, we have

$$4.4\times10^{12}\frac{\text{kWh}}{\text{y}}\frac{1\text{ D}}{2.32\dfrac{\text{kWh}}{\text{d cap}}}\frac{\text{C}}{2.23\text{ D}}\frac{2\dfrac{\text{lb}}{\text{d cap}}}{1\text{ C}}\frac{\text{kg}}{2.2\text{ lb}}\frac{\text{tonne}}{10^3\text{ kg}}\frac{\text{Gt}}{10^9\text{tonne}} = 0.77\frac{\text{Gt CO}_2}{\text{y}}$$

Thus, the reduction would be 0.67 Gt (1.435–0.77 Gt), or a 45% reduction in CO_2 emissions by using natural gas instead of coal.

(c) Using the given information for cement, we have as an estimate of energy to produce a kg of cement:

$$4.4\times10^{12}\frac{\text{kWh}}{\text{y}}\frac{\text{y}}{4.074\times10^9\text{ tonne}}\frac{1\text{ tonne}}{10^3\text{kg}} = 1.08\frac{\text{kWh}}{\text{kg}}$$

For concrete, we use the given ratio to calculate:

$$1.08\frac{\text{kWh}}{\text{kg cement}}\frac{1\text{ kg cement}}{6.5\text{ kg concrete}} = 0.17\frac{\text{kWh}}{\text{kg concrete}}$$

These estimates for energy assumed that only Portland cement was used, and it was combined with only water, sand, and gravel. The use of other cements or materials, as well as a mix of fuels, could be used to lower these estimates.

The above example shows that it is possible to reduce CO_2 emissions by about half (from 1.435 to 0.77 Gt) just by switching from coal to natural gas. Further reductions of 5–10% in energy use might be possible with natural gas due to more efficient use of the gaseous fuel than coal. However, even with these reductions the amount of CO_2 emissions would remain very large due to emissions both from the fuel and the CO_2 released from the limestone. Thus, further changes are needed to reduce overall energy consumption and CO_2 emissions.

Reducing Energy Use and Using Other Materials to Produce Concrete

Improving the energy efficiency of the cement and concrete manufacturing processes is important, but overall ir can produce only modest reductions in CO_2 emissions compared to other options (Fig. 13.2). Switching from other fossil fuels to natural gas is perhaps the easiest change, and one that is occurring today, as this can immediately impact CO_2 emissions as noted in the above example. Capture and storage of carbon released at these point source sites could be quite effective in avoiding the release of CO_2 into the atmosphere, but such capture processes are not currently used in cement production. Furthermore, increasing the durability of concrete could extend the lifetime of some structures, for example from 50 to 100 years, which would translate to a reduced need for maintaining and rebuilding these infrastructures in the coming years.

Substitution of other materials for Portland cement can greatly reduce overall CO_2 emissions by avoiding release of CO_2 from the limestone (Scrivener et al., 2016). The most promising materials for reducing the amount of Portland cement used in concrete, and their relative upper percentages in cement replacement, include calcined clays (30%), pulverized limestone (15%), fly ash from coal fuel power plants (40%), and blast furnace slag from steel production (typically 40%, but up to 70%) (Favier et al., 2018). Calcined clay and pulverized limestone show the most promise for widespread impact on use as they have a key advantage of wide availability among these different materials. Both calcined clay and fillers such have an estimated availability at 6 Gt/y (Scrivener et al., 2016), which is higher than the current use of Portland Cement (4.6 Gt/y). Fly ash also has relatively good availability of 0.90 Gt/y but the quality is highly variable, the source (coal-fired

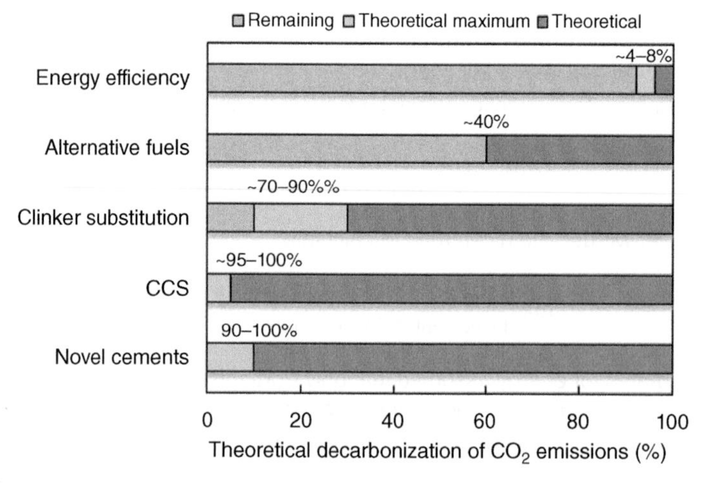

Figure 13.2 Possible reduction in CO_2 emissions due to a variety of approaches. *Source*: Adapted from Lehne et al. (2018).

power plants) is a large emitter of CO_2, and a reduction in coal use is envisioned in order to reduce global carbon emissions. Blast furnace slag has an availability of 0.33 Gt/y but ~90% of the available source is already used for cement production, and thus, slag is not expected to be useful for further reductions in CO_2 emissions. Other materials that can be substituted for Portland dement include natural pozzolans (primarily volcanic ashes), biomass ashes, waste glass, and silica fume (amorphous or noncrystalline silica dioxide, SiO_2), but these have relatively limited availability. The greatest changes in carbon emissions rely upon either the use of novel cements (>90% estimated reduction possible) or clinker substitutions (up to 90% reductions) (Lehne and Preston, 2018).

Limestone Calcined Clay Cement, or LC^3, is a commercially available alternative to traditional Portland cement (see https://lc3.ch). This cement with less clinker content was developed in Switzerland is claimed to achieve 40% less CO_2 emissions than traditional cement (95% clinker and 5% gypsum), comparable strength and performance, and improved durability compared to Portland cement. LC^3 is composed of 50% clinker, 30% calcined clay, 15% limestone, and 5% gypsum. Although the calcined clay is also prepared at a high temperature of 800°C, this is much lower than that currently used for the clinker (1450°C) and it does not release CO_2. Limestone and gypsum are used without heat treatment. Various structures have now been built at modest scales using this new LC^3 material that include a Swiss Embassy building in India, several houses, and pavements.

Other novel cements include geopolymer-based materials that do not use limestone, and materials that can absorb CO_2 while they harden. The geopolymer materials are novel as they avoid the use of clinker but they have not yet gained traction in commercial applications. Other materials are being developed to take up CO_2, such as those by Solidia and Cemex (Vertua Ultra). These are indicated to reduce CO_2 emissions by up to 70% while also reducing water consumption. The uptake rate of CO_2 by the materials made by Solidia are very low when exposed to air. To substantially improved the amount of CO_2 that is incorporated into the material, it must be cured inside CO_2 incubators which limits the use of this material to very specialized conditions.

Alternative materials to Portland cement have so far not appreciably impacted the market. Creating the next generation of concrete could take many years and proving its ease of use, low cost, and durability will take even longer (Montiero et al., 2017). Changes in energy use and materials have reduced emissions per mass of concrete produced over the past decades. However, the greatly increased production rates of concreate have completely overshadowed gains based on the CO_2 per mass produced. Therefore, greater efforts need to be directed at reducing the amount of concrete used in building construction as well further reducing CO_2 emissions by using new materials (Scrivener et al., 2018).

Recycling Concrete

Given the high carbon footprint of CO_2 from cement production, it would make sense to recycle as much concrete as possible. It was estimated in 2010 that 140 million ton/y of concrete was recycled for use in the aggregates market, with 84% of states recycling concrete as an aggregate base (Portland Cement Association, 2010). In 2017, it was similarly found that 85% of 342 million tons, or 280 million tons of concrete and other aggregates was recycled with the balance landfilled (Townsend and Anshassi, 2017). If concrete is to be economically recycled, then it must be used close to the point of production. The recycling process consists of crushing large chunks into smaller sizes, sorting, and cleaning (to remove brick, wood, rebar, and other impurities). The recycled aggregate can then be used for roads (base material), ready mix concrete for foundations, slabs, and walkways, or for soil stabilization and landscaping among other uses. Energy use for crushing concrete, based on that for crushing rocks, is relatively small (~0.003 kWh/kg) (Ciezkowski et al., 2017), and thus, it is only a fraction of the energy needed to produce the concrete (0.16 kWh/kg).

Sand

The most used natural resource in the world is water, but number two on the list is sand primarily due to its use in concrete (Torres et al., 2017). Sand and gravel are the largest mass inputs for concrete, with a typical volume ratio of 1:1.7:3.3 for cement:sand:aggregates, with nearly 29 Gt of sand and gravel used in 2010. You might think that the global supply of sand is essentially unlimited, but its availability is a balance between where the sand is located and where it is needed. Sand mining has therefore been intense in areas of high development where the sand needs to be obtained close to the point of use to reduce construction costs. This extraction of sand nearby to very active areas of construction has resulted in severe environmental impacts, for example in the Mekong Delta where it has impacted water supplies, and Sri Lanka where sand mining was claimed to be responsible for contributing to extensive crop losses in the region (Logan, 2018).

Another complicating issue in sand use and location is that not all sand is the same, with river sand the most desired of many different sources. This extraction of river sand has had unintended consequences. For example, in India, extraction from the Yamuna river has caused declines in dolphins and terrapins in the Ganges River. Illegal sand mining is also frustrating efforts to protect many endangered species and habitats. While used concrete can be recycled for aggregates in new concrete, sand remains a precious resource for new concrete. Thus, sand availability is becoming limited especially in Asia due to the high growth rate of concrete-based construction. While sand seems to be an infinite resource, unfortunately it is not (Larson, 2018). Crushing quarried aggregates to sizes comparable to sand can produce a material that works well in concrete, but this "manufactured sand" so far has not been shown to be as popular as river sand.

13.3 STEEL

Steel is used in a variety of ways, most notably in our daily lives in new buildings. The large buildings around the world could not be built without steel, and when you look at large-scale construction sites, you will inevitably find classic steel structures (Fig. 13.3). Approximately half of the steel produced is used for the construction industry (Fig. 13.4). Other major uses of steel are mechanical equipment (15%), automobile production (12%), and metal products (11%). In 2019, the production of steel was estimated at 1.87 billion tonnes (1.87×10^{12} kg), with an increase for 2021 predicted to be 1.9 billion tonnes (Statistica, 2020).

CO_2 emissions from steel manufacturing are only slightly less than those from cement production. In 2017, global CO_2 emissions from steel plants were estimated at 2.1 Gt, compared to 2.3 Gt estimated for the cement industry (Fig. 13.5) (Pales et al., 2019). The steel industry has become slightly more energy efficient over the past few decades, with an 8.3% reduction in energy per mass of steel, but the amount of energy consumed remains large as it required 5500 kWh/ton of steel produced (in 2017). Despite the reduction in the energy per mass of steel, the total energy consumption by this industry doubled between 2000 and 2017 reaching 9060 TWh in 2017. China accounted for nearly half of global steel production in 2017.

Figure 13.3 An example of a new steel building being constructed in State College, PA. *Source*: Photograph by B.E. Logan.

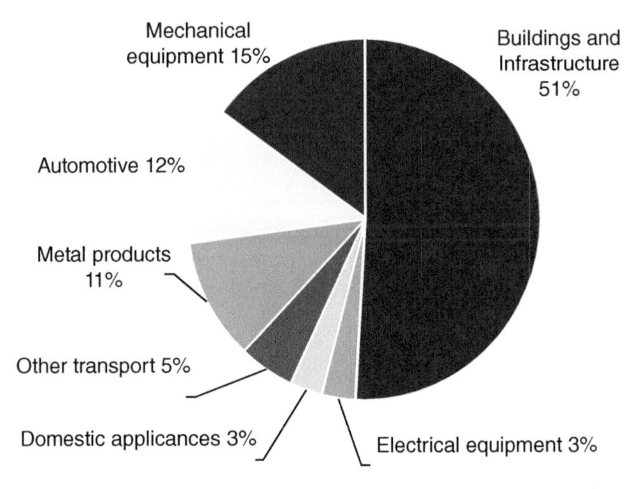

Figure 13.4 Steel use globally based on sector. *Source*: Data from Statistica (2020).

The main reason for such high CO_2 emissions is that the iron and steel industries primarily use coal in the manufacturing and production processes, with ~77% of energy used derived from coal (Fig. 13.6). In conventional steel production, a source of carbon must be provided for the reduction of the raw material (hematite, or Fe_2O_3) to solid iron, which produces a large amount of CO_2 via

$$2Fe_2O_3 + 3C \rightarrow 4Fe + 3CO_2$$

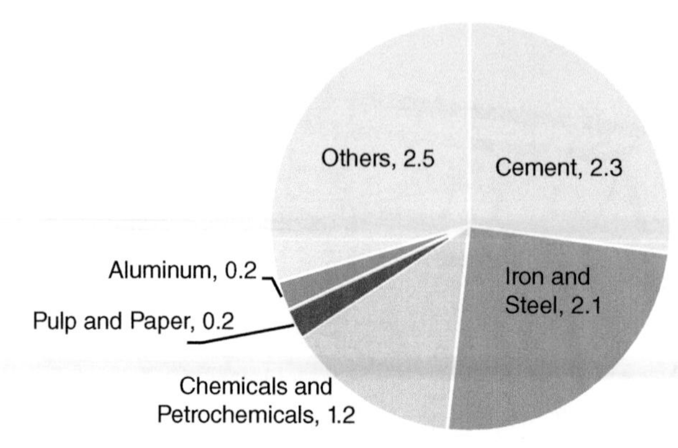

Figure 13.5 Distribution of global CO_2 annual emissions (Gt) for different industries in 2018. *Source*: Data from Pales et al. (2019).

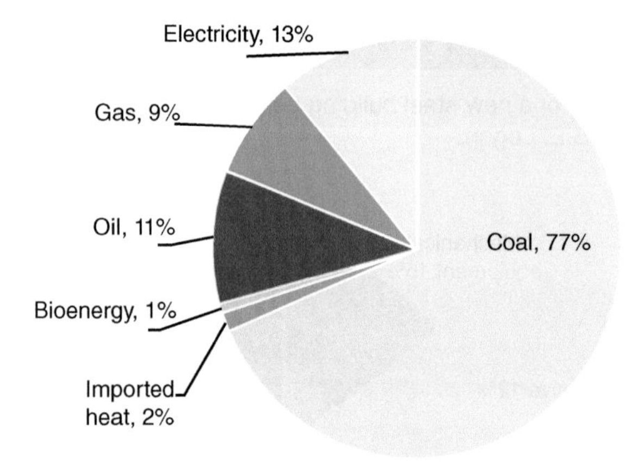

Figure 13.6 Distribution of energy US based on percentage for a total annual use of 9060 TWh in 2017. *Source*: Data from Pales et al. (2019).

The amount of coal used for steel production has even slightly increased relative to the total proportion of energy sources, with 75.9% of all energy used in 2000 compared to 77.3% in 2017 (Pales et al., 2019). Large decreases in CO_2 emissions from the steel production will require the industry to move away from coal and develop innovative technologies that could include new smelting techniques, direct iron reduction, and repurposing the CO_2 released by its incorporation into bio-based products or carbon capture and sequestration.

Assuming a global population of 7.55 billion people in 2017, for an energy production of 9060 TWh (Terra-watt hours), energy use for the iron and steel industry would be ***equivalent to 1.42 D***. With the associated 2.1 Gt of CO_2 emissions, this would be ***0.84 C*** if we normalized to every person on the planet.

Emerging Technologies for Reducing CO_2 Emissions from the Steel Industry

The United Nations has established sustainable development goals for the steel industry, and the International Energy Agency has formulated a Sustainable Development Scenario (SDS) to meet those goals by 2030 (Pales et al., 2019). Energy efficiency alone is not sufficient, and therefore, meeting these goals will require large changes to the steel industry. For example, it was estimated that the electric arc furnaces would need to increase to become 40% of the market by 2030 (Pales et al., 2019). To date, there have not been sufficient changes in the operations of most steel plants. The reliance on coal by this industry, based on percentage of energy used, for example, has not appreciably changed while the mass of steel produced has doubled.

There are options for reducing CO_2 emissions from steel production but some of these will require relatively massive changes to the industry such as shifting from coal to electricity, and carbon recovery. In a study on how steel plants could reduce carbon emissions, Flores-Granobles (2020) estimated that a standard steel plant currently emitted a total of 1.9 t of CO_2 per tonne of liquid steel produced (tCO_2/t), with 50% of all CO_2 emissions from the onsite power plant. Therefore, one immediate way to reduce CO_2 emissions by the steel industry would be to purchase electricity from the grid rather than to producing it on site. Most steel plants produce electricity primarily from coal with power plants having with lower thermal efficiencies than the larger power plants supplying electricity to the grid. In the EU, electricity production emits 0.42 t CO_2/MWh, while coal plants have nearly double that (0.8 t/MWh) due to lower efficiencies and use of renewables (Flores-Granobles and Saeys, 2020). Switching to grid electricity could therefore reduce overall CO_2 emissions by 22% and using natural gas on site for heat could reduce CO_2 emissions another 7%.

When steel is manufactured, the chemicals in the gases that are produced in the overall process contain energy that can be recovered (Flores-Granobles and Saeys, 2020). Three sources combined contain 8199 MJ for every tonne of liquid steel produced (Table 13.1). CO is a toxic gas, and it cannot be released so instead it is burned to produce CO_2 and electricity. However, this gas could be used for other purposes and thus avoid these gas emissions. Two suggested products from steel plant gases are methanol, which is a mature technology, and ethanol, which is currently being investigated at the pilot scale in a steel plant to evaluate overall economic feasibility. CO_2 capture from the emissions of the blast furnace alone could only reduce overall CO_2 emissions by 17%, and thus, greater changes are needed than just carbon capture. With full implementation of using gases

Table 13.1 Composition, volume, lower heating value, and total energy for typical steel mill gases.

Category	Components	Coke oven	Blast furnace	Basic oxygen furnace
Composition (v/v, %)	CH_4	23.04	—	—
	H_2	59.53	2.63	2.64
	CO	3.84	22.34	56.92
	CO_2	0.96	22.10	14.44
	N_2	5.76	48.77	13.83
	O_2	0.19	—	—
	H_2O	3.98	3.15	12.16
	Other	2.69	—	—
Volume (m^3/t)		166	1467	82
Lower heating value (MJ/m^3)		17.33	3.21	7.47
Total energy (MJ/t)		2877	4709	613

Units indicate t = ton of liquid steel, m^3 = normal cubic meter of gas at 1 atm and 0°C.
Source: Adapted from Flores-Granobles and Saeys (2020).

produced in the existing process, carbon capture, and electricity from the grid (in Europe), a steel plant could reduce CO_2 emissions by 55%. The greatest reduction that could be achieved would be to completely alter the processes by reducing the iron using H_2 if it is produced using a fossil-free approach, for example by a water electrolyzer, followed by an electric arc furnace powered by renewable electricity (Krüger et al., 2020). Using this approach and fully renewable energy sources for the electricity grid, and H_2 gas produced using renewable electricity, CO_2 emissions could decrease by 85% (Flores-Granobles and Saeys, 2020). However, this would require a complete replacement of most of the processes and thus would not seem to be likely due to the high costs involved for these changes. No matter which of these new and innovative processes are used large changes in CO_2 emissions from steel plants are possible based on technologies available today.

Steel Recycling

Recycling steel and iron is essential to further reducing energy consumption and CO_2 emissions from these industries. Steel scrap can be reused by processing in electric arc furnaces, which reduce CO_2 emissions by 58% compared to blast furnaces. Arc furnaces were also used for producing 28% of crude steel in 2017. Globally, over 85% of steel scrap was reused with 50% reuse in structural reinforcement steel and 97% in industrial equipment. In the European Union, over 90% of stainless steel is collected and recycled into new products, with an estimated €20 billion saved by using steel scrap in 2018 (EuRIC, 2020).

Other Metals and Materials

Steel and iron production have one of the highest contributions to global CO_2 emissions among many different types of materials, but these emissions from other metals and materials are substantial (Fig. 13.7). Around 20% of global CO_2 emissions can be attributed to production

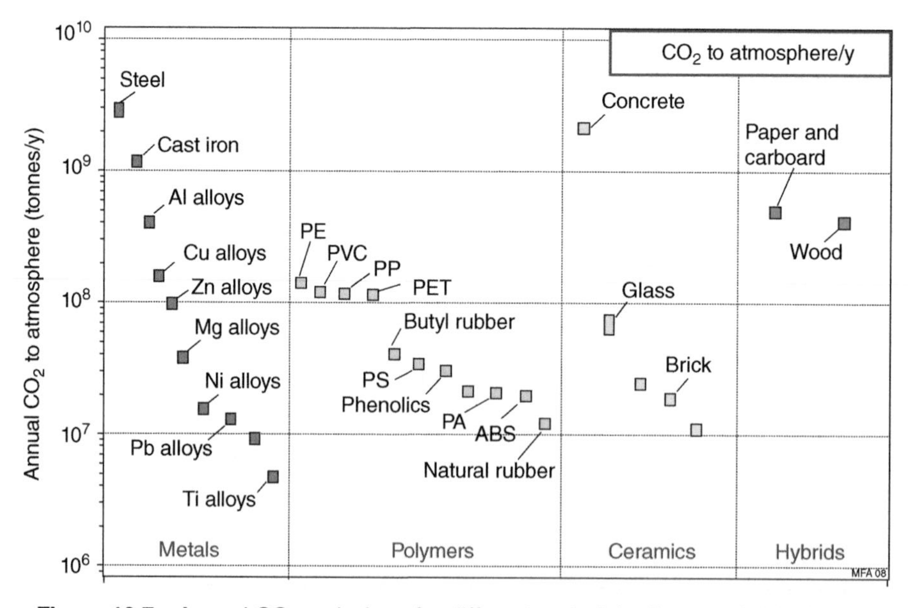

Figure 13.7 Annual CO_2 emissions for different materials. *Source*: Ashby (2009).

of these metals and materials, with steel, iron, aluminum, concrete, and paper and cardboard accounting for more emissions than all the others combined (Ashby, 2009).

13.4 DRAWDOWN SOLUTIONS FOR CEMENT AND STEEL

The Project Drawdown analysis of alternatives to cement suggests that 8.0–16.1 Gt of CO_2 emissions could be reduced by decarbonizing limestone, ranking those changes as 32 on their list of the areas with greatest potential to reduce global CO_2 emissions (Wilkinson, 2020). Translated to a period of 30 years (2020–2050), that would be equivalent to a reduction of 0.09 C. Given the huge impact of cement on global CO_2 emissions, it seems that more need to be done to reduce emissions from this source.

Project Drawdown does not mention steel production or how CO_2 emissions could be reduced relative to those used today by the steel industries. That omission is surprising given the enormous contribution to iron and steel production around the world, and the potential for greatly reducing CO_2 emissions from steel plants. Recycling steel is also helpful for avoiding the current use of coal-based furnaces and thus reducing CO_2 emissions due to coal emissions. Hopefully, the topic of steel production will receive more attention by the Project Drawdown group, and others, as there already exist effective methods to reduce carbon emissions from steel production. However, given the lack of any appreciable reductions in coal use in the past two decades for steel production, there are no indicators that this route to reduce CO_2 emissions is being pursued by the iron and steel industries.

References

ANDREW, R. M. 2018. Global CO_2 emission from cement production. *Earth System Science Data,* 10, 195–217.

ASHBY, M. F. 2009. *Materials and the environment: Eco-informed material choice*, Butterworth-Heinemann.

CARBONBRIEF.ORG. 2020. *Why cement emissions matter for climate change* [Online]. Available: https://www.carbonbrief.org/qa-why-cement-emissions-matter-for-climate-change [Accessed June 18 2020].

CIEZKOWSKI, P., MACIEJEWSKI, J. & BAK, S. 2017. Analysis of energy consumption of crushing processes – Comparison of one-stage and two-stage processes. *Studia Geotechnica et Mechanica,* 39, 17–24.

EURIC. 2020. *EuRIC Metal recycling factsheet* [Online]. EURIC. Available: https://www.euric-aisbl.eu/position-papers/item/335-euric-unveils-metal-recycling-brochure [Accessed November 24 2020].

FAVIER, A., DE WOLF, C., SCRIVENER, K. & HABERT, G. 2018. A sustainable future for the European Cement and Concrete Industry: Technology assessment for full decarbonisation of the industry to 2050, ETH Zurich. https://doi.org/10.3929/ethz-b-000301843 [Accessed June 25 2020].

FLORES-GRANOBLES, M. & SAEYS, M. 2020. Minimizing CO_2 emissions with renewable energy: A comparative study of emerging technologies in the steel industry. *Energy & Environmental Science,* 13, 1923–1932.

KRÜGER, A., ANDERSSON, J., GRÖNKVIST, S. & CORNELL, A. 2020. Integration of water electrolysis for fossil-free steel production. *International Journal of Hydrogen Energy,* 45, 29966–29977.

LARSON, C. 2018. Asia's hunger for sand takes toll on ecology. *Science,* 359, 964–965.

LEHNE, J. & PRESTON, F. 2018. *Making concrete change: Innovation in low-carbon cement and concrete*Chatham House. https://www.chathamhouse.org/2018/06/making-concrete-change-innovation-low-carbon-cement-and-concrete [Accessed June 18 2020].

LOGAN, B. E. 2018. Waste not, want it. *Environmental Science & Technology Letters,* 5, 301–301.

MONTIERO, P. J. M., MILLER, S. A. & HORVATH, A. 2017. Towards sustainable concrete. *Nature Materials,* 16, 698–699.

OLIVIER, J. G. J., JANSSENS-MAENHOUT, G., MUNTEAN, M. & PETERS, J. A. H. W. 2016. Trends in global CO_2 emissions: 2016 Report.

PALES, A. F., LEVI, P. & VASS, T. 2019. *Tracking Industry 2019: Iron and steel* [Online]. IEA. Available: https://www.iea.org/reports/tracking-industry-2019/iron-and-steel [Accessed July 26 2020].

PORTLAND CEMENT ASSOCIATION. 2010. *Recycled concrete* [Online]. Available: https://www.cement.org/docs/default-source/th-paving-pdfs/sustainability/recycled-concrete-pca-logo.pdf?sfvrsn=2&sfvrsn=2#:~:text=The%20Construction%20Materials%20Recycling%20Association,year%20in%20the%20U.S.%20alone [Accessed June 26 2020].

SCRIVENER, K. L., JOHN, V. M. & GARTNER, E. M. 2016. Eco-efficient cements: Potential, economically viable solutions for a low-CO_2, cement-based materials industry.

SCRIVENER, K. L., JOHN, V. M. & GARTNER, E. M. 2018. Eco-efficient cements: Potential economically viable solutions for a low-CO_2 cement-based materials industry. *Cement and Concrete Research,* 114, 2–26.

STATISTICA. 2020. *Distribution of steel end-use worldwide in 2018 by sector* [Online]. Available: https://www.statista.com/statistics/1107721/steel-usage-global-segment/ [Accessed October 4 2020].

TORRES, A., BRANDT, J., LEAR, K. & LIU, J. 2017. A looming tragedy of the sand commons. *Science,* 357, 970–971.

TOWNSEND, T. & ANSHASSI, M. 2017. Benefits of construction and demolition debris recycling in the United States.

WILKINSON, K. 2020. *The drawdown review: Climate solutions for a new decade*, Project Drawdown.

ZENODO. 2020. *Cement emissions data* [Online]. Available: https://zenodo.org/record/3380081#.XuzwREVKhaR [Accessed June 19 2020].

CHAPTER 14

ASSESSMENT AND OUTLOOK

14.1 ADDRESSING CLIMATE CHANGE WILL REQUIRE BOTH RENEWABLE ENERGY AND CARBON CAPTURE

Many people, communities, states, businesses, and nations have the desire to minimize the impacts of climate change by reducing CO_2 and other greenhouse gas (GHG) emissions. Unfortunately, there is not complete consensus of all these different groups on how to move forward and reduce GHGs, and the best ways to achieve drawdown of GHGs have not been firmly established. Leaders of most nations stand behind the concept of reducing GHGs. A total of 197 countries signed the 2015 Paris Agreement on climate change, and 179 have provided climate proposals that have received formal approval (including the United States). In 2017, the United States announced that it would pull out of the Paris Agreement and did that in November of 2020. While the United States rejoined in 2021, this lack of continuous commitment to the goals of the Paris agreement had a destabilizing impact on efforts to combat climate change. This unwillingness to participate in the global effort to reduce CHGs reflected the desire of the US administration during 2017–2021 to aggressively not promote activities that would address solutions needed to avoid climate change. Other countries have also had internal conflicts which have impeded progress in addressing global climate change issues. Therefore, even for the countries that signed the Paris Agreement, the importance of addressing climate change will still be debated and subjected to the goals of different administrations. Each country will continue to debate the advantages of taking steps to address climate change relative to other national and international challenges and needs.

The participation of global businesses is essential to addressing climate change, and these businesses are not part of the Paris Agreement. Unless large businesses can agree on the advantages of reducing GHGs (or be provided sufficient incentives), it is not clear that needed changes can occur. Fortunately, a group of businesses has recognized the need for corporate change and participation in climate change activities. Business members of the organization The Climate Pledge have come together to provide their support to reach the goal of near-zero carbon emissions by 2040 or 10 years sooner than that goal set by the Paris Agreement for 2050. In early 2021, there were over 50 businesses spanning 12 countries that had agreed to join The Climate Pledge including Amazon, IBM, Microsoft, and Siemens.

Some solutions for drawing down CO_2 represent new business opportunities in the form of jobs and manufacturing for these new technologies including zero-emission vehicles, wind and solar power systems, and urban infrastructure changes that can reduce GHG emissions. Some countries are embracing opportunities for new markets in renewable energy through new business revenues.

Daily Energy Use and Carbon Emissions: Fundamentals and Applications for Students and Professionals, First Edition. Bruce E. Logan.

For example, Saudi Arabia is planning a new city (Neom) that will invest US $5 billion in producing H_2 from renewable energy sources (wind and solar). While this production provides a new renewable energy source for the world, officials unfortunately did not announce plans to use any of this H_2 to reduce the use of fossil fuels in their own country. Thus, this business-oriented project by that country could provide renewable and green H_2 to the world but it would not impact overall fossil fuel consumption by the rest of the country.

The transition to a renewable and carbon-neutral energy infrastructure will require massive legislative efforts and large amounts of government spending. Although these anticipated expenditures are significant in size, they are not nearly as large as that expected to be needed to address the global economy relative to the 2020 COVID-19 pandemic. Altering the current path of the US energy infrastructure that is dependent on fossil fuels to low-carbon energy solutions could cost US $1.4 trillion annually between 2020 and 2024 (Andrijevic et al., 2020). However, the COVID-19 pandemic has shattered the economies of many nations and the larger efforts at economic recovery appear to be directed to a COVID-19 stimulus plan with (at the time of this writing) little money directed toward annual energy investments. Still, it was estimated that just 10% of the anticipated COVID-19 investments could address these needed low-carbon energy technologies. In addition to the development of new energy technologies, addressing climate change will also require divestment in high-carbon fossil fuels of US $280 billion per year over the same period of time (Andrijevic et al., 2020). These estimates only address the energy infrastructure and do not include activities that additionally are needed for addressing climate change such as refrigerants, food waste, and many others that are listed by Project Drawdown.

An analysis of reductions in CO_2 emissions based on current pledges indicates that we will not achieve the Paris goal of less than a 1.5°C increase (50% probability) unless we also implement actions that achieve negative carbon emissions (Anderson and Peters, 2016) (Fig. 14.1). Even with projected negative emissions and all the project activities possible, the world would not achieve zero net emissions in their scenario until 2070, with emissions in 2030 possibly only 25% lower than 2020. By 2050, it was estimated that CO_2 could be reduced by about two-thirds but that would require all global participants to achieve their pledges to the Paris agreement.

The impacts of climate change continue to be evaluated and predicted, but climate change is occurring, and its impacts can be felt around the world. The past 6 years have been the warmest on record, and overall, the planet has warmed up by 1.25°C compared to temperatures prior to the industrial age (Voosen, 2021). However, land temperatures are nearly 60% higher than global averages and now average 1.98°C warmer. Average global temperatures underestimate changes to land temperatures as the ocean temperatures are lower than those that occur on land. The results of these hotter land temperatures are now clearly seen by extraordinary numbers of extreme events, with Australia having record-setting combined drought and heat waves that produced catastrophic fires in 2020. The state of California set a record in 2018 for the most fires (1.8 million acres burned), a record that was broken two years later with more than double that in 2018 (4 million acres) (Wigglesworth and Serna, 2020). Parts of Siberia were 7°C warmer than during pre-industrial times, and 2020 was the warmest year on record in Russia, Europe, and Asia, and South America tied its previous record (Voosen, 2021). Heating of the subtropical Atlantic Ocean produced a record-breaking number of hurricanes. It is likely that these records will continue to be broken as the world continues to heat up because of continued releases of GHGs.

The impacts of climate change are already profoundly impacting the planet, and it does not seem likely that all pledges to reduce GHG emissions will be realized especially given the disruptive nature of the COVID-19 pandemic to the economies of many countries. Therefore, we need greater local efforts, in addition to global activities, to produce electricity from renewable sources

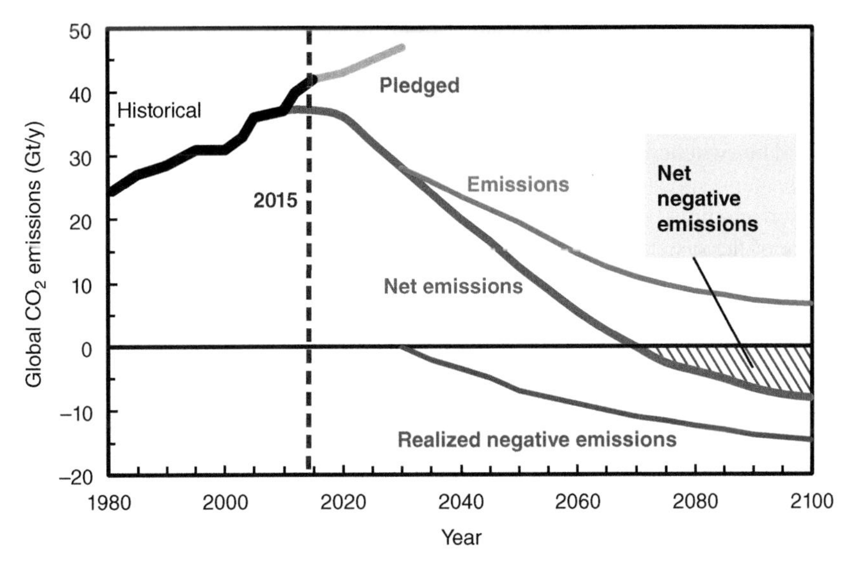

Figure 14.1 Carbon reduction scenario that suggests by 2050 we will not be at negative CO_2 emissions. *Source*: Redrawn using data and information in Anderson and Peters (2016).

and implement CO_2 capture and sequestration processes. The assessment of our own energy use suggests that individuals can do a lot to reduce energy use and CO_2 emissions. However, reducing energy use and using renewable electricity is not enough to curb climate change. The CO_2 already emitted into the air from stored carbon sources in the form of fossil fuels, peat, and soil carbon, and emissions resulting from land use changes (LUC) must also be captured and put into the ground.

14.2 ASSESSING POSSIBLE CHANGES TO OUR OWN DAILY ENERGY CONSUMPTION

The two greatest impacts you could make to your own energy consumption are installing solar panels on your home and reducing your gasoline consumption. While 60–70% of people in the United States are living in homes they own, the rest of the population pays rent to an owner of the property, especially in large cities. Therefore, unless you own your home, you are likely unable to make major changes to the outer infrastructure, such as installing solar photovoltaics to produce electricity, adding insulation, or replacing windows. If you do own a home, many of these and other changes can provide a quick payback in only a few years, such as changing light-bulbs to all be LEDs, insulation around windows, and fixing windows and attic insulation (see Chapter 9). If (or when) you do buy a home, on average people, you will live in that home for about 13 years. Therefore, long-term investments such as solar photovoltaics make sense from both an economical point of view and for reducing CO_2 emissions. The average payback period for solar panels is around 11–15 years depending on location and the availability of energy discounts (Chapter 6). It will take many years, or even decades, to greatly reduce the reliance of the existing electrical grid on fossil fuels. However, it takes only a few months to switch your house to electricity from solar energy especially if your current roof is in a good condition. Otherwise, you would likely want to replace the roof shingles or fix the existing roof before covering it with solar panels.

If you have sufficient solar panels to power your home, you can replace your existing heating systems with electrical devices, for example, by using heat pumps for heating your home. Stovetops and ovens could be switched over to electricity as well if they are currently running on natural gas, and the water heater could similarly be switched to a heat pump (if in a useful location for drawing heat from its surroundings) or to an electric heater. The timing of these changes should take into account the age of the systems. For example, if you have a heating system that is more than 15–20 years old, it will likely need to be replaced soon. Planning to find a good company in your area to do that work is helpful so that you can make the most informed decision on what type of system to install in your home.

The vehicles you own and how much you drive can also have a large impact on your daily energy use and carbon emissions. Using one gallon of gasoline (or petrol) a day (15.2 D) consumes slightly more electrical energy than an average home in the United States. As discussed in Chapter 11, you can greatly reduce your energy consumption and carbon emissions by choosing a highly efficient hybrid or plug-in vehicle or going to a completely electric vehicle especially if you can charge it using renewable electricity from your solar home.

How much can you reduce your consumption of fossil fuels without other changes to the energy infrastructure? In Chapter 3, an energy analysis showed that for a typical US lifestyle that a person might consume 57.1 D of energy based on fossil fuels, with 10.6 D that was derived from non-CO_2 emitting energy sources. That result is shown in Figure 14.2 along with an estimate of how much you could reduce your fossil fuel energy by the following activities:

- Producing electricity from 37 photovoltaic panels, each rated at 320 W, with annual solar irradiance equal to 4.35 peak hours of sunlight (average of locations in the upper northern eastern and mid-western states of 4.5 and 4.2 peak hours, see Figure 6.1).
- Replacing an oil or natural gas furnace with a high-efficiency heat pump.
- Using an electric car that avoids using 1.2 gallons a day of gasoline (average in the *United States*) and instead uses electricity produced by your solar panels.

With the changes shown in Figure 14.2 based on no net energy consumption for your home due to solar panels, you could reduce your fossil fuel use to 25.1 D_{ff} and still take one trip a year by air, for example, from the United States to Europe (assuming Chicago to Frankfurt). The remaining energy consumption would be related to your food consumption. The energy for food production, harvesting, transport, and delivery is assumed to be unchanged from the previous scenario in Figure 14.2a. However, some of the energy associated with the food system is due to your own activities (such as driving to the store and preparing the food). Thus, your actual D_{ff} would be lower due to your use of an electric car and house. As the rest of the country transitioned to renewable energy sources, the 25.1 D_{ff} could be further reduced as the food systems become less dependent on fossil fuels. Skipping international or other travel that consumes aviation fuel could further reduce your D_{ff}.

If you can reduce your energy consumption to 25.1 D_{ff}, then your carbon emissions are now a function of the fuels used for the US electricity grid for food production and jet fuel (Fig. 14.3). These two sources now produce a total of 15.3 C or only 51% of the original C emissions (29.8 C). Avoiding the international flight would further reduce this to 11 C. Thus, while fossil fuel continues to be used for electricity generation, and it is likely that some always will be going even 50 years from now, the importance of carbon capture and sequestration remains an important aspect of achieving negative carbon emissions.

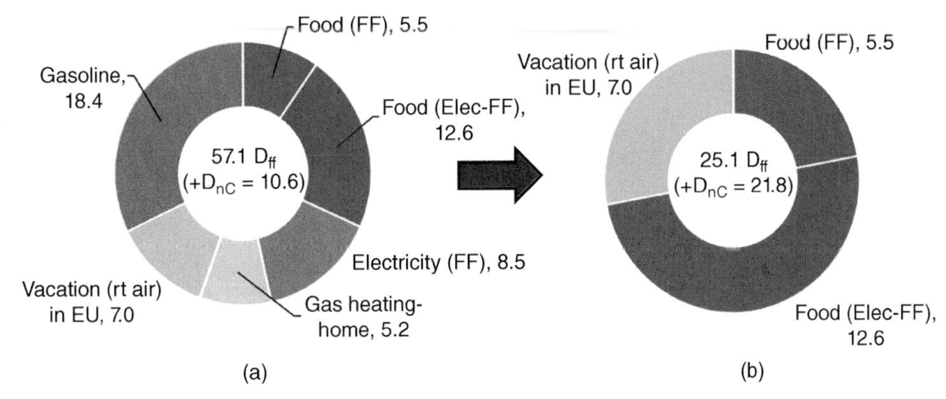

Figure 14.2 (a) Hypothetical D footprint for one person based on the primary energy uses shown (see Figure 3.3b. Numbers may not sum due to rounding in calculations). (b) Revised for a home with solar panels providing 21.8 D_e for home electricity (37 panels with 0.6 D each) for one person (although the average house would have 2.52 people). Natural gas hating was eliminated by using a heat pump powered by solar panels (see Table 9.1 where 32.1 D of natural gas is equivalent to 4.7 D_e for a high efficiency heat pump). Gasoline (18.4 D) for a combustion engine vehicle was replaced with 9.2 D_e provided by the solar panels for an electric vehicle. The total energy (46.9 D) is reduced compared to 67.1 by eliminating fossil fuel energy uses for all sources except as indicated (used in the food system or for a vacation).

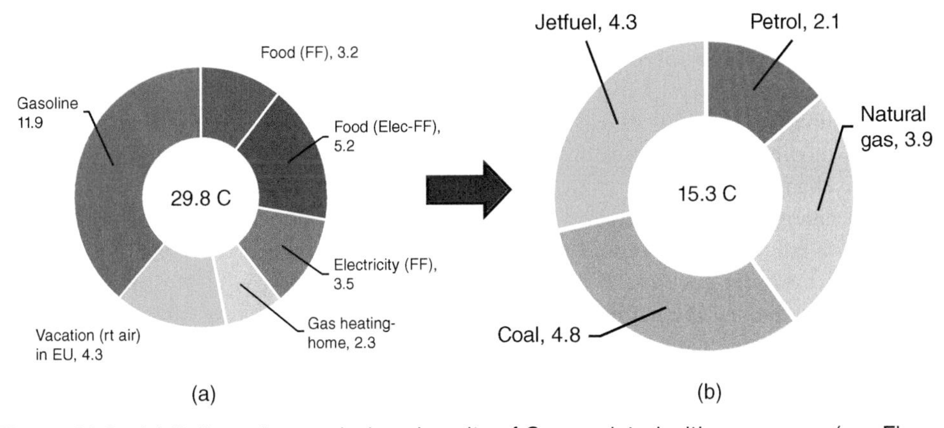

Figure 14.3 (a) Daily carbon emissions in units of C associated with energy use (see Fig. 4.1) and (b) revised carbon emissions based on the assumptions in Figure 14.2 (numbers may not sum due to rounding in calculations.

Energy Consumption Changes Beyond Our Direct Control in Our Daily Lives

There are many other aspects of energy consumption that we cannot directly control in the different categories of residential, commercial, industrial, transportation, and power generation. Electricity and thus fossil fuels are used in our towns for many different things outside of our own direct control, such as use in commercial and municipal buildings and for maintaining roads, lighting cities and streets, providing water and treating used water, and many other community services.

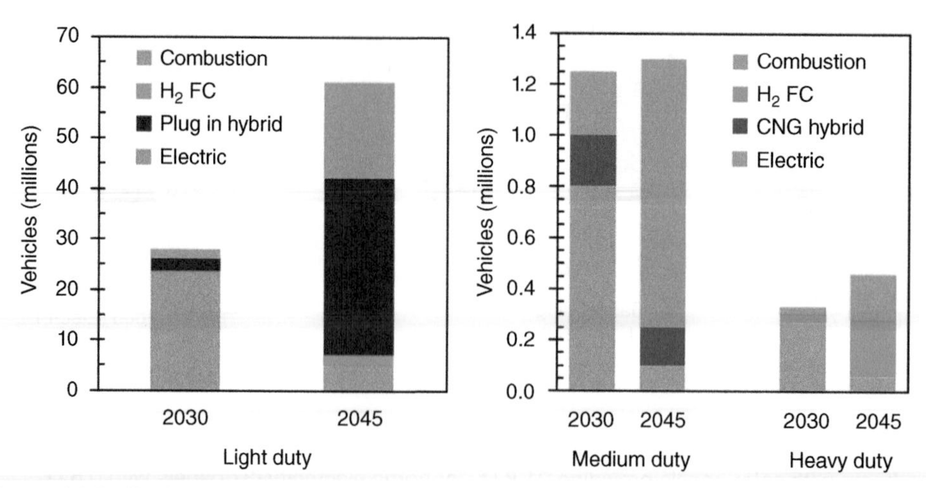

Figure 14.4 Fuel sources for light-, medium-, and heavy-duty vehicles in the plans for California for 2030 and 2045. *Source*: Data from Mahone et al. (2020).

Through participation in public meetings and councils, you can help to promote activities by your own town or city to take all possible steps to decarbonize their activities and minimize energy expenditures and carbon emissions to the maximum possible extent.

Many industries are already very efficient relative to energy consumption that goes into services compared to lost energy as waste heat (see Fig. 3.6). The motivation for most industries is to save money and maximize profits and unfortunately not to focus on reducing CO_2 emissions. However, if consumers shift their purchases to companies that are more responsible than others relative to addressing climate change, then collectively consumer spending can drive changes in energy consumption and CO_2 emissions. A global perspective must be used when evaluating companies as moving carbon emissions outside the borders of your country does not address global climate change.

There are energy expenditures that will be particularly difficult to address because they are so intertwined with international functions. For example, the airline industry, especially as it relates to international travel, is not tied to the decisions of any one country. Global aviation, which includes both national and international travel of people and freight, has contributed to 1.5% of CO_2 emissions, with international shipping accounting for 1.83% in 2017 (Wikipedia, 2020). Steel and cement production are global industries that contribute 13% of total CO_2 emissions (Chapter 13). While steel can be recycled with greatly reduced fuel use and emissions, shifting concrete production away from reliance on heating calcium carbonates, which directly release CO_2, to other materials will remain both an enormous need and challenge.

Transportation remains one of the largest challenges due to the extent of global consumption of fuels, especially due to the energy densities of fuels needed to move heavy-duty vehicles (trucks) and planes. In the plan by the California Air Resources Board, transportation energy sources shift from primarily combustion-based light-duty vehicles to hybrid and electric vehicles by 2045 (Fig. 14.4). For the larger vehicles, the shift is more pronounced to electric for medium-duty vehicles and to H_2 fuel cell vehicles for the heavy-duty vehicles.

The types of vehicles used in the above scenario and their associated fuels in this report do not fully convey the magnitude of the changes needed in California by 2045 because it does not detail the origin of the fuels for transportation as well as other uses that include buildings, industry,

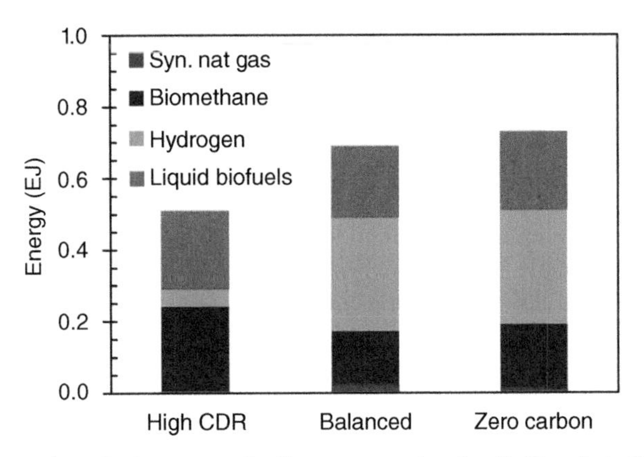

Figure 14.5 Low-carbon fuel sources for three scenarios for California in 2045: Zero-carbon emissions, a reduced need for zero-carbon fuels based on assuming high levels of carbon dioxide sequestration (High CDR), and an intermediate plan (Balanced). *Source*: Data from Mahone et al. (2020).

agriculture, and electricity generation. In the zero-carbon energy scenario in 2045, approximately 30% of fuels will be liquid biofuels such as ethanol and biodiesel and 25% from biomethane, for a total of 55% of the fuels derived from biological processes (Fig. 14.5). Hydrogen gas would provide the single largest percentage of fuels (44%). Hydrogen gas would be produced by water electrolysis using renewable wind, solar, and zero-carbon energy sources as opposed to using fossil fuels as mostly done now. In the high carbon dioxide reduction (CDR) approach, less total energy is needed as it is assumed carbon capture and sequestration can provide the needed CO_2 reductions, and so, fossil fuels can be used as a part of the energy infrastructure. In this high CDR approach, there is greatly reduced production of renewable H_2, and liquid biofuels provide the largest component of the fuels. There is also an intermediate or "balanced" plan that allows for a lower amount of carbon sequestration through a greater reliance on H_2 production (Fig. 14.5).

The analyses conducted by California highlight that liquid biofuel production is a required component of the further low-carbon emission plans. Also, two of the scenarios recognize that it might not be realistic to assume that the whole state could completely move to zero carbon. The changes for 2020–2030 are relatively modest in the plan, but the changes for the period of 2030–2045 for transportation are immense given only a 15-year window for those changes (Fig. 14.4). Thus, it seems prudent to anticipate a high reliance on methods to capture and sequester CO_2 from the atmosphere to meet a zero-carbon net balance.

Embedded Carbon

The focus of this book is on energy consumption due to fossil fuels and the resulting carbon emissions. However, a broader view of energy consumption should include more than just direct use of fossil fuels or electricity. Everything we buy and use contributes to our total energy use and carbon emissions because it takes energy to make, package, and transport those items. We saw for food how the 1 D of energy we consume was magnified by 24.5 times due to the energy to put that amount of food on our table. In the same way, there is energy consumption embedded in every cellphone, piece of clothing, and chair that we buy and use, and then, there is the energy to collect our garbage and used items, as well as to bury (which can result later in the release methane from landfills) or incinerate them. Reducing our carbon footprint therefore requires reducing consumption, recycling, and reuse as much as possible.

How could we track our embedded carbon use? Dieter Helm, a Professor of Economic Policy at the University of Oxford, suggests that we keep a dairy of all our energy use, whether it is in the form of food, gasoline, or even toilet paper (Helm, 2020). However, it would be difficult to quantify energy use or carbon emissions unless everything we bought was stamped with quantities of D and C. In the United States, packaged food items are labeled with nutritional information such as Calories and percent of daily "values" for sodium, cholesterol, fat, carbohydrates, and protein. Imagine if that list included not only Calories or better yet energy in units of D both in the item and to produce the item, but also the carbon emissions related to the production and delivery of that item in units of C, and how much water was used in a unit of w! We could immediately know how our food choices impacted our carbon and water footprints. Taking this concept further, items in other stores could use a similar label. A new television, for example, would not only list the D value for typical electricity use but also the D, C, and w for the production of that item, from the time it was made to a "typical" energy point in the United States. When we buy a car on the United States, there is a "delivery charge" to get the car to our dealer. If that charge can be calculated, then it should be just as easy to say how much energy was used (on average) to make and get the car to that dealer.

There have been many proposals to put a carbon tax on fossil fuel use. By extension, this should mean that every item made in a factory would have a trail of D, C, and w values that would allow a final tag on the item to contain this information. With such a direct and itemized accounting, consumers could make better choices. Have you ever looked at a menu with the Calories next to the items and changed your mind about what you were going to have for your meal? Even if some consumers ignored listed D, C, and w information, there might be enough people that would pay attention and make choices that would greatly reduce fossil fuel use and carbon emissions.

14.3 HOW MUCH CO_2 CAN WE CAPTURE INTO BIOMASS AND THE DEEP SUBSURFACE?

Increasing the use of renewable energy and carbon-neutral energy production alone appears unlikely to solve climate change, so additional solutions will be needed to reduce atmospheric levels of CO_2 and other GHGs. Many solutions to reduce warming of the planet, collectively referred to as geoengineering solutions, have been proposed to address climate change. In this section, we consider two approaches: natural climate solutions, based on sequestering CO_2 in trees and soils; and active carbon capture and CO_2 sequestration in the deep subsurface. Other more controversial solutions, such as stratospheric aerosol injection, marine cloud brightening, and ground-based albedo modification, will not be discussed due to uncertainties in their effectiveness relative to possible adverse outcomes from such large-scale activities.

Different units are used to describe the captured carbon that depend on the context of the calculation. Carbon in soil or in biomass is usually reported in mass of the element carbon, with global masses in terms of Peta grams of carbon (Pg C). Note that this C is carbon, based on the use of units, and not the daily carbon unit of C which is unitless. Carbon dioxide concentrations can be expressed as CO_2 equivalent in terms of Pg CO_2-C or Pg CO_2. When the dash is used, the CO_2 is expressed based on the molecular weight of carbon (12), and when the dash is omitted, it is expressed in terms of the molecular weight of CO_2 (44). Thus, 1 Pg C is equivalent to 1 Pg CO_2-C or 3.67 Pg CO_2 where the number 3.67 is the ratio of the molecular weights of CO_2 and C (3.67 = 44/12).

Carbon Release from the Soil

While CO_2 emissions from fossil fuels are a huge concern, carbon released from soils has also been an ongoing challenge in controlling GHG emissions. Over the past 12,000 years, 133 Gt of soil carbon may have been lost from the top 2 m of soil cover around the world, which is a period of time that spans human land use (Sanderman et al., 2017) (Fig. 14.6). Not surprisingly, given the increase in farming and in the global population, the rate of soil loss has been rapidly increasing over the past 200 years. By some optimistic estimates, two-thirds of this carbon could be reclaimed by adopting best management practices. Thus, returning 88 Gt of carbon into the soil would capture 322 Gt CO_2 from the air (Sanderman et al., 2017). However, more likely estimates are that only 10–30% of this carbon could be captured into soil over a period of 20 years. If 20% were captured, that would amount to 3.2 Gt of CO_2 per year.

Natural Climate Solutions

One way to sequester carbon is through restoring CO_2 into the soil and additional measures of natural capture into biomass, referred to as natural capture solutions (NCS), that include restoration, conservation, and land management practices that either increase carbon storage or avoid GHGs or do both. These NCS are achieved by land changes and practices related to forests, grasslands and agricultural lands, wetlands, and coastal regions (Fig. 14.7). Most of the carbon would be sequestered in forests (10.1 Gt/y or 68.3% of the total). Examples of changes related to forests include reforestation, avoided forest conversion into other uses, natural forest management, avoided wood fuels, and fire management. For agriculture fields and grasslands, several changes relate to grazing, producing biochar, and avoided grassland conversion. Wetland changes include coastal and peat restoration and avoided impacts on these two systems. Overall, the global impact of these changes could enable 23.8 Gt/y of CO_2 equivalent storage in 2030 (Griscom et al., 2017), which is about 72% of global fossil fuel-derived CO_2 emissions assuming the IEA estimate of 33.3 Gt in 2019 (IEA, 2020). It was estimated that about half of this amount of CO_2 (11.3 Gt) could be achieved at a cost of <\$100 US per tonne by 2030, which is sufficient to achieve 37% of the needed CO_2 reduction by 2030 to achieve a <2°C rise (versus 1.5°C based on the Paris agreement pledges).

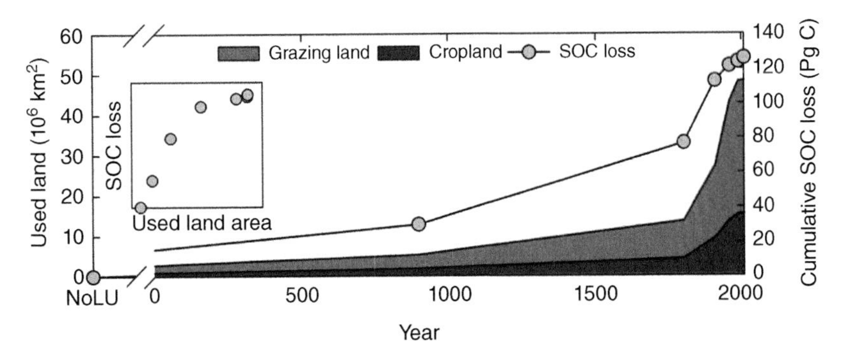

Figure 14.6 Impact of land use on soil organic carbon (SOC) over a period of time representing human impacts on cropland and grazing lands. Units of SOC in Pg of carbon, where 1 Pg = 1 Gt of carbon, equivalent to 3.67 Gt of CO_2 in the atmosphere. Inset: Pg of carbon lost per of 10^6 km^2 of used land area. *Source*: Sanderman et al. 2017.

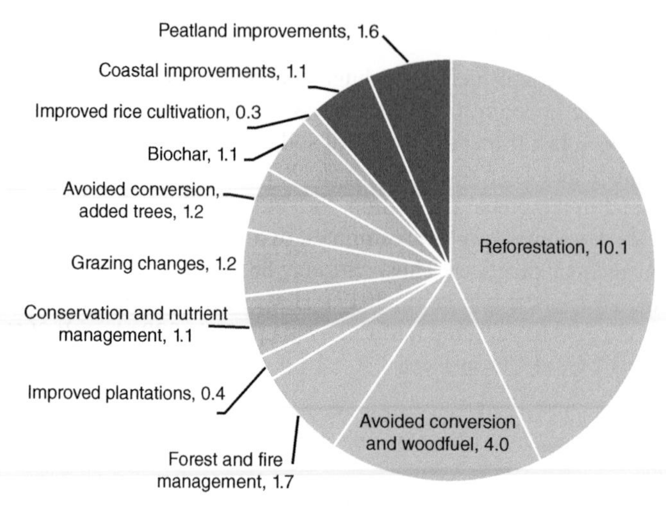

Figure 14.7 Maximum possible climate mitigation solutions for CO_2 capture (Gt/y) related to forests (green), agriculture and grasslands (orange), and wetlands (blue), with an overall total of 23.8 Gt/y by 2030. *Source*: Data from Griscom et al. (2017).

Reductions in CO_2 emissions possible through NCS are impressive but full of challenges in terms of establishing acceptable methods. Estimates of future carbon capture vary depending on assumptions and methods used by different research groups. The amount of carbon captures by NCS, based on a study by Griscom et al. (2017) summarized above, was 30% larger than several previous estimates and differences in assumptions made could reduce capture depending on specific decisions made for how much land could be reclaimed in the future. Griscom et al. (2017) assumed that grazing land in their defined ecoregions was fully reforested. However, if 75% of that land was not reforested, then it would reduce the maximum mitigation potential of reforestation by 31%, as this was the largest category for carbon capture in their study. Other NCS estimates using machine learning and canopy cover have produced unrealistically high estimates, such as those by Bastin et al. (2019), with claims that 205 Gt of CO_2 could be captured into trees. The magnitude of this potential capture was denounced as being completely unreasonable when a more detailed analysis of carbon flows and budgets was considered (Friedlingstein et al., 2019, Lewis et al., 2019).

Estimates of future carbon capture must also consider disruptions that may already be altering existing trends in carbon sequestration by forests. During a twenty-year period (1990–2010), it was estimated that 0.162 Gt of carbon was sequestered in forests and wood products and that rates would continue going into the future (Woodbury et al., 2007). Similarly, there was an expectation that there would be net carbon sequestration in forests in California. However, in 2014, a study on California forests found that between 2001 and 2008, there was a loss in net carbon storage (i.e. a release) of 14 million tonnes due to a reduction in storage per area (61%) with the balance due to a reduction in available area (Battles et al., 2014). This finding of net release was in sharp contrast to previous reports suggesting that past trends may not be reliable going into the future especially as these future conditions will be increasingly impacted by climate change.

Patterns in forest storage capacity are showing changes now due to large-scale land disturbances, many of which are increasing in frequency due to climate change. Thus, predicted expectations for NCS are being overturned by increasingly disruptive activities which are leading to greater net carbon releases than in the past. While elevated CO_2 levels can be shown to increase growth rates of plants and forests, this advantage is being offset by disturbances due to wildfires, increased drought, windthrow (devastation of forest cover due to severe storms and high winds), biotic attack (disease and insects), and LUC (McDowell et al., 2020). As a result of these many different impacts, tree mortality rates in mature forest ecosystems have doubled in Europe and the Americas over the past forty years. Wildfires can produce extensive ecosystem damage and release CO_2 from wood and subsurface soils due to increased erosion and loss of soil stability. The frequency of fires is increasing due to climate change, land use, and management practices, leading to a reduction in carbon storage in these affected systems. When the intervals between fires become shorter (reduced from 250 to 100 years), more carbon is released into the air (Smithwick et al., 2007). A study on the Yellowstone forests in the United States predicted that large wildfires would be increased from a historical frequency of 100–300 years to <30 years due to climate change. Better forest management practices could reduce the impact of fires (Westerling et al., 2011). For example, thinning of the lower woody materials can reduce the release of carbon from live tree biomass by 98% (Hurteau et al., 2008). However, it is not likely that forest thinning could be practiced on as large a scale as that needed to dramatically alter the impact of fires on carbon storage in forests.

Global commerce has caused the introduction of non-native insects and pathogens into a range of ecosystems. Dutch elm disease, which caused the demise of nearly 77 million trees between 1930 and 1970, is perhaps one of the best-known reasons for a rapid decline in trees due disease. In the 1950s in the United States, elm trees lined the streets of many small towns, with their large branches spreading over the streets providing both beauty and shade. However, the introduction of the fungus responsible for Dutch Elm disease wiped out these stands of trees as they were often planted together, which enabled the fungus to move between trees through the roots or aided by transfer of beetles. Tree mortality due to biotic agents that include insects, pathogens, insect–pathogen combinations, and many species-specific diseases has been increasing in severity, frequency, and extent of impact in recent years (McDowell et al., 2020). For example, the hemlock woolly adelgid has devastated trees eastern Hemlock trees in Pennsylvania, along with the spotted lanternfly damaging crops and trees. Such tree mortalities could limit the effectiveness of carbon sequestration particularly with insufficient tree diversity and ecosystem management.

Another concern for NCS as a solution for drawing down CO_2 is permanence as lands converted for NCS could revert to other uses that might increase carbon releases. Land changes are subjected to political decisions, and so changes in officials that can control regulations and commerce, and economic downturns that impact the public's opinions on the importance of controlling climate change versus other needs, can drive land use in new directions. For example, Brazil was successful for many years in slowing deforestation through regulations, but recent changes in political leadership is removing those obstacles and spurring additional losses of forests as the land is converted for other uses.

NCS is an essential component to carbon sequestration. For this approach to be most effective, all countries will need to participate not only in re-planting forests, but also in maintaining these ecosystems. Not only can this growth of forests enhance carbon capture, but it can also help to better maintain the biodiversity of these ecosystems all over the planet.

Example 14.1

In Example 4.2, we estimated that 15 trees were needed to capture 1 C (based on 48 lb CO_2/y/tree) and that this was the equivalent of 104 trees per acre based on capturing 2.5 ton of CO_2 per acre. (a) How many acres of trees would be needed to capture the carbon from a person in the United States based on daily CO_2 emissions of 48.9 C? (b) If PA has 12.8 million people emitting 48.9 C, how many acres are needed to take up this CO_2? (c) The state of PA is 46,055 mi² with 58% of the state covered in forests. What percentage of the population of PA is there sufficient forested land to capture their CO_2?

(a) Using 15 trees = 1 C, we have

$$\frac{15 \text{ trees}}{1 \text{ C}} 48.9 \text{ C} = 734 \text{ trees}$$

$$\frac{734 \text{ trees}}{104 \dfrac{\text{trees}}{\text{ac}}} = 7.1 \text{ ac}$$

(b) The amount of forested land in PA needed for the state population is:

$$7.1 \frac{\text{ac}}{\text{cap}} \; 12.8 \times 10^6 \text{ cap} = 90.9 \times 10^6 \text{ ac}$$

(c) The percentage of population is calculated from the forested acres as:

$$46 \text{ mi}^2 \frac{640 \text{ acre}}{\text{mi}^2} (0.58) = 17.1 \times 10^6 \text{ ac}$$

$$\text{Ratio of available to needed:} \; \frac{17.1 \times 10^6 \text{ ac}}{90.9 \times 10^6 \text{ ac}} \times 100\% = 19\%$$

This result indicates that only 19% of the CO_2 emitted from a person based on the national average can be sequestered by existing forests in the state. This calculation does not consider energy use in the state versus energy exports. PA is third largest net supplier of energy to other states and the third largest generator of electricity.

Simultaneous Biofuel Production and Carbon Storage

While capturing CO_2 into forests and other biomass is needed to draw down CO_2, this approach eliminates using the land for other purposes. As was shown in Figure 14.4, an increased production rate of biofuels will be needed for transportation, especially for heavy-duty vehicles. Therefore, some land will be needed to grow the biomass to produce that fuel, which would eliminate its use purely for natural CO_2 sequestration. This suggests that biofuel production could reduce overall carbon storage which could offset the benefits of using biofuels. Traditional biofuel production, for example, ethanol production from corn, can have a neutral to marginal impact on reducing CO_2 emissions (Chapter 6). Conversion of additional non-agricultural land for production of a biofuel can also result in a large loss of existing carbon storage and thus increase CO_2 emissions into the environment.

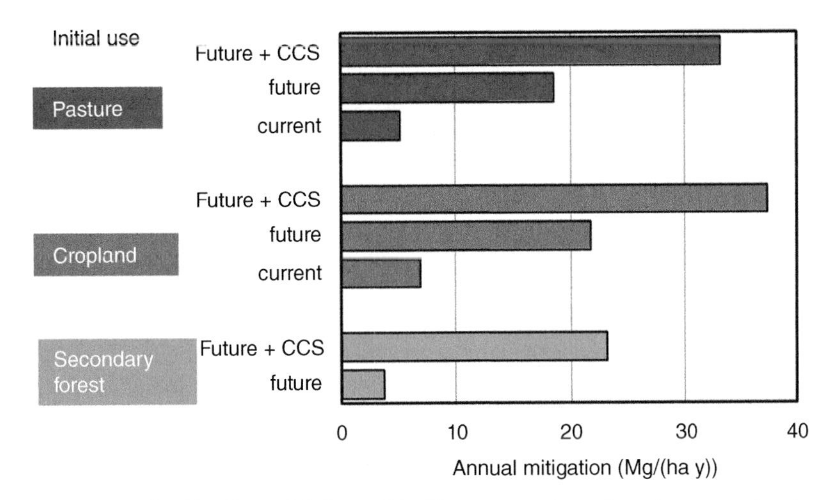

Figure 14.8 Annual mitigation of CO_2 enabled by conversion of pasture, croplands, or secondary forests into production of cellulosic biofuels using Miscanthus based on current technologies, or future technologies in the absence and presence of carbon capture and storage (CCS). *Source*: IEA (2021), Field et al. (2020).

By using cellulosic biomass, it may be possible to simultaneously produce biofuels and achieve carbon capture using certain types of lands (Field et al., 2020). There are substantial amounts of land that either have low productivity for crops or have been abandoned for this purpose, as well as lands no longer used due to dietary shifts away from their original uses. These lands, as well as low production pastures, offer opportunities for their conversion into land used for biofuel production. Cellulosic biomass production is much more efficient in terms of biomass production and reduced use of fertilizers than growing crops such as corn or canola. For example, with current farming and cellulosic conversion rates, it is possible to produce 456 gallons of ethanol using corn, compared to ~1200 gallons using Miscanthus. The greatest opportunities for achieving carbon capture, while growing biofuels, results from conversion of existing agricultural lands with an existing amount of low-carbon storage into production of Miscanthus. An analysis of land currently used for pasture, cropland (agricultural food production), and secondary forests was examined for biofuel production using current technologies (ethanol production) three sites in the United States in the states of New York, Iowa, and Louisiana. With their conversion to biofuel production, it was possible to achieve an annual CO_2 capture of 5 Mg/(ha y) for sites originally used for pasture and 7 Mg/(ha y) for cropland production (Fig. 14.8). Increased carbon capture would not be possible for secondary forest land as conversion of forests into biofuel production could result in a net loss of CO_2 storage −12 Mg/(ha y), compared to a positive capture of 8 Mg/(ha y) with continued forest growth (Field et al., 2020).

Increased benefits of land conversion occur by using next-generation or "future" biorefinery technology conversion and production methods based on hybrid conversion processes to produce ethanol and Fischer–Tropsch liquids. For this production scenario, there is net carbon capture for all three types of land, with the highest amount of 22 Mg/(ha y) for land originally used as cropland (Field et al., 2020). If active carbon capture and storage (CCS) can be implemented at locations sufficiently near these sites, so that the CO_2 from the production sites is captured, compressed, and injected into the deep subsurface, then even greater amounts of net carbon capture are possible at all three types of sites. Using advanced or future technologies of bioenergy fuel productio18n combined with CCS, referred to as bioenergy carbon capture and storage (BECCS), could achieve 37 Mg/(ha y) and thus achieve a dual benefit of producing biofuels needed for the transportation infrastructure while simultaneously capturing carbon. To put these benefits in perspective, transitioning of this land for cellulosic biofuel production could sequester 4 times more carbon than forest restoration and up to 15 times more carbon than grassland restoration (Field et al., 2020).

The benefits of cellulosic biofuel production, especially on certain repurposed lands, have not been sufficiently recognized by policymakers or the public. The widespread negative perception of the production of biofuels in general is based on analyses originally applied to the production of biofuels from food crops. However, while cellulosic production and conversion approaches have advanced, this prior perception has been maintained, and thus, the opportunities for BECCS cannot be realized until these greater benefits can change the existing paradigm. Increased attention to the benefits of cellulosic biofuels from the perspective of climate change offers an opportunity for a new paradigm in carbon capture and bioenergy conversion technologies.

Example 14.2

Field et al. (2020) concluded that continued forest growth captures 7.52 Mg CO_2 eq/ha-y. Convert a previous estimate of 2.5 ton/ac-y used above in Example 4.1 to the same units.

Using the given capture rate and some unit conversions, we have

$$\frac{2.5\text{ ton}}{\text{ac y}}\ \frac{2000\text{ lb}}{\text{ton}}\ \frac{1\text{ kg}}{2.2\text{ lb}}\ \frac{1\text{ Mg}}{10^3\text{ kg}}\ \frac{2.47\text{ ac}}{\text{ha}} = 5.61\frac{\text{Mg CO}_2}{\text{ha y}}$$

The estimate of Field et al. is based on "forest growth," while the 2.5 ton/ac-y is based on biomass in trees. Converting the estimate of Field into English units, we would obtain 3.34 tons of CO_2 equivalent per year or 134 trees/acre if using 48 lb/y/tree based on the calculations in Example 14.1.

Simultaneous Renewable Electricity Production and Engineered Carbon Capture for Sequestration

There are many different ways to capture CO_2 but one of the oldest engineering methods is to use amines to extract the CO_2 from concentrated sources (like flue gas) and then to reverse the process to produce a high concentration of CO_2 (Rochelle, 2009). The chemicals used for relatively concentrated sources like flue gases have been modified to enable CO_2 capture directly from air. One of the challenges in making the process economical is moving enough air through the absorber to produce reasonable rates of capture, but fans and blowers that need to be built and operated can make the overall process expensive. In addition, heat is used for CO_2 release. Estimated costs for direct air capture are on the order of \$200–700 USD per tonne of CO_2, with about 75% of this due to capital expenditures. Most of the remaining costs are due to the heat used in the operation of the system and ventilation. Another promising method is to use an aqueous KOH sorbent, which was estimated to be able to achieve costs of \$94–230 USD per tonne of CO_2 (Keith et al., 2018).

Geothermal systems used for heating and electricity production provide an excellent opportunity to adapt an existing infrastructure to also extract CO_2 from the air using amine-based methods, with only minimal capital expenditures and greatly reduced operational expenses. Fans are already used to provide cooling for these geothermal systems, and therefore, little additional expenses are needed to adapt these processes to achieve air flow through CO_2 absorbers. The waste heat from electricity generation using geothermal systems can also be used to release the CO_2 from the chemical absorbents, and thus, there are little costs associated with this heat above that already being spent on geothermal energy extraction. At a site operated by the company Vito, in Mol, Belgium, a pilot process is underway to adapt a geothermal system used there for heating the industrial facility and generating electricity to additionally capture CO_2. It is estimated that the cost can be reduced to ~60€/tonne of CO_2 (~$72 USD/t).

14.4 MAJOR CHANGES TO THE WATER INFRASTRUCTURE WITH RENEWABLE ENERGY

The shift to renewable energy sources that are not dependent on heat generation from fuels could greatly alter water use in the United States and elsewhere. A total of 48% of the water in the United States are used for thermoelectric power plant cooling. Eliminating water use for cooling water purposes therefore has the potential to cut the current overall water use of 1300 w by about half. There is also an energy expenditure for pumping this cooling water at power plants which is a part of the 0.53 EJ of energy lost from the 14.22 EJ of total electricity generation (Table 2.8). The energy loss for water pumping at a power plant is therefore a relatively small energy loss. However, water intakes at some power plants can have negative impacts on fish and other aquatic organisms through their impingement on screens or destruction when the water is heated. When cooling water is pulled into the plant, screens are used to keep large objects in the water from entering the plant. Some fish pulled against this screen will be unable to escape and will die. This is particularly a problem in the winter when fish are less active and more susceptible to impingement. There is also entrainment of fish eggs and biota small enough to pass through those screens which will be killed or inactivated, which can adversely impact the productivity and stability of fish populations in the lakes or rivers.

Avoiding the need for cooling water could also avoid adverse effects of climate change on power production relative to water needed for thermoelectric cooling. For example, between 2000 and 2015, there were an estimated total of 43 times that power plants either curtailed or shut down power generation due to water-related situations. Most of the plants involved were nuclear power plants (McCall et al., 2016), with a majority of the remaining plants fueled using coal. Plants must shut down when the effluent they produce will be too warm for the receiving water body as it can adversely impact fish and other aspects of the aquatic ecosystem. While warming ambient temperatures can raise the intake water pressure, there is also a problem when there is insufficient water in the receiving water body to mix with the heated plant effluent. The lack of water is particularly a problem during severe draughts, a situation which will become increasingly exacerbated with climate change as more drastic swings in wet and dry cycles occur. In more recent years, these shutdowns have become more frequent. For example, the Hoover dam electricity generation plant shut down in June of 2014 due to a series of droughts, and the City of Atlanta was in danger in 2020 of having power plants that serve that city having to shut down due to draughts in that area as well.

Water shortages around the world are becoming more frequent, and excessive pumping of groundwater has depleted freshwater sources in many parts of the word. The increased use of brackish water and seawater desalination will increase energy use for water services (Chapter 5). It is possible to power water desalination plants using solar energy, but so far, there has been little commercial activity in connecting these water plants to renewable energy, especially in the fossil fuel-rich Middle Eastern countries. Long-term studies on smaller brackish water desalination plants (416 m^3/d) suggest that with appropriate operating procedures, it would be possible to provide dependable desalination plant performance with intermittent energy sources (Ruiz-García and Nuez, 2020). An increasing number of used water treatment plants are purchasing power from renewable energy sources and therefore, it may be possible to achieve energy neutrality of the water treatment processes. Such efforts, when combined with water pumping that is also powered by renewable energy, would greatly aid the sustainability of our water infrastructure.

14.5 HOW MUCH CAN THE WORLD REDUCE ENERGY CONSUMPTION AND CARBON EMISSIONS?

What Technology Solutions Are Needed?

The main goal in reducing energy consumption is reducing carbon emissions and minimizing climate change. So how far can technological solutions drive reductions in CO_2 emissions? The International Energy Agency (IEA) conducted a study on the technologies needed to achieve a global decrease of 2°C (2DS) by 2060 (IEA, 2017). Instead of starting with the "business as usual" baseline in their assessment, they established a reference technology scenario (RTS) that incorporated current and announced policies and commitments. This baseline was also assumed in the analysis by Anderson and Peters (2016) (Fig. 14.1). The RTS predicted CO_2 emissions would not peak until 2060, reaching around 40 Gt/y, or a 16% increase in emissions in 2060 compared to those in 2014 (Fig. 14.9). The RTS scenario further predicts a 2.7°C rise in average global temperatures by 2100, with temperatures continuing to rise into the next century unless aggressive mitigation technologies were used between now and the end of the century to keep that rise at 2°C or less.

The 2DS goal requires a cumulative reduction in CO_2 emissions of 760 Gt between 2015 and 2060 (IEA, 2019). CO_2 emissions were predicted by Anderson and Peters (2016) to decrease from ~34 Gt/y in 2020 to ~20 Gt/y in 2050, with net emissions of 12–13 Gt/y by 2050 in order to limit the temperature rise to 2°C. For the IEA study (Fig. 14.9), there is a similar reduction in the 2DS plan by 2050 with a further reduction to a net of 9 Gt/y by 2060 for the 2DS.

All of these emissions scenarios to limit temperature rises, whether it is to 1.5°C (Hausfather and Peters, 2020) or 2°C (Anderson and Peters, 2016, IEA, 2017), all require substantial carbon sequestration. If the world emitted ~33.3 Gt/y of CO_2 due to energy use (IEA, 2020), a total of ~1000 Gt of CO_2 would be emitted by 2030 (not including other GHGs). However, the world population is expected to increase from 7.80 billion in 2019 to 9.4 billion by 2050. Therefore, these scenarios are also highly dependent on population increases and CO_2 emissions resulting from mitigation strategies. With pledged reductions by countries (RTS scenario) and population growth, the cumulative emissions reach ~570 Gt by 2030 and 1350 Gt by 2050. The 2DS goal of 760 Gt therefore is achieved through reductions in CO_2 emissions due to increasing use of renewables (33%) and nuclear energy (6%), fuel switching (primarily from coal to natural gas) (7%), and efficiency gains (40%), reducing emissions from the RTS scenario from 40 Gt/y to ~9 Gt/y of CO_2 in the 2DS scenario (Fig. 14.9). Coal use would be reduced by 65%, and all coal plants would have CCS, while half of the gas-fired power plants would have CCS. Bioenergy is expected to play a central role in renewables, with twice the energy obtained from biofuels used today or 145 EJ of energy in 2060.

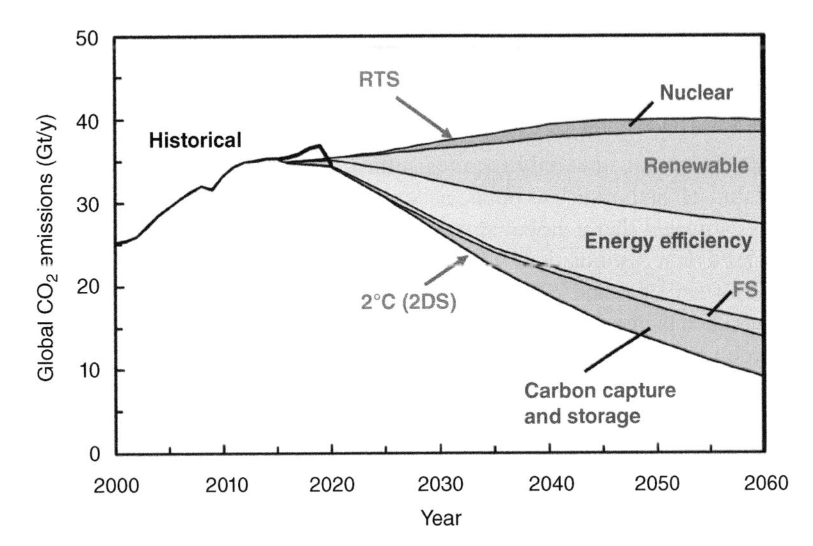

Figure 14.9 Global CO_2 emissions predicted by the IEA for years 2015–2060 based on existing pledges from countries defined as the reference technology scenario (RTS), historical data (Hanna et al., 2020; IEA, 2021), and reductions that could lead to an average global rise in temperature of 2°C (2DS) (IEA, 2017) (FS indicates reductions based on fuel switching). *Source*: Modified from a figure by Rahimi (2021).

The total reduction in CO_2 reduction in the 2DS plan due to CCS is 14%, with around 142 Gt captured between 2015 and 2060. This requires the use of CCS to rapidly increase from 30 Mt today to 1 Gt in 2030 and 6.8 Gt in 2060. CCS is mostly assumed to come from power plants (48%), with a balance between industry (26%) and fuel transformation (26%). The plan does not include direct air capture of CO_2.

The EIA 2DS plan requires large changes for the origins of energy use and the resulting CO_2 emissions. For example, in 2014, heating and cooling in buildings and industry were responsible for 40% of global energy use, with 27% of emissions for transportation and with 65% of this energy for these uses derived from fossil fuels. CO_2 emissions were mainly from power production (40%), with the rest from industry (24%), buildings (8%), and transport (22%). By 2060, the power industry is essentially completely decarbonized through a combination of renewables, fuel switching, and CCS, accompanied by a 57% reduction in CO_2 emissions from industry and 36% for transportation.

The IEA also presented a more aggressive scenario for reductions in CO_2 emissions called beyond the 2DS scenario (B2DS), with the goal of keeping the global average temperatures to a maximum rise of 1.75°C. In this case, CO_2 emissions fall to zero in 2060, compared to 9 Gt/y as shown in Figure 14.9. This reduction would be achieved through increases of 34% in additional CCS, 34% in further advances in efficiencies, 18% due to fuel switching, 15% in renewables, and a minor contribution (1%) by nuclear power. However, *this B2DS goal seems unlikely to occur*. For the period of 2015 to 2019, the RTS predicted a nearly constant amount of CO_2 emissions, but actual data subsequently obtained for that period showed about a 4% increase in emissions in 2019 (Fig. 14.9). The COVID-19 pandemic produced ~6% decrease in CO_2 emissions from 33.4 Gt in 2018 to 31.5 Gt in 2020 (IEA, 2021), temporarily reducing emissions to essentially the same IEA prediction for that year. However, an analysis of historical data showed that after global disruptions that included two oil crises, two financial crashes, and the collapse of the USSR, CO_2 emissions quickly rebounded to levels consistent with those predicted before those events (Fig. 4.4) (Hanna et al., 2020). Thus, the RTS plan is already several years behind in these goals for just the 2DS plan, making it highly unlikely the goals of the B2DS plan can be met.

Successful carbon storage not only requires sufficient underground capacity, but also consideration of non-technical factors such as public acceptance, property leasing, regulations, siting, and financing. It was estimated that a global storage capacity of 10,000 Gt of CO_2 could be reduced to only 400 Gt based on a combination of technical, economic and political factors (Zahasky and Krevor, 2020). Between 1996 and 2020, CO_2 storage increased at a rate of 8.6% per year. If that rate were constant, then the total storage would reach only 441 Gt by 2100, which would be below the amounts needed for the pathway to reduce global warming to <2°C by 2100 of 865 Gt. To reach that target, the cumulative storage capacity will need to be doubled every 8.4 years for the next 60 years (Zahasky and Krevor, 2020).

Technological and Societal Considerations for Reducing CO_2 Emissions

Project Drawdown estimated that it is possible to reduce GHG emissions by a total of 997 Gt by 2050, and CO_2 is the major component of these GHGs. The IEA 2DS projections total about half of this by 2050 or 470 Gt. Thus, there is quite a difference in the amount of GHG reductions estimated to be possible in the next 30 years in these two approaches. The 2DS scenario is more narrowly focused on technical considerations, while Project Drawdown has a much broader view that considers social change and other factors as well as GHGs other than CO_2.

What is assumed in the integrated assessment models used by the IPCC, the IEA, Project Drawdown, as well as many other model simulations, is that *these proposed solutions are not only technically and economically feasible but also that they are socially acceptable*. As shown by the United States withdrawing from the Paris Treaty in 2020, there is no guarantee that the pledged reductions will occur or that the nations will all work together to draw down CO_2 and other GHG emissions. There are uncertainties in the models used to predict the extent of the warming due to CO_2 and other GHGs, but the uncertainties associated with social acceptance of the need for these reductions and social and political forces over the next 30 years are likely the most difficult to model. Global agreement on both reductions of emissions and sequestration are both essential components of a drawdown plan. Even with pledged reductions in emissions without substantial negative carbon emissions due to sequestration, there is little to suggest in these scenarios of the future that we can avoid global warming below levels set at the Paris Agreement.

Energy Consumption in Units of D and Carbon Emissions in Units of C in 2050

Where will we be by 2050? In the 2DS scenario, it is estimated that there will be 13 Gt/y of CO_2 emissions, or assuming a population of 9.4 billion people on average, the daily carbon emissions for one person would be 4.2 C. Assuming a D:C ratio of 1.6 (for petroleum) that would be equivalent to a daily energy use of 6.7 D. This result for D would be about half the 15.3 D estimated for a person in the United States that invests completely in renewable energy for their personal lifestyle, but a person in the United States could reduce this by reductions in international travel and by avoiding consumption of foods with high amounts of energy input (such as beef and other meats). However, these personal reductions do not consider the energy used for global activities, such as steel and concrete production, international travel, and all the other activities that occur in our lives that we cannot directly control by changing our own personal lifestyles.

Can we achieve the needed energy and CO_2 emissions required to minimize the impact of climate change? As we discussed in Chapter 2, the goal of the 2000 W society (equivalent to 20.7 D) is to use only 500 W of fossil-based energy (5.2 D), with the remaining 1500 W supplied by carbon-neutral energy. That use of 500 W is equal to an average of 12,000 kWh per day of fossil fuel energy or 5.2 D_{ff}. This goal is only slightly less than the 6.7 D estimated based on the 2DS plan, and it does not include any carbon sequestration. These D values are comparable to those for Africa (5.0) and India, but are still much higher than those of people in less developed nations where D values approach 1 D. People in developing countries want to raise their standard of living and therefore their energy use, which could substantially increase their D and C values depending on the energy sources. Thus, if there is to be a more equitable sharing in the burden of achieving net global carbon reductions, the industrialized countries and regions, like the United States, China, and the EU, will need to substantially reduce their own energy consumption to allow greater energy use and carbon emissions by the developing nations. The willingness of these countries to lower energy use and reduce CO_2 emissions, while simultaneously enabling the increase in energy consumption by others, will be a challenge going into the future that no modeling scenario at this time can predict.

Shifting the Blame

One problem with each country setting a cut in emissions is that this activity can shift the location of the energy use and carbon emissions, but not change the total amounts of use or emissions. For example, if the United Kingdom wants to increase its use of renewables and it uses wood pellets produced in the United States, then it reduces its carbon footprint by not consuming fossil fuels while it produces heat or electricity from that material. However, this just shifts energy use and emissions to the United States for cutting down the trees and harvesting biomass, preparing the pellets, and shipping the biomass to the United Kingdom. Pledges to reduce carbon emissions therefore will not work unless all players participate, and the embedded carbon is shared across national borders.

Many global companies have shifted manufacturing to certain countries over the past few decades, notably to China in recent years, which can reduce their carbon emissions but increase those of the other country. In 1990, China was a relatively poor country where labor was very cheap and environmental laws were either non-existent or not enforced, but due to its large population, it produced 2.4 Gt/y of GHG emissions or about 10.7% of the world total (Ritchie and Roser, 2020). Fast forward to today where energy use has increased to 10.1 Gt/y or 29.7% of the total global emissions, so that its GHG emissions are the highest in the world for a single country (although not on a per person basis). Shifting steel production from places like the United States or the United Kingdom to China resulted in less emissions for the United States and the United Kingdom, but this shift did not result in any decrease in net global energy use or emissions. This shift in CO_2 and other gas emission has resulted from all the new factories in China and the economic growth of the country. The gross domestic product (GDP) of China has increased by 38 times in just the past three decades (from US $0.36 to $13.6 trillion) because of the rise in manufacturing in that country and advances as shown by this economic development (Helm, 2020).

One approach to fully incentivize reducing carbon emissions for countries and businesses, rather than just shifting the point of production, is to have a global carbon tax as suggested by Helm (2020). Therefore, whether the goods are made in China, Japan, or the United States, a carbon tax is levied in proportion to the emissions. As noted by Helm, if an item shipped to a country is not stamped with proper approvals indicating a paid carbon tax, it would be sent back to the original country, or the carbon tax would be paid by the shipper to the country receiving the materials. This payment would provide incentive for the companies to both minimize carbon emissions and pay the tax in the country of origin rather than giving that money to another country.

14.6 REDUCING CO_2 EMISSIONS FROM FOSSIL FUELS WILL NOT BE ENOUGH

The focus of this book has been on energy use and CO_2 emissions from fossil fuels from the perspective of our daily lives. While we do have some control over our own energy use and CO_2 emissions as we have discussed, there are still fossil fuel-derived emissions that we cannot control, such as those from global concrete and steel production, fossil fuels used for the food processing system, and international travel. For these larger-scale challenges, countries will need to collectively engage in a global effort to reduce CO_2 emissions and to capture and sequester emissions from fossil fuels through natural and engineered CCS technologies. But fossil fuel sources of CO_2 are only a part of the total amount of CO_2 emissions, and other gases contribute to climate change as well.

The different factors that are contributing to climate change due to GHG emissions are well summarized by the United Nations (UN) Environment Programme, in their Emissions Gap Reports, with the most recent report in 2020 for emissions up to 2019 (United Nations Environment Programme, 2020). These comprehensive reports are inclusive of all GHG emissions that occur from using fossil fuels, emissions due to other sources, and emissions those related to LUC. The most recent UN report indicated that 38.0 ± 1.9 Gt of CO_2 was released from fossil fuels and 14.4 Gt from other gases or a total of 52.4 ± 5.2 Gt of GHGs (Table 14.1). These other gases were primarily methane, mostly derived from animals (fermentation, gasification, and manure management), in addition to nitrous oxide and fluorinated chemicals. Overall, fossil fuel-derived CO_2 emissions contributed to 72.5% of this total.

An often overlooked portion of GHG emissions is the contribution (or capture) of gases from LUC, collectively defined as "emissions and removals of greenhouse gases resulting from direct human induced land use, land use change and forestry activities" (United Nations Environment Programme, 2020). When CO_2 emissions from LUC are included with fossil fuel emissions, the

Table 14.1 Global greenhouse gas emissions for 2019 based on CO_2 equivalents.

Emissions	CO_2e emissions (Gt/y)
CO_2 fossil fuel gas emissions	38.0 ± 1.9
Other gases	14.4
Methane (CH_4) = 9.8	
Nitrous oxide (N_2O) = 2.8	
Fluorinated gases = 1.7	
Subtotal	52.4 ± 5.2
Land use, land use changes (LUC), and forestry	6.7
CO_2 = 6.3	
CH_4 and N_2O = 0.5	
Total	59.1 ± 5.9

Numbers may not sum as given due to rounding.
Source: United Nations Environment Programme (2020).

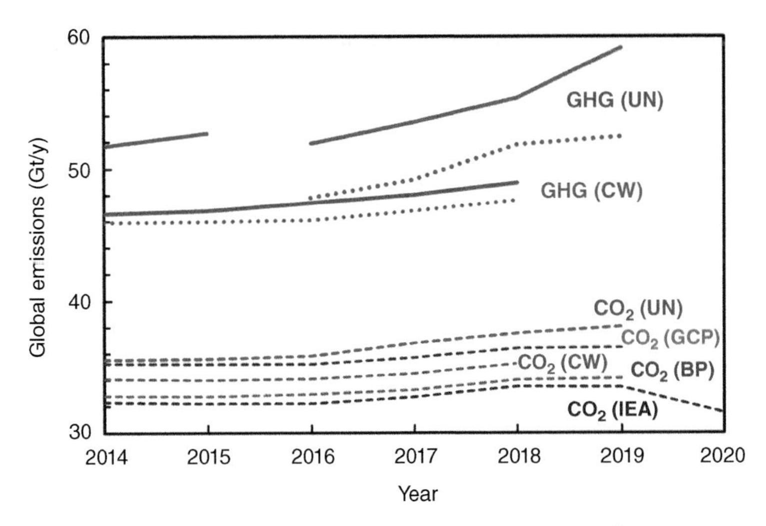

Figure 14.10 Global GHG or CO_2 emissions reported by several different agencies: all GHG plus LUC, solid lines; all GHG with no LUC, dotted lines; and only CO_2 emissions, dashed lines. The break in the line for the UN data reflects a change in how it was reported. *Source*: United Nations Environment Programme (2020); Climate Watch (2021); GCP, BP, Dudley (2019); IEA (2021).

total for just CO_2 emissions increases from 38.0 to 44.3 Gt for 2019 (Table 14.1). LUC are important because switching completely to renewable fuels does not necessarily change LUC sources of CO_2 emissions. The additional contribution of the two main LUC gases, CH_4 and NO_2, increases the contribution of "other gases" to a total of 14.9 Gt CO_2-e or 25% of all GHG emissions. With LUC included CO_2 emissions account for a smaller overall percentage of total GHG emissions (64%). The impact of LUC using data from the UN report is much larger than some other estimates. For example, the Climate Watch Program database reported 1.3 Gt due to LUC for 2018 (total of 48.9 Gt), compared to 3.5 for LUC reported by the UN in that year (55.3 Gt total) (Fig. 14.10). In 2019, the additional LUC emissions increased to 6.7 Gt (Table 14.1) (data were not yet available from the Climate Watch Program for 2019).

The estimated annual release of CO_2 from fossil fuels varies among many available reports on global emissions. For example, the UN report indicated a total of 38.0 ± 1.9 Gt for CO_2 from fossil fuels in 2019 (Fig. 14.10), compared to estimates of 37.1 Gt (Wikipedia, 2020), 36.1 Gt (IEA, 2021), and 34.0 Gt (Dudley, 2019) reported by others. These higher estimates likely arise from differences in activities included in the emissions analyses.

What Does a Pledge to Cut 50% of "Carbon Emissions" Really Mean?

The pledge of a nation to cut "carbon emissions" is often assumed by many people to be CO_2 emissions associated with combustion of fossil fuels, but it could also be a pledge to reduce total GHG emissions. In April of 2021, the United States announced a plan to cut GHG emissions, based on 2005 levels, by as much as 50–52% by 2030. While President Biden in a press conference indicated that the United States would cut "greenhouse gases," some reports stated that it would be a reduction in carbon emissions (Sullivan and Liptak, 2021) and therefore only CO_2 emissions. Certainly, the magnitude of this change of 50% or more is an ambitious goal that can help reduce

CO_2 emissions from fossil fuels, but the terminology needs to be clear to accurately report on what exactly these cuts apply to.

Consider how the different numbers for GHGs would change depending on what applied to the 50% cut. The most recent US EPA report indicated emissions at a total of 6.58 Gt of CO_2e for 2019 by the United States, but with LUC in the United States (net negative), this is reduced to 5.79 Gt (EPA, 2021). The total CO_2 emissions from fossil fuel combustion were 4.89 Gt, but if other sources (such as cement and ammonia production, among others) are included, the total for all CO_2 emissions increases to 5.27 Gt. Thus, it is not clear whether the United States would reduce by 50% the total that included (in 2019) all GHGs (6.58 Gt), only CO_2 (5.27 Gt), or only CO_2 from fossil fuel combustion (4.89 Gt). For the United States, it is assumed that LUC would be included in reductions as LUC is a net negative for GHG emissions, with – 0.79 Gt in 2019. For other countries in the world, the LUC could add to GHG emissions. Further clarification is therefore needed on whether these total cuts in gas emissions include CO_2 from all sources or just fossil fuel combustion. The exact number has a large impact on how much the total GHG emissions would be reduced.

The baseline chosen by the United States for this 50% reduction is the year 2005, for which the United States had one of the very highest amounts of GHG emissions in its history, compared to lower emissions in more recent years such as 2015 (the year of the Paris Agreement) or 2019 (pre-COVID 19). Based on all sources and sinks, the United States has already reduced its net GHG emissions in 2019 compared to 2005 by 13% or from 6.64 Gt/y in 2005 to 5.79 Gt/y (with LUC) (EPA, 2021). For all CO_2 emissions (no LUC), that reduction was 14% (from 6.14 to 5.27 Gt), or even slightly larger at 15% by only considering fossil fuel combustion as sources (from 5.75 to 4.89 Gt). The reduction in all these GHG emissions occurred primarily as a result of switching from coal to natural gas as total energy use in the United States was essentially the same in 2005 and 2019 (not corrected for renewables in those years) (Lawrence Livermore National Laboratory, 2020).

The terms “carbon emissions” and “GHG emissions” are often loosely used and interchanged, and therefore, news articles often lack clarity on whether the topic is only CO_2 or all GHGs. Scientists and others reporting on the amounts of these emissions need to help continually reinforce the distinctions between CO_2, especially as it relates to only fossil fuels or also from land use, and the other GHGs. However, we often try to simplify our statements to focus on other points we are trying to make, rather than dive into the details every time we talk about the magnitude of GHG emissions. For example, the well-known climate scientist Michael Mann recently wrote in his book on climate change that “*we are currently generating the equivalent of roughly 55 billion tons per year of carbon dioxide through fossil fuel burning and other human activities*” (Mann, 2021). Does this amount include all GHGs and those from LUC? An expert like Mann knows his intent was to include all GHGs by saying emissions that were “equivalent” to CO_2, but others reading this statement might assume that these emissions of 55 Gt were *only* due to CO_2 since the statement stressed fossil fuels. It might have been better to write “carbon dioxide associated with fossil fuels and other GHGs” for clarity, but this revision lacks the brevity of the original statement. Also, did his estimate include LUC? Mann uses the word “roughly,” perhaps intending to avoid this issue of a specific number that might better inform us on LUC. The source for his statement is the same one used for data in Table 14.1, but the number would be either 52.4 Gt/y (all gases without LUC) or 59.1 Gt/y (with LUC), so 55 Gt/y is roughly between these numbers (and would not change the point he was making). It is a challenge for all writers about how to clearly distinguish CO_2 derived from fossil fuels from CO_2 from other sources, for example, from materials used to make cement. It gets more complicated if you want to include GHGs other than CO_2 or distinguish direct emissions from other sources such as those from LUC. There are perhaps places in this book where the source of the emissions is also not clear due to brevity (although the focus in previous chapters has been emissions related to energy use).

The main point to make when addressing GHGs is that eliminating CO_2 emissions from fossil fuels prevents only a part of the problem or only 38 Gt/y of all CO_2 equivalent emissions (based on the UN report). Eliminating CO_2 from fossil fuel combustion and using only renewable energy sources would still leave the world with 21.1 Gt/y (without LUC) to 44.3 Gt (with LUC) of CO_2, with a remaining 14.8 Gt associated with other gases that cannot be removed from the atmosphere in the same way as CO_2.

This concept of clarifying what amount of GHGs we can capture is also relevant to statements made in a recent book by Bill Gates on climate change. The overall thesis of Gates is that technology can address climate change through capturing CO_2 emissions and reducing other GHG emissions primarily using current and future technologies. Gates is well informed on the differences between CO_2 and other gases relative to climate change. However, in discussing possible solutions, Gates estimated "*the cost of absorbing emissions using DAC*" (direct air capture) for 51 Gt/y of emissions (Gates, 2021). The magnitude of this number, which is similar to the UN estimate of 52.4 Gt/y without LUC, indicates that these emissions included all GHG gases (CO_2, CH_4, N_2O, and fluorinated chemicals) but not LUC. However, DAC methods being developed only capture CO_2 and not other gases. Thus, the statement leads the reader to conclude that DAC can capture all annual emissions of 51 Gt/y as if they are all CO_2, rather than just capture a part of the gases emitted in a year (38 Gt without LUC or 44.3 Gt with LUC). The cost of capturing fossil fuel emissions at \$100 per ton of CO_2 was concluded to be \$5.1 trillion United States per year rather than \$3.8 trillion for only CO_2. It would have been better to reinforce that only CO_2 could be removed by DAC and that additional CO_2 (beyond that associated with fossil fuel combustion) would need to be removed to compensate for both CO_2 *and other GHGs*. A clearer distinction at this point relative to his discussion of needed solutions would help to reinforce that DAC itself does not "solve" non-CO_2-related GHG emissions.

The Largest Emitters can Make the Biggest Changes

The United States must lead by example as its citizens have both a very high per capital use of energy, and the United States ranks second in the world in total emissions. China has at least two challenges: It is the number one largest emitter of GHGs due in large part to its high reliance on coal, which is the most carbon-intensive fuel; and it is a country with the highest population in the world. China still considers itself to be a developing nation, and thus, a quest to further modernize could jeopardize all global climate action plans. In April of 2021, Chinese President Xi Jinping pledged that China would become carbon neutral by 2060, but he also indicated that emissions by China would continue to increase its emissions as he did not pledge for the country to reach a peak in emissions until 2030 (Sullivan and Liptak, 2021). This decade long delay is in stark contrast to all climate action plans calling for declining, and not increasing, emissions between 2020 and 2030.

The top four GHG emitters in the last 10 years, China, the United States, EU27+UK, and India, accounted for 55% of all GHG emissions (United Nations Environment Programme, 2020). If you add in the Russian Federation and Japan and include emissions from international transport, this increases the contributions of this group to 65% of all emissions. It is especially critical that these participants all cooperate in a plan to reduce these emissions. Agreement and action by the

G20 member countries could be a very effective catalyst for change, as they collectively account for 78% of GHG emissions. Even this group of countries cannot accomplish limiting the effects of climate change, however, if the rest of the world continues to increase its use of fossil fuels and not address changes in land use (such as the destruction of forests) that increase GHGs. The solution to climate change is dependent on the actions of individuals, cities, countries, and businesses. Success going forward is dependent on a complicated mix of technologies, as well as politics and beliefs, but the most important ingredients for a solution are determination and the will of the world address the greatest ever challenge to humanity of global climate change.

References

ANDERSON, K. & PETERS, G. 2016. The trouble with negative emissions. *Science,* 354(6309), 182–183.

ANDRIJEVIC, M., SCHLEUSSNER, C.-F., GIDDEN, M. J., MCCOLLUM, D. L. & ROGELJ, J. 2020. COVID-19 recovery funds dwarf clean energy investment needs. *Science,* 370(6514), 298–300.

BASTIN, J.-F., FINEGOLD, Y., GARCIA, C., MOLLICONE, D., REZENDE, M., ROUTH, D., ZOHNER, C. M. & CROWTHER, T. W. 2019. The global tree restoration potential. *Science,* 365(6448), 76–79.

BATTLES, J. J., GONZALEZ, P., ROBARDS, T., COLLINS, B. M. & SAAH, D. S. 2014. California forest and rangeland greenhouse gas inventory development final report, Final report, California Air Resources Board Agreement 10-778. https://ww3.arb.ca.gov/cc/inventory/pubs/battles%20final%20report%2030jan14.pdf.

Climate Watch. 2021. ClimateWatch. https://www.climatewatchdata.org/ [Accessed April 28 2021].

DUDLEY, B. 2019. *BP statistical review of world energy 2019*, Pureprint Group Limited UK. https://www.bp.com/content/dam/bp/business-sites/en/global/corporate/pdfs/energy-economics/statistical-review/bp-stats-review-2019-full-report.pdf.

EPA. 2021. Inventory of U.S. Greenhouse Gas Emissions and Sinks, 1990-2019. https://www.epa.gov/ghgemissions/inventory-us-greenhouse-gas-emissions-and-sinks-1990-2019.

FIELD, J. L., RICHARD, T. L., SMITHWICK, E. A. H., CAI, H., LASER, M. S., LEBAUER, D. S., LONG, S. P., PAUSTIAN, K., QIN, Z., SHEEHAN, J. J., SMITH, P., WANG, M. Q. & LYND, L. R. 2020. Robust paths to net greenhouse gas mitigation and negative emissions via advanced biofuels. *Proceedings of the National Academy of Sciences,* 117(36), 21968–21977.

FRIEDLINGSTEIN, P., ALLEN, M., CANADELL, J. G., PETERS, G. P. & SENEVIRATNE, S. I. 2019. Comment on "The global tree restoration potential". *Science,* 366(6463), eaay8060.

GATES, B. 2021. *How to avoid a climate disaster*, Knopf, New York.

GRISCOM, B. W., ADAMS, J., ELLIS, P. W., HOUGHTON, R. A., LOMAX, G., MITEVA, D. A., SCHLESINGER, W. H., SHOCH, D., SIIKAMÄKI, J. V., SMITH, P., WOODBURY, P., ZGANJAR, C., BLACKMAN, A., CAMPARI, J., CONANT, R. T., DELGADO, C., ELIAS, P., GOPALAKRISHNA, T., HAMSIK, M. R., HERRERO, M., KIESECKER, J., LANDIS, E., LAESTADIUS, L., LEAVITT, S. M., MINNEMEYER, S., POLASKY, S., POTAPOV, P., PUTZ, F. E., SANDERMAN, J., SILVIUS, M., WOLLENBERG, E. & FARGIONE, J. 2017. Natural climate solutions. *Proceedings of the National Academy of Sciences,* 114(44), 11645–11650.

HANNA, R., XU, Y. & VICTOR, D. G. 2020. After COVID-19, green investment must deliver jobs to get political traction. *Nature,* 582, 178–180.

HAUSFATHER, Z. & PETERS, G.P. 2020. Emissions – The 'business as usual' story is misleading. *Nature,* 577, 618–620.

HELM, D. 2020. *Net Zero: How we stop causing climate change*, HarperCollins Publishers.

HURTEAU, M. D., KOCH, G. W. & HUNGATE, B. A. 2008. Carbon protection and fire risk reduction: toward a full accounting of forest carbon offsets. *Frontiers in Ecology and the Environment,* 6(9), 493–498.

IEA. 2017. Energy technology perspectives 2017 – Catalysing energy technology transformations. https://www.iea.org/reports/energy-technology-perspectives-2017.

IEA. 2019. Projections: Energy policies of IEA countries 2019 edition. https://webstore.iea.org/ods-projections [Accessed August 7 2020].

IEA. 2020. Global CO_2 emissions in 2019. https://www.iea.org/articles/global-co2-emissions-in-2019 [Accessed October 18 2020].

IEA. 2021. Global Energy Review: CO_2 Emissions in 2020. https://www.iea.org/articles/global-energy-review-co2-emissions-in-2020 [Accessed March 14 2021].

KEITH, D. W., HOLMES, G., St. ANGELO, D. & HEIDEL, K. 2018. A process for capturing CO_2 from the atmosphere. *Joule,* 2(8), 1573–1594.

LAWRENCE LIVERMORE NATIONAL LABORATORY. 2020. Energy flow charts. https://flowcharts.llnl.gov/commodities/energy [Accessed April 18 2021].

LEWIS, S. L., MITCHARD, E. T. A., PRENTICE, C., MASLIN, M. & POULTER, B. 2019. Comment on "The global tree restoration potential". *Science,* 366(6463), eaaz0388.

MAHONE, A., SUBIN, Z., MANTEGNA, G., LOKEN, R., KOLSTER, C. & LINTMEIJER, N. 2020. Achieving carbon neutrality in California: PATHWAYS scenarios developed for the California Air Resources Board (Draft, August 2020). https://ww2.arb.ca.gov/sites/default/files/2020-08/e3_cn_draft_report_aug2020.pdf.

MANN, M. 2021. *The new climate war*, PublicAffairs.

MCCALL, J., MACKNICK, J. E. & HILLMAN, D. 2016 Water-related power plant curtailments: an overview of incidents and contributing factors. https://www.nrel.gov/docs/fy17osti/67084.pdf [Accessed October 31 2021].

MCDOWELL, N. G., ALLEN, C. D., ANDERSON-TEIXEIRA, K., AUKEMA, B. H., BOND-LAMBERTY, B., CHINI, L., CLARK, J. S., DIETZE, M., GROSSIORD, C., HANBURY-BROWN, A., HURTT, G. C., JACKSON, R. B., JOHNSON, D. J., KUEPPERS, L., LICHSTEIN, J. W., OGLE, K., POULTER, B., PUGH, T. A. M., SEIDL, R., TURNER, M. G., URIARTE, M., WALKER, A. P. & XU, C. 2020. Pervasive shifts in forest dynamics in a changing world. *Science,* 368(6494), eaaz9463.

RAHIMI, M. 2021. Reductions in global CO_2 emissions based on an IEA report, Personal communication.

RITCHIE, H. & ROSER, M.. 2020. CO_2 and greenhouse gas emissions. https://ourworldindata.org/co2-and-other-greenhouse-gas-emissions [Accessed March 15 2021].

ROCHELLE, G. T. 2009. Amine scrubbing for CO_2 capture. *Science,* 325(5948), 1652–1654.

RUIZ-GARCÍA, A. & NUEZ, I. 2020. Long-term intermittent operation of a full-scale BWRO desalination plant. *Desalination,* 489, 114526.

SANDERMAN, J., HENGL, T. & FISKE, G. J. 2017. Soil carbon debt of 12,000 years of human land use. *Proceedings of the National Academy of Sciences,* 114(36), 9575–9580.

SMITHWICK, E. A. H., HARMON, M. E. & DOMINGO, J. B. 2007. Changing temporal patterns of forest carbon stores and net ecosystem carbon balance: The stand to landscape transformation. *Landscape Ecology,* 22(1), 77–94.

SULLIVAN, K. & LIPTAK, K.. 2021. Biden announces US will aim to cut carbon emissions by as much as 52% by 2030 at virtual climate summit. https://www.cnn.com/2021/04/22/politics/white-house-climate-summit/index.html [Accessed April 25 2021].

UNITED NATIONS ENVIRONMENT PROGRAMME. 2020. Emissions gap report 2020. https://wedocs.unep.org/bitstream/handle/20.500.11822/34438/EGR20ESE.pdf?sequence=25.

VOOSEN, P. 2021. Global temperatures in 2020 tied record highs. *Science,* 371(6527), 334–335.

WESTERLING, A. L., TURNER, M. G., SMITHWICK, E. A. H., ROMME, W. H. & RYAN, M. G. 2011. Continued warming could transform Greater Yellowstone fire regimes by mid-21st century. *Proceedings of the National Academy of Sciences,* 108(32), 13165–13170.

WIGGLESWORTH, A. & SERNA, J. 2020. California fire season shatters record with more than 4 million acres burned, Newspaper: Los Angeles Times, October 4, 2020.

WIKIPEDIA. 2020. List of coutnries by carbon dioxide emissions. https://en.wikipedia.org/wiki/List_of_countries_by_carbon_dioxide_emissions, [Accessed June 12 2020].

WOODBURY, P. B., SMITH, J. E. & HEATH, L. S. 2007. Carbon sequestration in the U.S. forest sector from 1990 to 2010. *Forest Ecology and Management,* 241(1), 14–27.

ZAHASKY, C. & KREVOR, S. 2020. Global geologic carbon storage requirements of climate change mitigation scenarios. *Energy and Environmental Science,* 13(6), 1561–1567.

APPENDIX 1: CONVERSION FACTORS

Table A1.1 Conversion factors based on electrical grid energy and primary or fossil fuel energy.

Conditions	Calculation	D:D	$(D:D)^{-1}$
Primary energy (D_p) for electricity generation for US grid electricity (D_e)	$D_p/D_e = 35.01/14.21$ (Table 3.6)	2.46	0.41
Fossil fuel energy (D_{ff}) for electricity generation for US grid electricity (D_e)	$D_{ff}/D_e = 23.24/14.21$ (Table 3.6)	1.64	0.61
Carbon-free energy (D_{nC}) for electricity generation for US grid electricity (D_e)	$D_{nC}/D_e = 11.77/14.21$ (Table 3.6)	0.83	1.21
Coal power plant primary energy (D_p, $\eta = 32.2\%$ efficiency) to electricity (D_e)	$D_p/D_e = 1/0.322$ (Example 4.4, η from Table 2.9 using ratio of primary to produced electricity)	3.11	0.32
Coal power plant primary energy (D_p, $\eta = 32.6\%$ efficiency) to electricity (D_e)	$D_p/D_e = 1/0.326$ (Example 4.4 method, η from Table 2.9 for EIA reported plant efficiency)	3.07	0.33
Petroleum power plant primary energy (D_p, 31.3% efficiency) to electricity (D_e)	$D_p/D_e = 1/0.313$ (Example 4.4 method, η from Table 2.9 based on ratio of primary to produced electricity)	3.19	0.31
Petroleum power plant primary energy (D_p, 30.8% efficiency) to electricity (D_e)	$D_p/D_e = 1/0.308$ (Example 4.4 method, η from Table 2.9 for EIA reported plant efficiency)	3.25	0.31
Natural gas power plant primary energy (D_p, 43.1% efficiency) to electricity (D_e)	$D_p/D_e = 1/0.431$ (Example 4.4 method, η from Table 2.9 based on primary to produced electricity)	2.32	0.43
Natural gas power plant primary energy (D_p, 43.6% efficiency) to electricity (D_e)	$D_p/D_e = 1/0.436$ (Example 4.4 method, η from Table 2.9 for EIA reported plant efficiency)	2.35	0.44
Primary energy for food system (D_p) to food system energy (D)	$D_p/D = 24.5/13.2$ (Table 8.1)	1.86	0.54

(continued overleaf)

Table A1.1 *(continued)*

Conditions	Calculation	D:D	(D:D)$^{-1}$
Fossil fuel energy for food system (D_{ff}) to food system energy (D)	D_{ff}/D = 18.1/13.2 (Table 8.1)	1.37	0.73
Carbon-free energy for food system (D_{nC}) to food system energy (D)	D_{nC}/D = 6.4/13.2 (Table 8.1)	0.79	1.27
Example of a person that uses primary energy with energy use (some as electricity) of 48.9 D	D_p/D = 67.6/48.9 (Figs. 3.2 and 3.3)	1.38	0.72
Example of a person that uses fossil fuel (D_{ff}) energy with energy use (some as electricity) of 48.9 D	D_{ff}:D = 57.1/48.9 (Figs. 3.2 and 3.3)	1.18	0.85

Values for D:C indicate D/C, and values for C:D are calculated as C/D.

Table A1.2 Conversion factors based on electrical grid energy and primary or fossil fuel energy.

Conditions		D:C	C:D
Total energy used by the US for all energy sources (D_p) to all CO_2 emissions (C):	D_p:C = 101.6 D/48.9 C (Example 4.5, based on total primary energy in Fig. 3.1 and total CO_2 emissions based on C in Fig. 4.1)	2.08	0.48
Total energy used by the US in all fossil fuels (D_p) to all CO_2 emissions (C):	D_p:C = 84.5 D/48.9 C (Total energy in Table 2.6, and CO_2 emissions based on C in Fig. 4.1 and Example 4.5)	1.73	0.58
Primary energy for electricity (D_p) to CO_2 emissions (C):	D_p:C = 35 D_p/14.59 C (Example 4.5, based on total primary electricity energy in Fig. 2.4 converted to D, and D:C ratios in Table 4.2)	2.40	0.42
Primary energy for electricity only for fossil fuels (D_{ff}) to CO_2 emissions (C):	D_p:C = 23.2 D/14.59 C (Example 4.5, based on primary energy in Table 2.7 and D:C ratios in Table 4.2)	1.59	0.63
Electricity energy (D_e) to CO_2 emissions (C):	D_e:C = 14.22 D_e/14.59 C (Table 3.6 for the D_e, and Example 4.5 for C from electricity)	0.97	1.03
Coal power plant electricity (D_e) to CO_2 emissions (C):	D_e/C = [C = D_e/($\eta \times$ D:C)] = [1/(0.322 $\times$ 1.20)] (Example 4.4, Table 4.3)	2.59	0.39
Petroleum power plant electricity (D_e) to CO_2 emissions (C):	D_e/C = [C = 1/($\eta \times$ D:C)] = [1/(0.313 $\times$ 1.63)] (Example 4.4, Table 4.3)	1.96	0.51

Table A1.2 (*continued*)

Conditions		D:C	C:D
Natural gas power plant electricity (D_e) to CO_2 emissions (C):	D_e/C = [C = 1/(η × D:C)] = [1/(0.431 × 2.23)] (Example 4.4, Table 4.3)	1.04	0.96
Primary energy for food system (D_p) to carbon emissions (C):	D_p:C = 24.5/10.9 (Figs. 8.2 and 8.5)	2.25	0.44
Fossil fuel energy for food system (D_{ff}) to carbon emissions (C):	D_{ff}:C = 18.1/10.89 (Figs. 8.2 and 8.3)	1.66	0.60
Example of a person that uses 67.6 D_p and produces 30.5 C:	D_p:C = 67.6/30.5 (Figs. 3.3 and 4.2)	2.22	0.45
Example of a person that uses 57.1 D_{ff}, producing 30.5 C:	D_{ff}:C = 57.1/30.5 (Figs. 3.3 and 4.2)	1.87	0.53
Coal, 13.6 lb/d (energy equal to 1 gal of gasoline)	D = 15.2, C = 12.6 (Table 4.2)	1.20	0.83
Gasoline, 1 gal/d	D = 15.2, C = 9.8 (Table 4.2)	1.55	0.65
Natural gas, 1.65 CCF/d (energy equivalent to 1 gal of gasoline)	D = 15.2, C = 6.82 (Table 4.2)	2.23	0.45

Values for D:C indicate D/C, and values for C:D are calculated as C/D.

APPENDIX 2: ENERGY RELATED TO ELECTRICITY GENERATION IN THE UNITED STATES

Table A2.1 Primary energy (EJ) used for electricity production in the United States.

Year	Solar	Hydro	Wind	Geoth	Total
1990	0.00	3.18	0.03	0.17	3.38
1995	0.01	3.32	0.03	0.15	3.51
2000	0.01	2.92	0.06	0.15	3.14
2001	0.01	2.33	0.07	0.15	2.56
2002	0.01	2.80	0.11	0.16	3.07
2003	0.01	2.90	0.12	0.15	3.18
2004	0.01	2.80	0.15	0.16	3.11
2005	0.01	2.82	0.19	0.16	3.17
2006	0.01	3.00	0.28	0.15	3.43
2007	0.01	2.56	0.36	0.15	3.08
2008	0.01	2.63	0.58	0.15	3.37
2009	0.01	2.80	0.76	0.15	3.72
2010	0.01	2.66	0.97	0.16	3.80
2011	0.02	3.25	1.23	0.16	4.66
2012	0.04	2.75	1.41	0.16	4.36
2013	0.09	2.67	1.69	0.16	4.60
2014	0.17	2.59	1.82	0.16	4.74
2015	0.24	2.44	1.87	0.16	4.71
2016	0.35	2.59	2.21	0.15	5.30
2017	0.51	2.90	2.47	0.16	6.04
2018	0.61	2.80	2.62	0.15	6.17
2019	0.69	2.62	2.88	0.15	6.33

For the energy without the energy added use data in Table A2.3.
Source: Data from US Environmental Protection Agency (2020).

Table A2.2 Primary energy (EJ) used for electricity production in the United States.

Year	Renewable	Biomass	Nuclear	Petrol	Coal	NatGas	Total
1990	1.11	0.33	6.44	1.36	17.16	3.49	29.89
1995	1.16	0.45	7.46	0.80	18.43	4.54	32.83
2000	1.05	0.48	8.29	1.21	21.33	5.58	37.95
2001	0.85	0.36	8.47	1.35	20.69	5.76	37.47
2002	1.03	0.40	8.59	1.01	20.87	6.08	37.99
2003	1.07	0.42	8.40	1.27	21.30	5.53	37.99
2004	1.06	0.41	8.68	1.27	21.42	5.90	38.74
2005	1.08	0.43	8.61	1.29	21.88	6.35	39.63
2006	1.18	0.43	8.67	0.67	21.59	6.73	39.27
2007	1.06	0.45	8.92	0.68	21.95	7.39	40.46
2008	1.17	0.46	8.89	0.48	21.64	7.20	39.85
2009	1.30	0.47	8.81	0.40	19.23	7.41	37.62
2010	1.33	0.48	8.90	0.39	20.19	7.94	39.23
2011	1.64	0.46	8.72	0.31	19.03	8.14	38.30
2012	1.56	0.48	8.51	0.23	16.69	9.80	37.26
2013	1.65	0.50	8.70	0.27	17.36	8.84	37.30
2014	1.70	0.56	8.80	0.31	17.33	8.82	37.52
2015	1.72	0.55	8.80	0.29	14.92	10.47	36.75
2016	1.96	0.53	8.89	0.26	13.71	10.87	36.22
2017	2.24	0.54	8.88	0.23	13.32	10.08	35.29
2018	2.31	0.52	8.90	0.27	12.72	11.51	36.24
2019	2.37	0.47	8.93	0.20	10.74	12.31	35.02

Renewable energy from Table A2.3.
Source: Data from US Environmental Protection Agency (2020).

Table A2.3 Electricity production (EJ) for the United States.

Year	Solar	Hydro	Wind	Geoth	Total
1990	0.00	1.04	0.01	0.06	1.11
1995	0.00	1.10	0.01	0.05	1.16
2000	0.00	0.98	0.02	0.05	1.05
2001	0.00	0.77	0.02	0.05	0.85
2002	0.00	0.94	0.04	0.05	1.03
2003	0.00	0.98	0.04	0.05	1.07
2004	0.00	0.95	0.05	0.05	1.06
2005	0.00	0.96	0.06	0.05	1.08
2006	0.00	1.03	0.10	0.05	1.18
2007	0.00	0.89	0.12	0.05	1.06
2008	0.00	0.91	0.20	0.05	1.17
2009	0.00	0.98	0.27	0.05	1.30
2010	0.00	0.93	0.34	0.05	1.33
2011	0.01	1.14	0.43	0.06	1.64
2012	0.01	0.99	0.51	0.06	1.56
2013	0.03	0.95	0.60	0.06	1.65
2014	0.06	0.93	0.65	0.06	1.70
2015	0.09	0.89	0.69	0.06	1.72
2016	0.13	0.96	0.82	0.06	1.96
2017	0.19	1.08	0.91	0.06	2.24
2018	0.23	1.05	0.98	0.06	2.31
2019	0.26	0.98	1.08	0.06	2.37

Source: Data from US Environmental Protection Agency (2020).

Table A2.4 Electricity generation (EJ) in the United States.

Year	Renewable	Biomass	Nuclear	Petrol	Coal	NatGas	Total
1990	1.11	0.07	2.08	0.43	5.66	1.11	10.46
1995	1.16	0.09	2.42	0.25	6.07	1.51	11.50
2000	1.05	0.11	2.71	0.38	7.00	1.86	13.11
2001	0.85	0.08	2.77	0.43	6.78	2.00	12.89
2002	1.03	0.08	2.81	0.32	6.88	2.19	13.31
2003	1.07	0.08	2.75	0.41	7.03	2.04	13.39
2004	1.06	0.08	2.84	0.41	7.05	2.26	13.70
2005	1.08	0.08	2.82	0.42	7.17	2.46	14.03
2006	1.18	0.09	2.83	0.21	7.09	2.64	14.05
2007	1.06	0.09	2.90	0.22	7.19	2.93	14.40
2008	1.17	0.09	2.90	0.15	7.09	2.89	14.29
2009	1.30	0.10	2.88	0.13	6.27	3.03	13.70
2010	1.33	0.10	2.91	0.12	6.58	3.24	14.29
2011	1.64	0.10	2.84	0.10	6.18	3.33	14.20
2012	1.56	0.10	2.77	0.07	5.40	4.08	13.98
2013	1.65	0.11	2.84	0.09	5.64	3.70	14.03
2014	1.70	0.12	2.87	0.10	5.65	3.72	14.16
2015	1.72	0.12	2.87	0.10	4.83	4.46	14.09
2016	1.96	0.11	2.90	0.08	4.43	4.61	14.09
2017	2.24	0.11	2.90	0.07	4.31	4.31	13.94
2018	2.31	0.11	2.91	0.09	4.11	4.92	14.45
2019	2.37	0.10	2.91	0.06	3.45	5.31	14.22

Renewable energy from Table A2.3.
Source: Data from US Environmental Protection Agency (2020).

Table A2.5 Average power plant efficiencies calculated from heat rates using different fuels.

Year	Nuclear	Petrol	Coal	NatGas	Average fossil fuel
1990	0.322	NA	NA	NA	0.328
1995	0.325	NA	NA	NA	0.331
2000	0.327	NA	NA	NA	0.334
2001	0.327	0.318	0.329	0.339	0.330
2002	0.327	0.321	0.331	0.358	0.335
2003	0.327	0.322	0.331	0.371	0.337
2004	0.327	0.323	0.330	0.395	0.341
2005	0.327	0.321	0.329	0.399	0.341
2006	0.327	0.316	0.330	0.403	0.344
2007	0.325	0.316	0.329	0.406	0.345
2008	0.326	0.310	0.329	0.411	0.346
2009	0.326	0.312	0.328	0.418	0.350
2010	0.326	0.311	0.328	0.417	0.350
2011	0.326	0.315	0.327	0.419	0.351
2012	0.326	0.310	0.325	0.424	0.359
2013	0.327	0.318	0.326	0.429	0.358
2014	0.326	0.316	0.327	0.432	0.359
2015	0.326	0.319	0.325	0.433	0.366
2016	0.326	0.316	0.325	0.434	0.370
2017	0.326	0.315	0.326	0.437	0.370
2018	0.326	0.308	0.326	0.436	0.375
2019	0.326	0.308	0.326	0.436	0.375
2020	0.326	0.308	0.326	0.436	0.375

The average given for the fossil fuels is the efficiency assumed for renewables (NatGas = natural gas).
Source: Data from US Environmental Protection Agency (2020).

Reference

US ENERGY INFORMATION ADMINISTRATION. 2020 July 2020 Monthly Energy Review. vol. 2020, DOE/EIA-0035(2020/7). https://www.eia.gov/totalenergy/data/monthly/pdf/mer.pdf [Accessed August 7, 2020].

APPENDIX 3: WORLD AND US POPULATION

Table A3.1 Population of the United States and the world from 1950 to 2019.

Year	US	World
1950	152.3	2557.60
1955	165.9	2782.10
1960	180.7	3043.00
1965	194.3	3350.80
1970	205.1	3713.50
1975	216	4089.10
1980	227.2	4445.40
1985	237.9	4849.30
1990	249.6	5285.70
1995	266.3	5691.90
2000	282.2	6081.80
2001	285.0	6158.70
2002	287.6	6236.00
2003	290.1	6313.80
2004	292.8	6390.90
2005	295.5	6468.70
2006	298.4	6548.40
2007	301.2	6630.20
2008	304.1	6713.30
2009	306.8	6796.30
2010	309.3	6877.80
2011	311.6	6958.90
2012	313.8	7040.10
2013	316.0	7122.30
2014	318.3	7204.20
2015	320.6	7285.20
2016	322.9	7365.70
2017	325.0	7445.40
2018	326.7	7524.50
2019	328.2	7604.70

Source: Data from US Energy Information Administration (2020).

Reference

US ENERGY INFORMATION ADMINISTRATION. 2020. July 2020 Monthly Energy Review. https://www.eia.gov/totalenergy/data/monthly/pdf/mer.pdf [Accessed 07 August].

Daily Energy Use and Carbon Emissions: Fundamentals and Applications for Students and Professionals, First Edition. Bruce E. Logan.

APPENDIX 4: WORLD ENERGY USE

Table A4.1 World annual total energy use (EJ) for select countries and all countries sorted into regions.

Country/region	2010	2012	2014	2016	2018	2019
Argentina	3.23	3.38	3.51	3.58	3.54	3.46
Australia	5.50	5.63	5.75	5.88	6.00	6.41
Brazil	10.98	11.69	12.4	11.92	12.13	12.4
China	104.28	117.05	124.2	126.95	135.77	141.7
Egypt	3.28	3.50	3.47	3.74	3.92	3.89
France	10.65	10.22	9.87	9.76	9.87	9.68
Germany	13.71	13.37	13.17	13.62	13.44	13.14
India	22.55	25.11	27.86	30.07	33.3	34.06
Iran	8.94	9.41	10.28	10.79	11.83	12.34
Japan	21.13	19.92	19.24	18.65	18.84	18.67
Russian Federation	27.99	28.98	28.71	28.76	30.04	29.81
Saudi Arabia	8.92	9.76	10.5	10.98	10.91	11.04
South Africa	5.29	5.14	5.22	5.30	5.30	5.40
South Korea	10.94	11.54	11.64	12.16	12.55	12.37
United Kingdom	8.94	8.55	8.02	8.01	7.96	7.84
US	92.97	89.69	93.05	92.02	95.60	94.65
North America	113.29	110.86	114.78	113.74	117.79	116.58
S&C America	26.16	27.93	28.76	28.5	28.53	28.61
Europe	88.69	86.32	82.01	83.90	84.76	83.82
CIS	35.28	37.04	36.74	36.73	38.81	38.68
Middle East	29.74	32.12	34.05	36.23	37.61	38.78
Africa	16.07	16.69	17.66	18.38	19.39	19.87
Asia Pacific	196.80	214.02	225.15	233.13	249.35	257.56
World	506.02	524.98	539.25	550.6	576.23	583.90

Source: Data from Dudley (2019).

Table A4.2 World annual electricity use (EJ) for select countries and all countries sorted into regions.

Country or region	2010	2012	2014	2016	2018	2019
Argentina	0.45	0.49	0.50	0.53	0.53	0.50
Australia	0.90	0.90	0.89	0.93	0.95	0.95
Brazil	1.86	1.99	2.13	2.08	2.17	2.25
China	15.15	17.96	20.86	22.08	25.80	27.01
Egypt	0.52	0.59	0.62	0.68	0.72	0.72
France	2.05	2.03	2.03	2.00	2.07	2.00
Germany	2.28	2.27	2.26	2.34	2.32	2.20
India	3.37	3.93	4.54	5.05	5.59	5.61
Iran	0.85	0.90	0.98	1.03	1.13	1.15
Japan	4.16	3.98	3.83	3.71	3.80	3.73
Russian Federation	3.74	3.85	3.83	3.93	3.99	4.03
Saudi Arabia	0.86	0.98	1.12	1.24	1.29	1.29
South Africa	0.93	0.93	0.92	0.91	0.92	0.91
South Korea	1.78	1.91	1.95	2.02	2.14	2.10
United Kingdom	1.38	1.31	1.22	1.22	1.20	1.17
United States	15.82	15.52	15.71	15.65	16.05	15.84
North America	19.00	18.88	19.13	19.19	19.65	19.53
S&C America	4.11	4.43	4.63	4.70	4.79	4.79
Europe	14.64	14.59	14.18	14.48	14.64	14.38
CIS	4.62	4.79	4.82	4.93	5.10	5.15
Middle East	3.15	3.41	3.78	4.12	4.41	4.55
Africa	2.42	2.59	2.76	2.89	3.04	3.13
Asia Pacific	29.73	33.40	37.20	39.41	44.31	45.69
World	77.65	82.10	86.51	89.72	95.95	97.22

Source: Data from Dudley (2019).

Table A4.3 Daily energy use per person per day, in units of D, for select countries and all countries sorted into regions.

Country or region	2010	2012	2014	2016	2018	2019
Argentina	25.9	26.6	27.0	27.0	26.2	25.4
Australia	81.4	80.6	79.9	79.5	79.0	83.4
Brazil	18.4	19.3	20.1	19.0	19.0	19.3
China	25.0	27.8	29.1	29.5	31.2	32.4
Egypt	13.0	13.3	12.6	13.0	13.1	12.7
France	55.6	52.8	50.5	49.5	49.8	48.7
Germany	55.6	54.2	53.0	54.4	53.0	51.6
India	6.0	6.5	7.1	7.4	8.1	8.2
Iran	39.7	40.9	43.5	44.5	47.4	48.8
Japan	53.9	50.9	49.2	47.9	48.6	48.3
Russian Federation	64.0	66.0	65.1	65.0	67.6	67.0

(*continued overleaf*)

Table A4.3 *(continued)*

Country or region	2010	2012	2014	2016	2018	2019
Saudi Arabia	106.7	109.8	111.4	111.0	106.2	105.6
South Africa	33.9	31.9	31.4	30.9	30.0	30.2
South Korea	72.4	75.6	75.4	78.2	80.4	79.2
United Kingdom	46.2	43.5	40.2	39.7	38.9	38.1
United States	98.7	93.7	95.8	93.5	95.9	94.3
North America	81.3	78.0	79.3	77.3	78.8	77.4
S&C America	18.0	18.8	19.0	18.4	18.1	18.0
Europe	44.2	42.7	40.3	40.9	41.1	40.5
CIS	49.9	51.8	50.6	50.0	52.1	51.7
Middle East	45.0	46.5	47.4	48.7	48.9	49.6
Africa	5.1	5.0	5.0	5.0	5.0	5.0
Asia Pacific	16.7	17.7	18.3	18.6	19.6	20.0
Total	23.8	24.2	24.2	24.2	24.8	24.8

Source: Data from Dudley (2019).

Table A4.4 Electricity generation (EJ) using renewables and other sources for select countries and all countries sorted into regions in 2019.

Country or region	Hydro	Renewables and biomass	Nuclear	Total	Others
Argentina	0.134	0.030	0.030	0.193	0.002
Australia	0.051	0.148	—	0.199	0.000
Brazil	1.437	0.424	0.058	1.920	—
China	4.571	2.636	1.255	8.463	0.203
Egypt	0.048	0.023	—	0.072	—
Germany	0.073	0.807	0.270	1.150	0.093
India	0.582	0.486	0.163	1.231	0.001
Iran	0.104	0.002	0.023	0.130	—
Japan	0.266	0.436	0.236	0.939	0.152
Russian Federation	0.700	0.006	0.752	1.459	0.015
Saudi Arabia	0.006	—	—	0.006	1.287
South Africa	0.003	0.045	0.051	0.099	0.017
South Korea	0.010	0.105	0.526	0.641	0.035
United Kingdom	0.022	0.408	0.202	0.632	0.028
United States	0.976	1.763	3.067	5.807	0.050
North America	2.437	2.077	3.469	7.983	0.130
S&C America	2.573	0.663	0.089	3.324	0.001
Europe	2.277	3.012	3.343	8.631	0.278
CIS	0.894	0.012	0.760	1.666	0.008
Middle East	0.120	0.048	0.023	0.191	—
Africa	0.478	0.162	0.051	0.691	0.010
Asia Pacific	6.421	4.126	2.330	12.878	0.413
World	15.200	10.100	10.066	35.365	0.841

Source: Data from Dudley (2019).

Table A4.5 Electricity generation (EJ) for renewables and C-neutral and fossil fuels for select countries and all countries sorted into regions in 2019.

Country or region	Renewables and C-neutral	Petrol	Coal	NatGas	Total
Argentina	0.193	0.010	0.003	0.296	0.502
Australia	0.199	0.021	0.538	0.196	0.954
Brazil	1.920	0.028	0.093	0.212	2.253
China	8.463	0.022	17.473	0.851	26.809
Egypt	0.072	0.102	—	0.549	0.722
Germany	1.150	0.018	0.616	0.328	2.112
India	1.231	0.030	4.095	0.256	5.611
Iran	0.130	0.297	0.002	0.718	1.147
Japan	0.939	0.161	1.174	1.305	3.578
Russian Federation	1.459	0.025	0.656	1.870	4.010
Saudi Arabia	0.006	0.742	—	—	0.748
South Africa	0.099	0.004	0.782	0.007	0.893
South Korea	0.641	0.027	0.859	0.543	2.070
United Kingdom	0.632	0.004	0.025	0.477	1.138
United States	5.807	0.072	3.793	6.123	15.795
North America	7.983	0.222	4.084	7.113	19.402
S&C America	3.324	0.311	0.267	0.882	4.784
Europe	8.631	0.186	2.515	2.765	14.098
CIS	1.666	0.031	0.951	2.495	5.143
Middle East	0.191	1.426	0.081	2.854	4.553
Africa	0.691	0.293	0.913	1.226	3.123
Asia Pacific	12.878	0.502	26.555	5.337	45.272
World	35.365	2.971	35.367	22.672	96.376

Source: Data from Dudley (2019).

Reference

DUDLEY, B. 2019. BP Statistical Review of World Energy 2019. 68 ed. Pureprint Group Limited UK. https://www.bp.com/content/dam/bp/business-sites/en/global/corporate/pdfs/energy-economics/statistical-review/bp-stats-review-2019-full-report.pdf [Accessed 24 April 2020].

APPENDIX 5: CO_2 EMISSIONS

Table A5.1 World annual CO_2 emissions (Gt) for select countries and all countries sorted into regions.

Country or region	2010	2012	2014	2016	2018	2019
Argentina	0.166	0.175	0.183	0.186	0.180	0.175
Australia	0.403	0.403	0.406	0.412	0.411	0.428
Brazil	0.398	0.443	0.504	0.450	0.442	0.441
China	8.143	9.001	9.240	9.138	9.507	9.826
Egypt	0.189	0.200	0.204	0.217	0.221	0.217
France	0.360	0.336	0.301	0.312	0.307	0.299
Germany	0.783	0.773	0.751	0.771	0.731	0.684
India	1.661	1.848	2.084	2.243	2.453	2.480
Iran	0.518	0.535	0.578	0.597	0.644	0.671
Japan	1.202	1.296	1.249	1.193	1.164	1.123
Russian Federation	1.492	1.569	1.531	1.505	1.548	1.533
Saudi Arabia	0.486	0.526	0.571	0.600	0.574	0.580
South Africa	0.477	0.464	0.469	0.471	0.470	0.479
South Korea	0.591	0.615	0.615	0.630	0.662	0.639
United Kingdom	0.530	0.512	0.458	0.416	0.397	0.387
United States	5.486	5.090	5.255	5.042	5.117	4.965
North America	6.458	6.090	6.268	6.049	6.149	5.976
S&C America	1.173	1.275	1.348	1.308	1.263	1.255
Europe	4.681	4.544	4.205	4.263	4.246	4.111
CIS	1.940	2.063	2.029	1.997	2.096	2.085
Middle East	1.739	1.854	1.954	2.066	2.106	2.164
Africa	1.101	1.137	1.198	1.231	1.285	1.309
Asia Pacific	13.99	15.31	15.80	16.02	16.86	17.27
World	31.09	32.27	32.80	32.94	34.01	34.17

Source: Data from Dudley (2019).

Table A5.2 GHG gas emissions by the US in CO_2 equivalents (Gt/y).

Year	Carbon dioxide	Methane	Nitrous oxide	Fluorinated gases	Total
1990	5.128	0.774	0.435	0.100	6.437
1991	5.079	0.779	0.425	0.091	6.373
1992	5.183	0.778	0.424	0.095	6.480
1993	5.283	0.766	0.452	0.095	6.597
1994	5.377	0.772	0.438	0.099	6.686
1995	5.439	0.765	0.449	0.118	6.771
1996	5.627	0.758	0.461	0.129	6.974
1997	5.704	0.742	0.447	0.137	7.029
1998	5.751	0.727	0.447	0.153	7.078
1999	5.830	0.709	0.436	0.150	7.126
2000	5.998	0.703	0.423	0.151	7.275
2001	5.900	0.696	0.439	0.138	7.172
2002	5.943	0.688	0.437	0.147	7.214
2003	5.992	0.688	0.437	0.138	7.255
2004	6.108	0.682	0.447	0.146	7.382
2005	6.132	0.680	0.433	0.148	7.392
2006	6.051	0.684	0.429	0.151	7.314
2007	6.128	0.686	0.440	0.162	7.416
2008	5.931	0.692	0.424	0.164	7.210
2009	5.491	0.680	0.422	0.160	6.754
2010	5.698	0.682	0.431	0.170	6.982
2011	5.565	0.656	0.422	0.177	6.821
2012	5.368	0.647	0.392	0.174	6.581
2013	5.514	0.642	0.439	0.174	6.770
2014	5.562	0.639	0.449	0.179	6.829
2015	5.412	0.638	0.444	0.182	6.676
2016	5.292	0.624	0.426	0.182	6.524
2017	5.254	0.630	0.421	0.183	6.488
2018	5.425	0.634	0.435	0.183	6.677

Fluorinated gases include hydrofluorocarbons (HFCs), perfluorocarbons (PFCs), hexafluoride (SF6), and nitrogen trifluoride (NF3).
Source: US Environmental Protection Agency (2019).

References

DUDLEY, B. 2019. BP Statistical Review of World Energy 2019. 68 ed. Pureprint Group Limited UK. Available: https://www.bp.com/content/dam/bp/business-sites/en/global/corporate/pdfs/energy-economics/statistical-review/bp-stats-review-2019-full-report.pdf [Accessed April 24 2020].

US ENVIRONMENTAL PROTECTION AGENCY. 2019. *Sources of greenhouse gas emissions*. Available: https://www.epa.gov/ghgemissions/sources-greenhouse-gas-emissions [Accessed August 30 2020].

APPENDIX 6: HOURS OF PEAK SOLAR IN THE UNITED STATES

Table A6.1 Hours of peak solar light estimated from maps produced by the National Renewable Energy Laboratory (NREL) by different solar companies.

State	City	Range	City	Average
Alabama	Birmingham	4.5–4.7	Montgomery	4.23
Alaska	Anchorage	2.4–2.9	Fairbanks	3.99
Arizona	Tucson	7.5–7.9	Tucson	6.57
Arizona	—	—	Phoenix	6.58
Arkansas	Fort Smith	4.9–5.0	Little Rock	4.69
California	San Diego	5.8–6.7	La Jolla	4.77
California	—	—	Davis	5.10
Colorado	Grand Junction	6.0–6.6	Grand Junction	5.86
Colorado	—	—	Boulder	4.87
Connecticut	Hartford	4.0–4.3	—	—
Florida	Tampa	5.5–5.7	Tampa	5.67
Florida	—	—	Gainesville	5.27
Georgia	Macon	4.7–5.0	Atlanta	4.74
Idaho	Boise	5.1–5.7	Boise	4.92
Illinois	Peoria	4.3–4.4	Chicago	3.14
Indiana	Indianapolis	4.1–4.3	Indianapolis	4.21
Iowa	Des Moines	4.2–4.4	Ames	4.40
Kansas	Wichita	5.2–5.8	Manhattan	4.57
Kentucky	Louisville	41.–4.4	Lexington	4.94
Louisiana	New Orleans	4.9–5.1	New Orleans	4.92
Maine	Portland	4.0–4.3	Portland	4.51
Maryland	Baltimore	4.2–4.5	Silver Hill	4.47
Massachusetts	Boston	4.3–4.5	Boston	3.84
Michigan	Lansing	3.8–4.0	E. Lansing	4.00
Minnesota	Minneapolis	4.3–4.5	St. Cloud	4.53
Mississippi	Jackson	4.9–5.1	—	—
Missouri	Springfield	4.9–5.0	—	—
Montana	Helena	4.5–5.0	Glasgow	5.15
Nebraska	Lincoln	4.9–5.2	Lincoln	4.79
Nevada	Reno	6.4–7.1	Las Vegas	6.41
New Hampshire	Concord	4.0–4.3	—	—
New Jersey	Atlantic City	4.0–4.3	Sea Brook	4.21

(continued overleaf)

Table A6.1 *(continued)*

State	City	Range	City	Average
New Mexico	Albuquerque	7.0–7.5	Albuquerque	6.77
New York	Syracuse	3.6–3.9	New York	4.08
North Carolina	Greensboro	4.6–4.8	Greensboro	4.71
North Dakota	Bismarck	4.5–4.7	Bismarck	5.01
Ohio	Columbus	3.9–4.1	Columbus	4.15
Oklahoma	Oklahoma City	5.5–6.0	Oklahoma City	5.59
Oregon	Portland	3.7–4.5	Corvallis	4.03
Pennsylvania	Harrisburg	4.0–4.2	State College	3.91
Rhode Island	Providence	4.3–4.5	Newport	4.23
South Carolina	Columbia	4.4–4.6	Charleston	5.06
South Dakota	Huron	5.0–5.3	Rapid City	5.23
Tennessee	Nashville	4.2–4.4	Nashville	4.45
Texas	Dallas	5.2–5.9	El Paso	6.72
Utah	Salt Lake City	5.7–6.6	Salt Lake City	5.26
Vermont	Burlington	3.8–4.0	—	—
Virginia	Richmond	4.6–4.8	Richmond	4.13
Washington	Seattle	3.3–3.9	Seattle	3.57
Wisconsin	Madison	4.3–4.4	Madison	4.29
Wyoming	Lander	5.2–5.7	Lander	6.06

Source: Adapted from Solar Reviews (2021), and Solar Direct (2020).

References

SOLAR DIRECT. 2020. *Solar electric system sizing, step 4- determine the sun hours available per day.* Available: https://www.solardirect.com/archives/pv/systems/gts/gts-sizing-sun-hours.html [Accessed October 28 2020].

SOLAR REVIEWS. 2021. *What are the average peak sun hours for my state?* Available: https://www.solarreviews.com/blog/peak-sun-hours-explained [Accessed April 4 2021].

INDEX

Daily Energy Use and Carbon Emissions: Fundamentals and Applications for Students and Professionals, First Edition. Bruce E. Logan.